Stochastic Modelling of Big Data in Finance

Stochastic Modelling of Big Data in Finance provides a rigorous overview and exploration of stochastic modelling of big data in finance (BDF). The book describes various stochastic models, including multivariate models, to deal with big data in finance. This includes data in high-frequency and algorithmic trading, specifically in limit order books (LOB), and shows how those models can be applied to different datasets to describe the dynamics of LOB, and to figure out which model is the best with respect to a specific data set. The results of the book may be used to also solve acquisition, liquidation and market making problems, and other optimization problems in finance.

Features

- Self-contained book suitable for graduate students and post-doctoral fellows in financial mathematics and data science, as well as for practitioners working in the financial industry who deal with big data
- All results are presented visually to aid in understanding of concepts

Dr. Anatoliy Swishchuk is a Professor in Mathematical Finance at the Department of Mathematics and Statistics, University of Calgary, Calgary, AB, Canada. He got his B.Sc. and M.Sc. degrees from Kyiv State University, Kyiv, Ukraine. He earned two doctorate degrees in Mathematics and Physics (PhD and DSc) from the prestigious National Academy of Sciences of Ukraine (NASU), Kiev, Ukraine, and is a recipient of NASU award for young scientist with a gold medal for series of research publications in random evolutions and their applications.

Dr. Swishchuk is a chair and organizer of finance and energy finance seminar 'Lunch at the Lab' at the Department of Mathematics and Statistics. Dr. Swishchuk is a Director of Mathematical and Computational Finance Laboratory at the University of Calgary. He was a steering committee member of the Professional Risk Managers International Association (PRMIA), Canada (2006-2015), and is a steering committee member of Global Association of Risk Professionals (GARP), Canada (since 2015).

Dr. Swishchuk is a creator of mathematical finance program at the Department of Mathematics & Statistics. He is also a proponent for a new specialization "Financial and Energy Markets Data Modelling" in the Data Science and Analytics program. His research areas include financial mathematics, random evolutions and their applications, biomathematics, stochastic calculus, and he serves on editorial boards for four research journals. He is the author of more than 200 publications, including 15 books and more than 150 articles in peer-reviewed journals. In 2018 he received a Peak Scholar award.

Chapman & Hall/CRC Financial Mathematics Series

Aims and scope:
The field of financial mathematics forms an ever-expanding slice of the financial sector. This series aims to capture new developments and summarize what is known over the whole spectrum of this field. It will include a broad range of textbooks, reference works and handbooks that are meant to appeal to both academics and practitioners. The inclusion of numerical code and concrete real-world examples is highly encouraged.

Machine Learning for Factor Investing: R Version
Guillaume Coqueret, Tony Guida

Malliavin Calculus in Finance: Theory and Practice
Elisa Alos, David Garcia Lorite

Risk Measures and Insurance Solvency Benchmarks: Fixed-Probability Levels in Renewal Risk Models
Vsevolod K. Malinovskii

Financial Mathematics: A Comprehensive Treatment in Discrete Time, Second Edition
Giuseppe Campolieti, Roman N. Makarov

Pricing Models of Volatility Products and Exotic Variance Derivatives
Yue Kuen Kwok, Wendong Zheng

Quantitative Finance with Python: A Practical Guide to Investment Management, Trading, and Financial Engineering
Chris Kelliher

Stochastic Modelling of Big Data in Finance
Anatoliy Swishchuk

Introduction to Stochastic Finance with Market Examples, Second Edition
Nicolas Privault

Commodities: Fundamental Theory of Futures, Forwards, and Derivatives Pricing, Second Edition
M.A.H. Dempster, Ke Tang

For more information about this series please visit: https://www.crcpress.com/Chapman-and-HallCRC-Financial-Mathematics-Series/book series/CHFINANCMTH

Stochastic Modelling of Big Data in Finance

Anatoliy Swishchuk
Department of Mathematics and Statistics,
University of Calgary, Calgary, Canada

CRC Press is an imprint of the
Taylor & Francis Group, an **informa** business
A CHAPMAN & HALL BOOK

First edition published 2023
by CRC Press
6000 Broken Sound Parkway NW, Suite 300, Boca Raton, FL 33487-2742

and by CRC Press
4 Park Square, Milton Park, Abingdon, Oxon, OX14 4RN

CRC Press is an imprint of Taylor & Francis Group, LLC

Library of Congress Cataloging-in-Publication Data

Names: Swishchuk, Anatoliy, author.
Title: Stochastic modelling of big data in finance / Anatoliy Swishchuk, Department of Mathematics and Statistics, University of Calgary, Calgary, Canada.
Description: 1 Edition. | Boca Raton, FL : Chapman & Hall, CRC Press, 2023. | Series: Chapman and Hall/CRC financial mathematics series | Includes bibliographical references and index.
Identifiers: LCCN 2022022045 (print) | LCCN 2022022046 (ebook) | ISBN 9781032209265 (hardback) | ISBN 9781032209289 (paperback) | ISBN 9781003265986 (ebook)
Subjects: LCSH: Finance--Mathematical models. | Stochastic models. | Big data.
Classification: LCC HG106 .S95 2023 (print) | LCC HG106 (ebook) | DDC 332.01/5195--dc23/eng/20220801
LC record available at https://lccn.loc.gov/2022022045
LC ebook record available at https://lccn.loc.gov/2022022046

ISBN: 978-1-032-20926-5 (hbk)
ISBN: 978-1-032-20928-9 (pbk)
ISBN: 978-1-003-26598-6 (ebk)

DOI: 10.1201/ 9781003265986

Typeset in CMR10 font
by KnowledgeWorks Global Ltd.

Publisher's note: This book has been prepared from camera-ready copy provided by the authors.

To My Family and My Motherland, Ukraine

Contents

Foreword

It is my pleasure to introduce my book in stochastic modelling of big data in finance (BDF). Big data has now become a driver of model building and analysis in many areas, including finance. One of the main problems here is: how to deal with big data arising in electronic markets for algorithmic and high-frequency (milliseconds) trading that contain two types of orders, limit orders and market orders. More than half of the markets in today's financial world now use a limit order book (LOB) mechanism to facilitate trade. There are hundreds of thousands of orders even for one stock just for one day: e.g., for CISCO data on November 3, 2014, we had 0.5 million price orders for that day, and that is only for a 1-level order book, with the limit orders sitting at the best bid and ask. And if we take hundreds of stocks on an exchange, and not only a 1-level order book, then we will get an example of really big data in finance! And the question is: can we model this mechanism of trading and describe this big data? And the answer is "Yes".

Thus, the main motivations for writing this book are: show how to study big data in finance; visualize these big data; describe the stochastic models for these big data; justify the proposed models; and show how to solve many problems with these big data in finance using such proposed models.

The book consists of three parts: Part I: Semi-Markovian Modelling of BDF, Part II: Modelling of BDF with Hawkes Processes, and Part III: Multivariate Modelling of BDF. The book also contains an Appendix: Basics in Stochastic Processes, where a reader can find all necessary basic definitions and facts in stochastic processes which we use in this book. Thus, I can say that the book is self-contained. The Preface gives a taste of stochastic models used in the book. The book also includes a large amount of figures that visualize our results and describe much of the data presented throughout the book.

The audience for the book are Master's and PhD students, as well as Post-Doctoral Fellows in financial mathematics, high-frequency and algorithmic trading, limit order books, stochastics, statistics and data science, and also practitioners in the financial industry dealing with big data in finance. This book can also be considered as a textbook for courses that involve data science, stochastic modelling in finance, and high-frequency and algorithmic trading/limit order books.

Preface

Manifesto: Stochastic Modelling of Big Data in Finance (BDF)-Full Description

Use data first, then propose a model based on the data. Do not fool yourself with the inverse approach. That is how our great journey with modelling of LOB has started: see [14, 12].

1. If the inter-arrival times (successive durations between price changes) follow *exponential distribution* (use, e.g., QQ-plot for the available data against exponential distribution), and arrivals of a new book event at the bid and the ask are *independent*, then use compound Poisson process for the stock price S_t:

$$S_t = S_0 + \sum_{k=1}^{N_t} X_k,$$

where N_t is a Poisson process (number of price changes on $[0, t]$) and $X_k, k = 1, 2, 3, ...,$ are the successive moves in the price. See [1].

2. If the inter-arrival times (successive durations between price changes) *do not follow exponential distribution* (use, e.g., QQ-plot for the available data against exponential distribution), but *arbitrary*, and arrival of a new book event at the bid and the ask are *independent*, then use the following process for the stock price S_t:

$$S_t = S_0 + \sum_{k=1}^{\nu_t} X_k,$$

where ν_t is a counting process (number of price changes on $[0, t]$) and $X_k, k = 1, 2, 3, ...,$ are the successive moves in the price. See [5].

3. If the inter-arrival times (successive durations between price changes) *do not follow exponential distribution* (use, e.g., QQ-plot for the available data against exponential distribution), but *arbitrary*, and arrivals of a new book event at the bid and the ask are *dependent*, and have *two different values*,

then use *Markov renewal process* for the stock price S_t:

$$S_t = S_0 + \sum_{k=1}^{\nu_t} X_k,$$

where ν_t is a counting process (number of price changes on $[0,t]$) and $X_k, k = 1,2,3,...,$ is a Markov chain with two states $\{+\delta, -\delta\}$(successive moves in the price), δ is a tick size. See [14].

4. If the inter-arrival times (successive durations between price changes) *do not follow exponential distribution* (use, e.g., QQ-plot for the available data against exponential distribution), but *arbitrary,* and arrivals of a new book event at the bid and the ask are *dependent*, and have *many different values*, then use the following *general Markov renewal process* for the stock price S_t:

$$S_t = S_0 + \sum_{k=1}^{\nu_t} a(X_k),$$

where ν_t is a counting process (number of price changes on $[0,t]$), $a(x)$ is a bounded function on $X = \{1,2,...,N\}$-space states for a Markov chain $X_k, k = 1,2,3,...,$ with N states (successive moves in the price changes). See [11].

5. If the inter-arrival times (successive durations between price changes) *do not follow exponential distribution* (use, e.g., QQ-plot for the available data against exponential distribution), but *arbitrary*, and exhibit *clustering effect*, and arrivals of a new book event at the bid and the ask are *independent*, then use *compound Hawkes process* for the stock price S_t:

$$S_t = S_0 + \sum_{k=1}^{N_t} X_k,$$

where N_t is the Hawkes process (number of price changes on $[0,t]$) and $X_k, k = 1,2,3,...,$ is i.i.d.r.v. (successive moves in the price). See [7], [8], [13].

6. If the inter-arrival times (successive durations between price changes) *do not follow exponential distribution* (use, e.g., QQ-plot for the available data against exponential distribution), but *arbitrary*, and exhibit *clustering effect*, and arrivals of a new book event at the bid and the ask are *dependent, and have two different values* then use *general compound Hawkes process* for the stock price S_t:

$$S_t = S_0 + \sum_{k=1}^{N_t} X_k,$$

where N_t is the Hawkes process (number of price changes on $[0,t]$) and

$X_k = \{+\delta, -\delta\}$ is a two-state Markov chain (successive moves in the price). See [13].

7. If the inter-arrival times (successive durations between price changes) *do not follow exponential distribution* (use, e.g., QQ-plot for the available data against exponential distribution), but *arbitrary,* and exhibit *clustering effect,* arrivals of a new book event at the bid and the ask are *dependent*, and *price changes can take more than two different values*, and price changes are not fixed to one tick, then use *general compound Hawkes process* for the stock price S_t:

$$S_t = S_0 + \sum_{k=1}^{N_t} a(X_k),$$

where N_t is the Hawkes process (number of price changes on $[0, t]$) and $X_k, k = 1, 2, 3, ...,$ is a Markov chain (successive moves in the price) with state space $X = \{1, 2, ..., N\}$. See [13], [6], [7].

8. If background intensity λ or self-exciting function $\mu(t)$ exhibit regime-switching behaviour, then use regime-switching general compound Hawkes processes. See [9, 10].

9. If arrival rates for market and limit orders are time-dependent, and they are different for bid and ask orders, then use our model in [2, 3].

10. All above-mentioned approaches are applicable to *MGCHP* for cases with multiple stock prices as well, i.e., for $(S_{1,t}, S_{2,t}, \cdots, S_{d,t}) = \vec{S_t}$. See [4]. We say $\vec{N_t} = (N_{1,t}, N_{2,t}, \cdots, N_{d,t})$ is a d-dimensional Hawkes process when the intensity function for each N_i is in the form of

$$\lambda_i(t) = \lambda_i + \int_{(0,t)} \sum_{j=1}^{d} \mu_{ij}(t-s) dN_{j,s},$$

where $\lambda_i \in R_+$ and the intensity $\mu_{ij}(t)$ is a function from R_+ to R_+. Let $\vec{N_t} = (N_{1,t}, N_{2,t}, \cdots, N_{d,t},)$ be a d-dimensional Hawkes process. Then the multivariate general compound Hawkes process (MGCHP) $\vec{S_t} = (S_{1,t}, S_{2,t}, \cdots, S_{d,t},)$ is defined as

$$S_{i,t} = S_{i,0} + \sum_{k=1}^{N_{i,t}} a(X_{i,k}),$$

where $X_{i,k}$ are independent ergodic continuous-time Markov chains.

11. In the most general case (when we do not have those mentioned above stylized properties for N(t)), use *multivariate general compound point process* (MGCPP) for the cases of one stock price or multiple stock prices, i.e., for

$(S_{1,t}, S_{2,t}, \cdots, S_{d,t}) = \vec{S}_t$. See [5]. Let $\vec{N}_t = (N_{1,t}, N_{2,t}, \cdots, N_{d,t},)$ be a d-dimensional point process. Then the MGCPP $\vec{S}_t = (S_{1,t}, S_{2,t}, \cdots, S_{d,t},)$ is defined as

$$S_{i,t} = S_{i,0} + \sum_{k=1}^{N_{i,t}} a(X_{i,k}),$$

where $X_{i,k}$ are independent ergodic continuous-time Markov chains for each i. Of course, in one-dimensional case we have a general compound point process (GCPP)

$$S_t = S_0 + \sum_{i=1}^{N(t)} a(X_i),$$

where $N(t)$ is a point process and X_i is an ergodic continuous-time Markov chain.

Remark 0.1 *All these models were introduced and proposed, and then studied after investigating many HFT data, presented in the book, which supported and justified the models above.*

Anatoliy Swishchuk
Department of Mathematics & Statistics
University of Calgary
Calgary, Alberta, Canada
April 26, 2022

Bibliography

[1] Cont, R. and de Larrard, (2013): Markovian modelliing of limit order markets. *SIAM J. Finan. Math..*, (2017).

[2] Chavez-Casillas, J., Elliott, R., Remillard, B. and Swishchuk, A. (2017): A level-1 limit order book with time dependent arrival rates. *Proceed. IWAP, Toronto.*

[3] Chavez-Cassilas, J., Remillard, B., Elliott, R. and Swishchuk, A. (2019): A level-1 limit order book with time dependent arrival rates. *Methodology and Computing in Applied Probability.* 21(3), 699–719.

[4] Guo, G. and Swishchuk, A. (2020): Multivariate General Compound Hawkes Processes and their Applications in Limit Order Books. *Risks*, (107), 45–51.

[5] Guo, Q., Remillard, B. and Swishchuk, A. (2020): Multivariate general compound point processes in limit order books. *Risks*, 8(3), 98.

[6] Swishchuk, A. and Huffman, A. (2020): General compound Hawkes processes in limit order books. *Risks*, 8(1), 28.

[7] Swishchuk, A. and He, Q. (2019): Quantitative and Comparative Analyses of Limit Order Books with General Compound Hawkes Processes. *Risks*, 7(4), 110.

[8] Swishchuk, A., Remillard, B. Elliott, R. and Chavez-Casillas, J. (2019): Compound Hawkes processes in limit order books. *Financial Mathematics, Volatility and Covariance Modelling.* Routledge: Taylor and Francis Group. v. 2, 1st ed., (Editors: Chevallier, J., Goutte, S., Guerreiro, D., Saglio, S. and Sanhaji, B.)

[9] Swishchuk, A. (2020): Modelling of limit order books by general compound Hawkes processes with implementations. *Meth. Comput. Appl. Probab.*, 23, 299–428 (2021).

[10] Swishchuk, A. (2020): Stochastic modelling of big data in finance. *Method. Comput. Appl. Probab.*, 30 pages.

[11] Swishchuk, A., Cera, K., Hofmeister, T. and Schmidt, J. (2017): General semi-Markov model for limit order books, *Inern. J. Theoret. Applied Finance*, 20, 20.

[12] Swishchuk, A. and Vadori, N. (2017). Semi-Markov modeliing of limit order markets. *SIAM J. Finan. Math.*

[13] Swishchuk, A. (2017): General compound Hawkes processes in LOB. arXiv:1706.07459 [q-fin.MF]

[14] Swishchuk. A. and Vadori, N. (2015): Semi-Markov model for the dynamics in limit order markets. *Poster: IPAM FMWS1, UCLA, CA*, March 23–27.

Symbols

Symbol Description

Symbol	Description
BDF	Big Data in Finance.
LOB	Limit Order Books.
SMM	Semi-Markovian Model.
GSMM	General Semi-Markovian Model.
HP	Hawkes Process.
RSHP	Regime-switching Hawkes Process.
CHP	Compound Hawkes Process.
CHPDO	Compound Hawkes Process with Dependent Order.
RSCHP	Regime-switching Compound Hawkes Process.
GCHP	General Compound Hawkes Process.
MHP	Multivariate Hawkes Process.
MGCHP	Multivariate General Compound Hawkes Process.
MGCPP	Multivariate General Compound Point Process.
NLCHPnSDO	Non-Linear Compound Hawkes Process with n-state Dependent Orders.
GCHPnSDO	General Compound Hawkes Process With N-state Dependent Orders.
GCHP2SDO	General Compound Hawkes Process with Two-State Dependent Orders.
GCHPDO	General Compound Hawkes Process with Dependent Orders.
LLN	Law of Large Numbers.
CLT	Central Limit Theorem.
FCLT	Functional Central Limit Theorem.
CDF	Cumulative Distribution Function.
$N(t)$ or N_t	Poisson or Hawkes Process.
S_t or s_t	Stock Mid-price.
FCLT	Functional Central Limit Theorem.
CDF	Cumulative Distribution Function.

Acknowledgements

I would like to thank Callum Fraser (the Editor of the Financial Mathematics books series (CRC Press/Taylor & Francis Group)), who invited me to write this book.

My thanks also go to Shashi Kumar (KnowledgeWorks Global Ltd. (KGL)), who helped me a lot with the LaTEX preparation of all these chapters. His help was so valuable and timely that I submitted my book well before the deadline.

From the bottom of my heart, I would also like to thank all my graduate students and participants of our finance seminars "Lunch at the Lab", "Lobster" and "Hawks", for their curiosity and dedication to these seminars' activities. As a result, all our joint research efforts during those seminar series have been converted into interesting results and published papers. I would like to mention Nelson Vadori (JP Morgan, NY), Katharina Cera (TUM, Munich), Julia Smith (TUM, Munich), Tyler Hoffmeister (Morgan Stanley, London), Jonathan Chavez-Casillas (URI, RI), Qiyue He (Ernst & Young, Calgary), Aiden Huffman (MSc, UWaterloo), and Qi Guo (PhD, UCalgary), for their curiosity, passion, and dedication. I would also like also to thank my dear collaborators, Robert Elliott (University of SA, Adelaide, Australia) and Bruno Remillard (HEC, Montreal, Canada), for sharing their ideas and great research experience. The book is based on many papers co-authored with all of them.

I likewise want to thank NSERC for their continuous support.

Finally, many thanks go to my dear family, including my daughter Julianna and charming granddaughter, Ivanka Maria, for their everlasting support and inspiration. Julianna and Ivanka Maria have persistently taught me not only how to convey the abstract moments of life into fairy tales but also into mathematics.

1

A Brief Introduction: Stochastic Modelling of Big Data in Finance

In this introductory chapter, we present an overview of stochastic modelling of big data in finance (including limit order books), based on price changes modelling associated with high-frequency and algorithmic trading. We introduce a big data in finance through the limit order books (LOB), and describe them by Lobster data, academic data for studying LOB, and Xetra and Frankfurt Markets stocks data (Deutsche Boerse Group). Numerical results, associated with Lobster and Xetra and Frankfurt Markets stocks data are presented, and explanation and justification of our method of studying of big data in finance are considered. We also describe various stochastic models for mid-price changes in the market and explain how to use these models in practice, highlighting the methodological aspects of using the models.

1.1 Introduction

Finance may be defined as the study of how people allocate scarce resources over time. The outcomes of financial decisions (costs and benefits) are usually spread over time and not known with certainty ahead of time, i.e subject to an element of risk. Decision makers must therefore be able to compare the values of cash-flows at different dates take a probabilistic/stochastic view.

Big data has now become a driver of model building and analysis in a number of areas, including finance. More than half of the markets in today's highly competitive and relentlessly fast-paced financial world now use a limit order book (LOB) mechanism to facilitate trade. Two types of trading in equities are widely practiced today: high-frequency (limit-order and market) trading and statistical arbitrage or market neutral (generalized) pairs trading. These types of trading account for well over two-thirds of the volume traded today. Main problem here is: how to deal with big data arising in electronic markets for algorithmic and high-frequency (milliseconds) trading that contain two types of orders, limit orders and market orders. It is not yet clear how to quantify the systemic risk, or the market instabilities generated by these types of trading. Systemic risk, or instabilities, occur in many complex systems: In

DOI: 10.1201/9781003265986-1

ecology (diversity of species), in climate change, in material behaviour (phase transitions), insurance, finance, etc.

Thus, of the areas of mathematics, namely, financial mathematics comes to the rescue and is based on the idea in making good decisions in the face of uncertainty. As long as uncertainty is involved, the probability theory and stochastic modelling are the main instruments in financial mathematics and in studying big data in finance.

Recently, many new directions appeared in finance:

- Energy Finance – use financial instruments to manage storage impact, seasonality, mean-reversion, illiquidity, decentralized energy markets;
- Environmental Finance – use of financial instruments to protect the ecological environment (climate exchanges for trading greenhouse gases (GHG) in Chicago, Europe, China, Canada, Australia);
- Carbon Finance – investment in GHG emission reduction projects and use financial instruments that are tradable on the carbon markets;
- Weather Derivatives – use financial instruments to reduce risk associated with adverse or unexpected weather conditions (derivatives are non-tradable);
- Renewable Energy Finance – use financial instruments to manage wind, solar, hydro & marine, water, energy, etc.;
- Systemic Risk (very recent);
- Big Data Science, in particular, Bid Data in Finance and Limit Order Books (LOB).

In this book we present a new approach to study big data in finance, specifically, in limit order books (LOB), based on stochastic modelling of price changes associated with high-frequency and algorithmic trading.

Interest in the modelling of limit order markets has recently increased significally. Some research has focused on optimal trading strategies in high-frequency environments: for example, Fodra et al. (2015a) studies such optimal trading strategies in a context where the stock price follows a semi-Markov process, while market orders arrive in the limit order book via a point process correlated with the stock price itself. Cartea et al. (2015a) develops an optimal execution strategy for an investor seeking to execute a large order using both limit and market orders, under constraints on the volume of such orders. The execution of portfolio transactions with the aim of minimizing a combination of volatility risk and transaction costs arising from permanent and temporary market impact considered in Almgren et al. (2001). The paper Avellaneda et al. (2008) studies the optimal submission strategies of bid and ask orders in a limit order book. In Fodra et al. (2015b), a semi-Markov model for the stock price is introduced: the price increments are correlated and equal to arbitrary multiples of the tick size. In Cont et al. (2010), the whole limit order book is modelled (not only the ask and bid queues) via an integer-valued continuous-time Markov chain. The survey paper Gould et al. (2013) highlights the insights that have emerged from the wealth of empirical and theoretical studies of limit order books. Semi-Markov modelling of limit order books was first considered in Swishchuk et al. (2017d) to model the mid-price change with two ticks. The general semi-Markovian modelling of limit order books with n ticks was considered in Swishchuk et al. (2017e). The general compound Hawkes process was first introduced in Swishchuk (2017a) to model the risk process in insurance, and it was studied in detailed in Swishchuk (2018). In the paper Swishchuk (2017b), we first obtained FCLTs and LLNs for so-called general compound Hawkes processes with dependent orders and regime-switching compound Hawkes process. We note that compound Hawkes processes were first applied to limit order books in Swishchuk et al. (2017c). The general compound Hawkes processes were considered in Swishchuk et al. (2017b) and Swishchuk et al. (2018). Multivariate general compound Hawkes pocesses in limit order books have been considered in Quo et al. (2020). For a more thorough literature on limit order markets, we refer to the above cited articles and book, and the references thereof.

The chapter is organized as follows. Section 1.2 introduces big data in finance, namely, limit order books and describes them by Lobster data-academic data for studying LOB, Xetra, and Frankfurt Markets (Deutsche Boerse Group) data on September 23, 2013 and CISCO Data on November 3, 2014. Section 1.3 describes various stochastic models for big data in finance via mid-price changes in the market. Section 1.4 introduces into numerical results associated with Lobster and other data, and explains, illustrates, and justifies our method of studying big data in finance. Methodological aspects of using the models are considered in Section 1.5. Section 1.6 concludes the paper.

1.2 Big Data in Finance: Limit Order Books

1.2.1 Description of Limit Order Books Mechanism

As we mentioned in the Introduction, more than half of the markets in today's highly competitive and relentlessly fast-paced financial world now use a limit order book (LOB) mechanism to facilitate trade.

Let us give a quick description of this mechanism. Orders to buy and sell an asset arrive at an exchange. There are three types of orders:

1. Market buy/sell order – specifies number of shares to be bought/sold at the best available price, right away;
2. Limit buy/sell order – specifies a price and a number of shares to be bought/sold at that price, when possible;
3. Order cancellation – agents who have submitted a limit order may cancel the order before it is executed.

With respect to these types of orders, we have:

- Market orders are executed immediately;
- Limit orders are queued for later execution, but may cancel; thus,
- The limit order book is the collection of queued limit orders awaiting execution or cancellation.

The *bid price* s_t^b is the highest limit buy order price in the book. It is the best available price for a market sell. The *ask price* s_t^a is the lowest limit sell order price in the book. It is the best available price for a market buy. We are interested in the *mid-price*:

$$S_t := \frac{s_t^b + s_t^a}{2}. \tag{1.1}$$

There are hundred thousands of orders for one stock just for one day: e.g., for CISCO data on November 3, 2014, we have 0.5 million price orders for that day, and that is only for 1-level order book, meaning the limit orders sitting at the best bid and ask. And if we take hundreds of stocks on an exchange and not only 1-level orders book, then we will get an example of really big data in finance!

And the question is: can we model this mechanism of trading and describe this big data in finance? And the answer is "Yes". Below we show and explain how to do this for LOBster and other data.

1.2.2 Big Data in Finance: Lobster Data

In this section we give description of Lobster data-academic data for studying LOB. LOBster Data may be found on the following website https://lobsterdata.com/info/DataSamples.php, that contains, in particular, sample files for Amazon, Apple, Google, Intel and Microsoft stocks on June 21, 2012. These sample files are based on the official NASDAQ TotalView-ITCH sample and looks like the one below, see Figure 1.1.

The actual LOBster files may be found on the website http://LOBSTER.wiwi.hu-berlin.de. LOBster generates a "message" and an "order book" file for each active trading day of a selected ticker. The "order book" file contains the evolution of the LOB up to the requested number of levels. The "message" file contains indicators for the type of event causing an update of the LOB in the requested price range. All events are timestamped to seconds after midnight, with decimal precision of at least milliseconds and up to nanoseconds depending on the requested period. For example, for Intel "Message" file is 3.3 MB, and "Order book" file is 4.9 MB. Both files are presented below, see Figure 1.2.

Snapshot of the "Message" file is illustrated on the figure below:

Snapshot of the "Order Book" file is illustrated on the figure below:

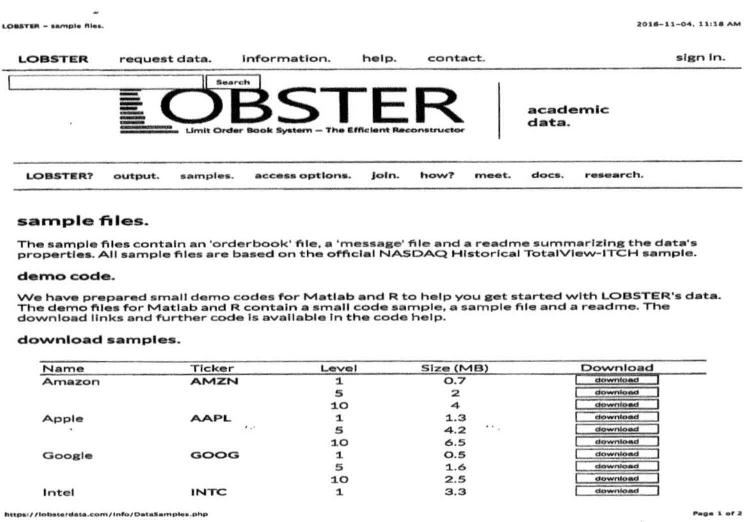

LOBSTER – sample files. 2016-11-04, 11:18 AM

LOBSTER request data. information. help. contact. sign in.

Search

LOBSTER
Limit Order Book System – The Efficient Reconstructor

academic data.

LOBSTER? output. samples. access options. join. how? meet. docs. research.

sample files.

The sample files contain an 'orderbook' file, a 'message' file and a readme summarizing the data's properties. All sample files are based on the official NASDAQ Historical TotalView-ITCH sample.

demo code.

We have prepared small demo codes for Matlab and R to help you get started with LOBSTER's data. The demo files for Matlab and R contain a small code sample, a sample file and a readme. The download links and further code is available in the code help.

download samples.

Name	Ticker	Level	Size (MB)	Download
Amazon	AMZN	1	0.7	download
		5	2	download
		10	4	download
Apple	AAPL	1	1.3	download
		5	4.2	download
		10	6.5	download
Google	GOOG	1	0.5	download
		5	1.6	download
		10	2.5	download
Intel	INTC	1	3.3	download

https://lobsterdata.com/info/DataSamples.php Page 1 of 2

FIGURE 1.1
Lobster Sample Files.

1.2.3 More Big Data in Finance: Xetra and Frankfurt Markets (Deutsche Boerse Group), on September 23, 2013 and CISCO Data on November 3, 2014

Of course, the five stocks (Amazon, Apple, Microsoft, Intel and Google) we have chosen are perhaps the most active (at least on the NASDAQ) and our numerical results might be misleading when considering more typical stocks.

However, we would like to point out that our assumptions about the non-Markovian behaviour of the limit order book and non-exponential distribution of inter-arrival events are valid not only for those five stocks but also for bunches of many others.

We used the financial instruments traded on the Xetra and Frankfurt markets (Deutsche Boerse Group), on September 23, 2013 (http://datashop.deutsche-boerse.com/1016/en).

The description of all instruments is presented in Figure 1.7: the first column gives the German security identification number, the second gives the international security identification number, the third gives the security name, and the last gives the one common name.

To study this set of big data in finance we divided 15 assets, presented in Figure 1.7, by three groups: (1) liquid assets (every 372–542 milliseconds (ms) in average an order arrives), (2) medium liquid assets (every 1392–1415 ms in average an order arrives), and (3) illiquid assets (every 8392–8467 ms in average an order arrives).

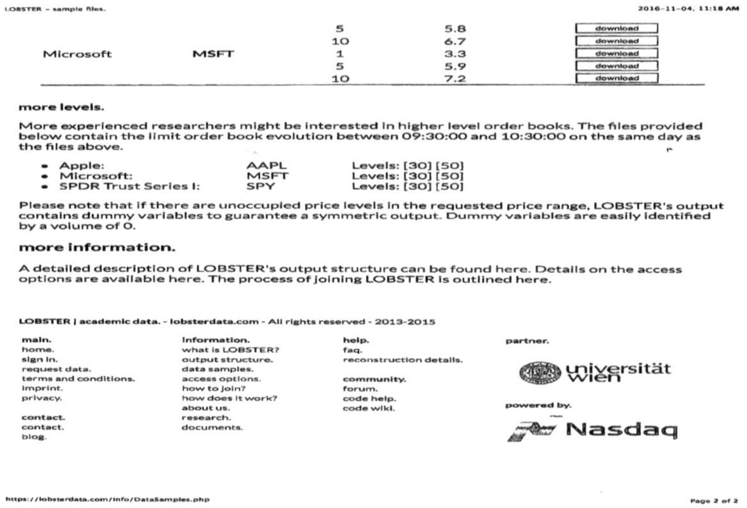

LOBSTER - sample files. 2016-11-04, 11:18 AM

		5	5.8	download
		10	6.7	download
Microsoft	MSFT	1	3.3	download
		5	5.9	download
		10	7.2	download

more levels.

More experienced researchers might be interested in higher level order books. The files provided below contain the limit order book evolution between 09:30:00 and 10:30:00 on the same day as the files above.

- Apple: AAPL Levels: [30] [50]
- Microsoft: MSFT Levels: [30] [50]
- SPDR Trust Series I: SPY Levels: [30] [50]

Please note that if there are unoccupied price levels in the requested price range, LOBSTER's output contains dummy variables to guarantee a symmetric output. Dummy variables are easily identified by a volume of 0.

more information.

A detailed description of LOBSTER's output structure can be found here. Details on the access options are available here. The process of joining LOBSTER is outlined here.

LOBSTER | academic data. - lobsterdata.com - All rights reserved - 2013-2015

main. home. sign in. request data. terms and conditions. imprint. privacy.

contact. contact. blog.

information. what is LOBSTER? output structure. data samples. access options. how to join? how does it work? about us. research. documents.

help. faq. reconstruction details.

community. forum. code help. code wiki.

partner. universität wien

powered by. Nasdaq

https://lobsterdata.com/info/DataSamples.php Page 2 of 2

FIGURE 1.2
Lobster Sample Files.

Moreover, we used even one more set of data, namely, CISCO on November 3, 2014, to show that inter-arrival times between limit orders at the best ask not follow an exponential distribution-see below.

1.3 Stochastic Modelling of Big Data in Finance: Limit Order Books (LOB)

In this section we present various models for the mid-price S_t (see (1), Section 2.1), and show how to study their evolutions. It includes semi-Markovian

TABLE 1.1
Apple Bid: Fitted Weibull and Gamma parameters. 95% confidence intervals in brackets. June 21, 2012.

Apple Bid	$H(1,1)$	$H(1,-1)$	$H(-1,-1)$	$H(-1,1)$
Weibull θ	75.9	180.9	31.5	78.2
	(71.6-80.5)	(172.6-189.7)	(29.5-33.6)	(73.4-83.3)
Weibull k	0.317	0.400	0.271	0.300
	(0.313-0.321)	(0.394-0.405)	(0.267-0.274)	(0.296-0.304)
Gamma θ	2187	1860	2254	2711
	(2094-2284)	(1787-1935)	(2157-2355)	(2592-2835)
Gamma k	0.206	0.276	0.168	0.196
	(0.202-0.210)	(0.271-0.282)	(0.165-0.171)	(0.192-0.199)

TABLE 1.2
Estimated probabilities for book event arrivals. June 21, 2012.

Microsoft	Bid	Ask
$P(1,1)$	0.63	0.60
$P(-1,1)$	0.36	0.41
$P(-1,-1)$	0.64	0.59
$P(1,-1)$	0.37	0.40
$P(1)$	0.49	0.51
$P(-1)$	0.51	0.49

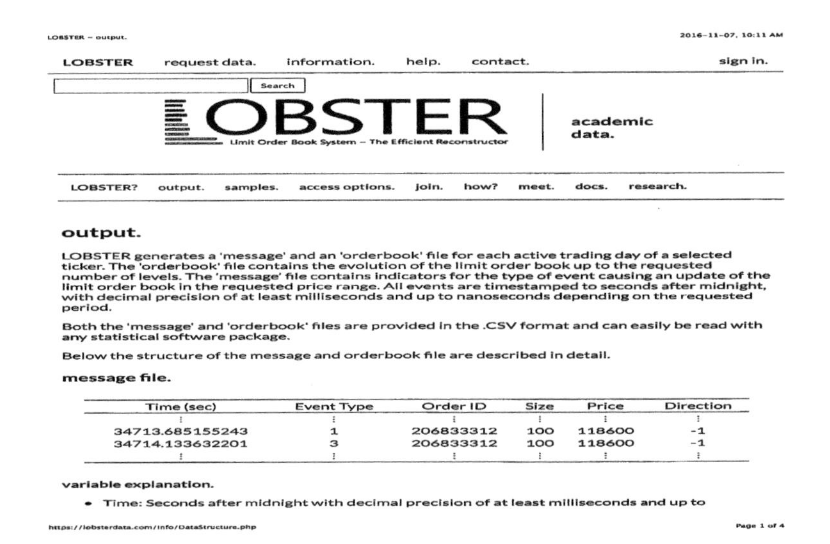

LOBSTER – output. 2016-11-07, 10:11 AM

LOBSTER request data. information. help. contact. sign in.

Search

LOBSTER
Limit Order Book System – The Efficient Reconstructor
academic data.

LOBSTER? output. samples. access options. join. how? meet. docs. research.

output.

LOBSTER generates a 'message' and an 'orderbook' file for each active trading day of a selected ticker. The 'orderbook' file contains the evolution of the limit order book up to the requested number of levels. The 'message' file contains indicators for the type of event causing an update of the limit order book in the requested price range. All events are timestamped to seconds after midnight, with decimal precision of at least milliseconds and up to nanoseconds depending on the requested period.

Both the 'message' and 'orderbook' files are provided in the .CSV format and can easily be read with any statistical software package.

Below the structure of the message and orderbook file are described in detail.

message file.

Time (sec)	Event Type	Order ID	Size	Price	Direction
⋮	⋮	⋮	⋮	⋮	⋮
34713.685155243	1	206833312	100	118600	-1
34714.133632201	3	206833312	100	118600	-1
⋮	⋮	⋮	⋮	⋮	⋮

variable explanation.

- Time: Seconds after midnight with decimal precision of at least milliseconds and up to

https://lobsterdata.com/info/DataStructure.php Page 1 of 4

FIGURE 1.3
Lobster Output Files.

model, general semi-Markovian model, compound Hawkes, general compound Hawkes and non-linear general compound Hawkes models.

1.3.1 Semi-Markov Modelling of LOB

We now specify formally the "state process", which is semi-Markov process, and which will keep track of the state of the limit order book at time t (stock price and sizes of the bid and ask queues), (see [23])

$$\widetilde{L}_t := (S_t, q_t^b, q_t^a),$$

LOBSTER - output. 2016-11-07, 10:11 AM

nanoseconds depending on the period requested
- Event Type:
  - 1: Submission of a new limit order
  - 2: Cancellation (partial deletion of a limit order)
  - 3: Deletion (total deletion of a limit order)
  - 4: Execution of a visible limit order
  - 5: Execution of a hidden limit order
  - 7: Trading halt indicator (detailed information below)
- Order ID: Unique order reference number
- Size: Number of shares
- Price: Dollar price times 10000 (i.e. a stock price of $91.14 is given by 911400)
- Direction:
  - -1: Sell limit order
  - 1: Buy limit order
  - Note: Execution of a sell (buy) limit order corresponds to a buyer (seller) initiated trade, i.e. buy (sell) trade.

order book file.

Ask Price 1	Ask Size 1	Bid Price 1	Bid Size 1	Ask Price 2	Ask Size 2	Bid Price 2	Bid Size 2	...
⋮	⋮	⋮	⋮	⋮	⋮	⋮	⋮	⋮
1186600	9484	118500	8800	118700	22700	118400	14930	...
1186600	9384	118500	8800	118700	22700	118400	14930	...
⋮	⋮	⋮	⋮	⋮	⋮	⋮	⋮	⋮

variable explanation.

- Ask Price 1: Level 1 ask price (best ask price)
- Ask Size 1: Level 1 ask volume (best ask volume)
- Bid Price 1: Level 1 bid price (best bid price)
- Bid Size 1: Level 1 bid volume (best bid volume)
- Ask Price 2: Level 2 ask price (second best ask price)
- Ask Size 2: Level 2 ask volume (second best ask volume)
- ...

further details.

level.

The term 'level' refers to occupied price levels. The difference between two levels in the LOBSTER output is not necessarily the minimum tick size.

https://lobsterdata.com/info/DataStructure.php Page 2 of 4

FIGURE 1.4
Lobster Output Files.

where $S_t := (s_t^a + s_t^b)/2$ is a mid-price,

$$S_t := s_0 + \sum_{i=1}^{N(t)} X_k, \tag{1.2}$$

$X_k = \{+\delta, -\delta\}$, δ-tick size, q_t^a, q_t^b are sizes of bid and ask queues, $N(t)$-number of price changes (renewal process). We note, that X_k takes here only two values: $X_k \in \{+\delta, -\delta\}$, thus X_k is a two-state Markov chain. In this setting, our process L_t is a semi-Markov process, meaning that:

- inter-arrival times between book events (limit orders, market orders, order cancellations) are not exponentially distributed, but may have arbitrary distribution,
- the arrival of a new book event at the bid or the ask is not independent from the previous events-they connected in a Markov chain.

In the context of many papers, including Cont and Larrard (SIAM J. Finan. Math., 2013), [7], this process $\widetilde{L}_t$ was proved to be Markovian. Here, we will need to "add" to this process the process (V_t^b, V_t^a) keeping track of the nature of the last book event at the bid and the ask to make it Markovian: in this sense we can view it as being semi-Markovian. The process:

$$L_t := (S_t, q_t^b, q_t^a, V_t^b, V_t^a)$$

is Markovian, where V_t^b, V_t^a are processes for events of increase or decrease

34200.0175	5	0	1	2238200	-1
34200.1896	1	11885113	21	2238100	1
34200.1902	4	11885113	21	2238100	1
34200.1902	4	11534792	26	2237500	1
34200.3728	5	0	100	2238400	-1
34200.3757	5	0	100	2238400	-1
34200.384	5	0	100	2238600	-1
34200.3858	5	0	100	2238600	-1
34200.3872	5	0	100	2239200	-1
34200.3885	5	0	100	2239300	-1
34200.3914	4	14585251	100	2239500	-1
34200.3914	4	3911376	20	2239600	-1
34200.3914	4	16202496	286	2239600	-1
34200.3935	4	2135294	100	2239900	-1
34200.3944	1	16207239	100	2239900	-1
34200.3985	5	0	100	2239700	-1
34200.3986	5	0	100	2239700	-1
34200.4003	5	0	100	2239700	-1
34200.4015	5	0	10	2239700	-1
34200.4015	4	16207239	90	2239900	-1
34200.4019	1	16208720	50	2239900	-1
34200.403	4	16207239	10	2239900	-1
34200.403	4	16208720	50	2239900	-1
34200.403	4	1365373	13	2240000	-1
34200.403	4	1847685	27	2240000	-1
34200.4079	4	1847685	73	2240000	-1
34200.4079	4	2051705	20	2240000	-1
34200.4079	4	3578212	4	2240000	-1
34200.4079	4	3581197	3	2240000	-1
34200.4111	4	3581197	7	2240000	-1
34200.4111	4	3591155	50	2240000	-1
34200.4111	4	3689544	43	2240000	-1
34200.4121	4	3689544	57	2240000	-1
34200.4121	4	3920363	6	2240000	-1
34200.4121	4	3920364	23	2240000	-1
34200.4121	4	3920367	6	2240000	-1
34200.4121	4	3920371	54	2240000	-1
34200.4121	4	3920373	54	2240000	-1
34200.4172	4	3920374	15	2240000	-1
34200.4172	4	4631442	76	2240000	-1
34200.4172	4	4631569	20	2240000	-1
34200.4172	4	4632895	15	2240000	-1

FIGURE 1.5
Message File.

the bid or ask queue by 1, respectively, meaning that we split the book events into 2 different types: limit orders that increase the size of the corresponding bid or ask queue, and market orders/order cancellations that decrease the size of the corresponding queue. (See [23] for more details.)

1.3.2 General Semi-Markov Modelling of LOB

The paper [24] considers a general semi-Markov model for limit order books with two states, which incorporates price changes that are not fixed to one tick. Furthermore, we introduced an even more general case of the semi-Markov model for limit order books that incorporates an arbitrary number of states for the price changes:

$$S_t := s_0 + \sum_{i=1}^{N(t)} a(X_k). \tag{1.3}$$

For both cases, the justifications, diffusion limits, where we used martingale approach (see [23]), implementations and numerical results are presented for different limit order book data: Apple, Amazon, Google, Microsoft, Intel on 2012/06/21 and Cisco, Facebook, Intel, Liberty Global, Liberty Interactive, Microsoft, Vodafone from 2014/11/03 to 2014/11/07.

2239500	100	2231800	100
2239500	100	2238100	21
2239500	100	2237500	100
2239500	100	2237500	74
2239500	100	2237500	74
2239500	100	2237500	74
2239500	100	2237500	74
2239500	100	2237500	74
2239500	100	2237500	74
2239500	100	2237500	74
2239600	306	2237500	74
2239600	286	2237500	74
2239900	100	2237500	74
2240000	1451	2237500	74
2239900	100	2237500	74
2239900	100	2237500	74
2239900	100	2237500	74
2239900	100	2237500	74
2239900	100	2237500	74
2239900	10	2237500	74
2239900	60	2237500	74
2239900	50	2237500	74
2240000	1451	2237500	74
2240000	1438	2237500	74
2240000	1411	2237500	74
2240000	1338	2237500	74
2240000	1318	2237500	74
2240000	1314	2237500	74
2240000	1311	2237500	74
2240000	1304	2237500	74
2240000	1254	2237500	74
2240000	1211	2237500	74
2240000	1154	2237500	74
2240000	1148	2237500	74
2240000	1125	2237500	74
2240000	1119	2237500	74
2240000	1065	2237500	74
2240000	1011	2237500	74
2240000	996	2237500	74
2240000	920	2237500	74
2240000	900	2237500	74
2240000	885	2237500	74

FIGURE 1.6
Order Book File.

1.3.3 Modelling of LOB with a Compound Hawkes Processes

The Hawkes process (HP) is so-called self-exciting point process which means that it is a point process with a stochastic intensity which, through its dependence on the history of the process, captures the temporal and cross sectional dependence of the event arrival process, as well as the self-exciting property observed in empirical analysis (see [11, 12]).

In the paper [22] we introduced two new Hawkes processes, namely, compound and regime-switching compound Hawkes processes, to model the price processes in limit order books:

$$S_t := s_0 + \sum_{i=1}^{N(t)} X_k, \tag{1.4}$$

where $N(t)$ is a Hawkes process. We note, that X_k takes here also only two values: $X_k \in \{+\delta, -\delta\}$, thus X_k is a two-state Markov chain.

We proved Law of Large Numbers and Functional Central Limit Theorems (FCLT) for both processes. The two FCLTs are applied to limit order books where we use these asymptotic methods to study the link between price volatility and order flow in our two models by using the diffusion limits of these price processes. The volatilities of price changes are expressed in terms of parameters describing the arrival rates and price changes. We also presented some numerical examples.

WKN	ISIN	INSTRUMENT NAME	COMMON NAME
		LIQUID ASSETS:	
A1JEAN	LU0665646815	UBS-ETF-MSCI EU.IN.2035 I	UBS-ETF MSCI Europe Infrastructure I
A1JVYM	IE00B7KMTJ66	UBS(I)ETF-SOL.G.P.GD IDDL	Solactive Global Pure Gold Miners UCITS ETF
A1JEAJ	LU0665646229	UBS-ETF-MSCI JA.IN.2035 I	UBS-ETF MSCI Japan Infrastructure I
A1JVYN	IE00B7KYPQ18	UBS(I)ETF-SOL.G.O.EQ.IDDL	Solactive Global Oil Equities UCITS ETF I
A1JVCB	IE00B7KL1H59	UBS(I)ETF-MSCI WORLD IDDL	MSCI World UCITS ETF I
		MEDIUM LIQUID ASSETS	
ETC057	DE000ETC0571	COMMERZBANK ETC UNL.	Coba ETC -3x WTI Oil Daily Short Index
ETC015	DE000ETC0159	COMMERZBANK ETC UNL.	Coba ETC -1x Gold Daily Short Index
ETC030	DE000ETC0308	COMMERZBANK ETC UNL.	Coba ETC 4x Brent Oil Daily Long Index
A0X8SE	IE00B3VWMM18	ISHSVII-MSCI EMU SC U.ETF	iShares MSCI EMU Small Cap UCITS ETF
A0JMFG	FR0010296061	LYXOR ETF MSCI USA D-EO	Lyxor UCITS ETF MSCI USA D-EUR
		ILLIQUID ASSESTS	
A1JB4P	DE000A1JB4P2	I.II-IS. D.J.G.S.S.UTS DZ	iShares Dow Jones Global Sustainability Screened UCITS ETF
630500	DE0006305006	DEUTZ AG O.N.	DEUTZ AG O.N.
A1T8GD	IE00B9CQXS71	SPDR S+P GL.DIV.ARIST.ETF	SPDR® S&P® Global Dividend Aristocrats UCITS ETF
851144	US3696041033	GENL EL. CO. DL -,06	General Electric STK
113541	DE0001135416	BUNDANL.V. 10/20	Bundesrepublik Deutschland 2,250% 9/2020 BOND

FIGURE 1.7
15 Stocks from Deutsche Boerse Group.

1.3.4 Modelling of LOB with a General Compound Hawkes Processes

We generalized the compound Hawkes process model in (4) for the general compound Hawkes process (see [17, 20]):

$$S_t := s_0 + \sum_{i=1}^{N(t)} a(X_k), \tag{1.5}$$

where $N(t)$ is a Hawkes process (see [11, 12]), X_k is n-state Markov chain, a(x) is a bounded function on $X = \{1, 2, ..., n\}$.

With regards to these general compound Hawkes processes, we proved a Law of Large Numbers (LLN) and a Functional Central Limit Theorems (FCLT) for several specific variations. We applied several of these FCLTs to limit order books to study the link between price volatility and order flow, where the volatility in mid-price changes is expressed in terms of parameters describing the arrival rates and mid-price process.

1.3.5 Modelling of LOB with a Non-linear General Compound Hawkes Processes

The latter general compound Hawkes process model in (5) was generalized in [17] for non-linear general compound Hawkes process:

$$S_t := s_0 + \sum_{i=1}^{N(t)} a(X_k), \tag{1.6}$$

where $N(t)$ is a non-linear Hawkes process (see [3]), X_k is n-state Markov chain, a(x) is a bounded function on $X = \{1, 2, ..., n\}$.

1.3.6 Modelling of LOB with a Multivariable General Compound Hawkes Processes

In the paper [16], we focused on various new multivariate Hawkes processes. We constructed multivariate general compound Hawkes processes (MGCHP) and investigate their properties in limit order books:

$$S_t^j := s_0^j + \sum_{i=1}^{N^j(t)} a(X_k^j), \tag{1.7}$$

where S_j is a set of j different assets, $N^j(t)$ are independent Hawkes processes, X_k^j are j independent n-state Markov chains, $a(x)$ is a bounded function on $X = \{1, 2, .., n\}$, $j = 1, 2, ..., m$.

For this model, we proved the Law of Large Number (LLN) and two Functional Central Limit Theorems (FCLT). The latter provides insights into the link between price volatilities and order flows in limit order books with several assets. Numerical examples with Intel and Microsoft trading data are also provided in this paper.

1.4 Illustration and Justification of Our Method to Study Big Data in Finance

In this section we give the justification of our semi-Markovian model for LOB, considered in Section 1.3.1, (2). Justifications for using other models for LOB in Section 1.3.2–1.3.6, (4)–(7), may be found in the following papers [16, 19, 18, 20, 22, 24].

Many papers, including R. Cont and A. de Larrard (SIAM J. Finan. Math, 2013), [7], introduced a tractable stochastic model, The Markovian model, to be precise, for the dynamics of a limit order book, computing various quantities of interest such as the probability of a price increase or the diffusion limit of the price process.

Among the various assumptions made in this article, we seek to challenge two of them while preserving analytical tractability:

- the inter-arrival times between book events (limit orders, market orders, order cancellations) are assumed to be independent and exponentially distributed;
- the arrival of a new book event at the bid or the ask is independent from the previous events.

As suggested by empirical observations, we extend [7] framework to:

1) *arbitrary distributions* for book events inter-arrival times (possibly non-exponential), and
2) both the nature of a new book event and its corresponding inter-arrival time *depend on the nature of the previous book event.*

We do so by stochastic modelling of the dynamics of the bid and ask queues by a stochastic process that is not-Markovian, as highlighted in Section 1.3 above.

We justify and illustrate our approach by calibrating our model to the five stocks Amazon, Apple, Google, Intel and Microsoft on June 21, 2012 (Courtesy: *https://lobster.wiwi.hu-berlin.de/info/DataSamples.php*), Xetra and Frankfurt Markets stocks (Deutsche Boerse Group), on September 23, 2013, and CISCO Data on November 3, 2014.

1.4.1 Numerical Results: Lobster Data (Apple, Google and Microsoft Stocks)

Regarding our first observation above (1), we calibrate the empirical CDF's $H^a(i,j,\cdot)$, $H^b(i,j,\cdot)$ to the Gamma and Weibull distributions (which are generalizations of the exponential distribution). We perform a maximum likelihood estimation of the Weibull and Gamma parameters for each one of the empirical distributions $H^a(i,j,\cdot)$, $H^b(i,j,\cdot)$ (together with a 95% confidence interval for the parameters).

When calibrating the empirical distributions of the inter-arrival times to the Weibull and Gamma distributions (Amazon, Apple, Google, Intel, and Microsoft on June 21, 2012), we find that the shape parameter is in all cases significantly different than 1 ($\sim$0.1 to 0.3), which suggests that the exponential distribution is typically not rich enough to capture the behaviour of these inter-arrival times, see Table 1.1 for Apple Bid.

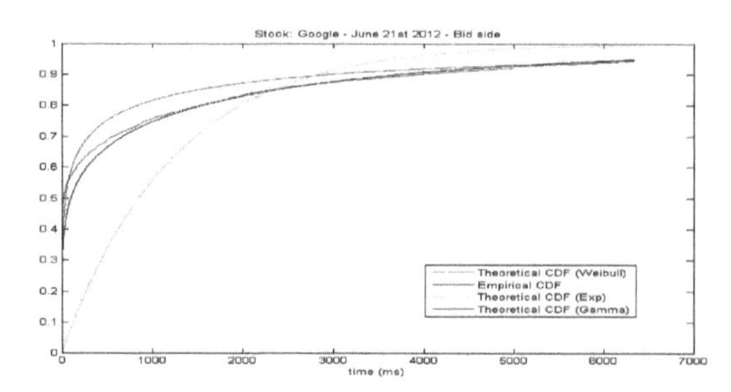

FIGURE 1.8
Comparison of CDFs for Empirical, Theoretical Weibull, Gamma, and Exponential Distributions for Google Stock.

Analogical results are valid for other Lobster data (see [23] for more details).

Comparison of CDFs for Empirical and Theoretical Weibull, Gamma, and Exponential distributions (stock-Google-June 21, 2012-Bid side) is presented on Figure 1.8.

Regarding our second observation above (2), we split the book events into 2 different types: limit orders that increase the size of the corresponding bid or ask queue, and market orders/order cancellations that decrease the size of the corresponding queue. Assimilating the former to the type "+1" and the latter to the type "−1", we find empirically that the probability to get an event of type "±1" is not independent of the nature of the previous event. Indeed, we present below the estimated transition probabilities between book events at the ask and the bid for the stock Microsoft on June 21, 2012. It is seen that the *unconditional* probabilities $P(1)$ and $P(-1)$ to obtain respectively an event of type "+1" and "−1" are relatively close to 1/2. Nevertheless, denoting $P(i,j)$ the *conditional* probability to obtain an event of type j given that the last event was of type i, we observe that $P(i,j)$ can significantly depend on the previous event i. For example, on the bid side, $P(1,1) = 0.63$, whereas $P(-1,1) = 0.36$:

1.4.2 Numerical Results: Xetra and Frankfurt Markets stocks (Deutsche Boerse Group), on September 23, 2013

Comparisons of ask PDF for the 5 Liquid assets, for the 5 Illiquid assets and for the 5 Medium Liquid assets (see Section 1.2.3, Table 1.7) show that the best fit for these set of assets gives the Burr type XII distribution $F(x) = 1-(1+x^c)^{-k}$, $(x>0, c>0, k>0$, both c and k are shape parameters, *not exponential.*

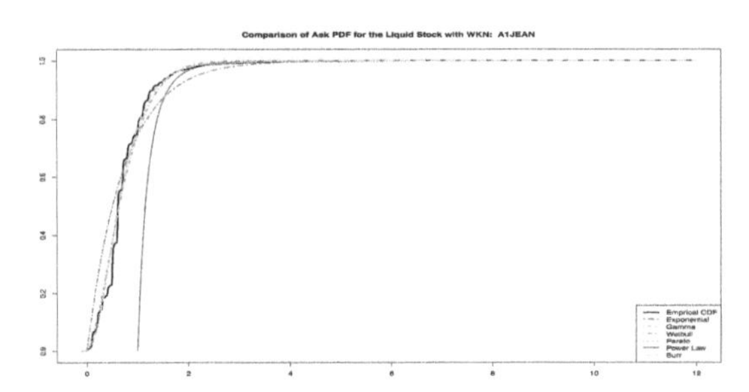

FIGURE 1.9
Comparison of Ask PDF for Liquid Stock with WKN: A1JEAN.

We note that all graphs contains comparison for empirical, exponential, Gamma, Weibull, Pareto, Power law, and Burr distributions (7 in total), see Figure 1.9.

1.4.3 Numerical Results: CISCO Data, November 3, 2014

Moreover, we used even one more set of data, namely, CISCO on November 3, 2014, to show that inter-arrival times between limit orders at the best ask does not follow an exponential distribution (see Figure 1.9).

Regarding the estimated probabilities (our observation 2)) for this set of data, CISCO data on November 3, 2014 and estimated the transition probabilities between book events at the best ask, using similar approach we have used for Microsoft on June 21, 2012 (see Table 1.2). We get the entries presented in Table 1.3.

As we can see, for the CISCO data, the probabilities $P(i, j)$ also simnifically depend on the previous event i $(i, j = 1, 2)$.

1.5 Methodological Aspects of Using the Models

Big data in science, including finance, usually bring challenges in handling the database. Here, we briefly discuss the methodology of handling the database in

TABLE 1.3
Estimated probabilities for book event arrivals for CISCO, November 3, 2014.

CISCO	Ask
$P(1, 1)$	0.599
$P(-1, 1)$	0.396
$P(-1, -1)$	0.604
$P(1, -1)$	0.401

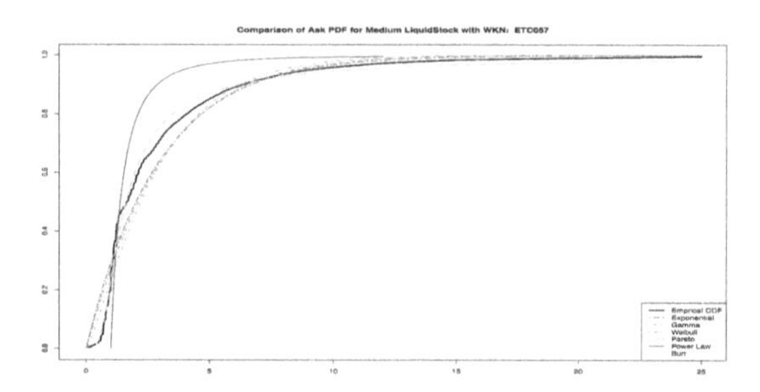

FIGURE 1.10
Comparison of Ask PDF for Medium Stock with WKN: ETC057.

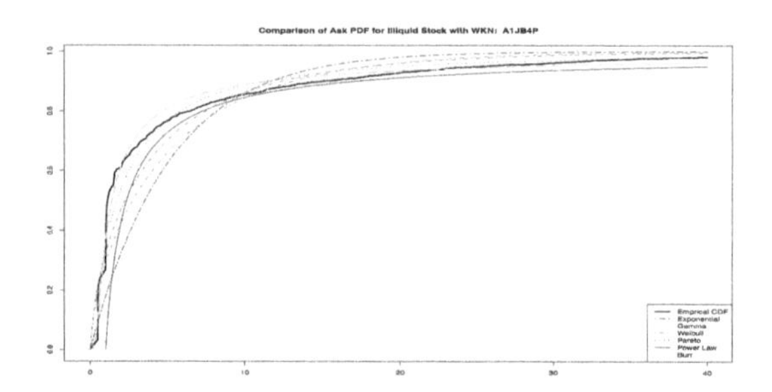

FIGURE 1.11
Comparison of Ask PDF for Illiquid Stock with WKN: A1JB4P.

LOB and details on statistical tests. To the author's opinion, the idea of linking the presented stochastic model to an existing and available LOB database is of great interest to practitioners or other researchers. Thus, below there is a description in details of how to handle the database in LOB, how to estimate the parameters of our models, and how to apply the model to some existing examples. Of course, the idea of our methodology is to check our main suggestions 1) and 2) in Section 4 with respect to both 1) arbitrary distributions for book events inter-arrival times (possibly non-exponential), and 2) dependency of both the nature of a new book event and its corresponding inter-arrival time on the nature of the previous book event, respectively.

i) Handling the Database in LOB/Checking the Suggestion 2), Section 1.4: split the book events into 2 different types: limit orders that increase the size of the corresponding bid or ask queue, and market orders/order cancellations that decrease the size of the corresponding queue. Assimilating the former to the type "+1" and the latter to the type "−1", find empirically using law of large numbers, the probability to get an event of type "±1". Denoting $P(i,j)$ the *conditional* probability to obtain an event of type j given that the last event was of type i, observe that $P(i,j)$ can significantly depend on the previous event i $(i,j = \pm 1)$.

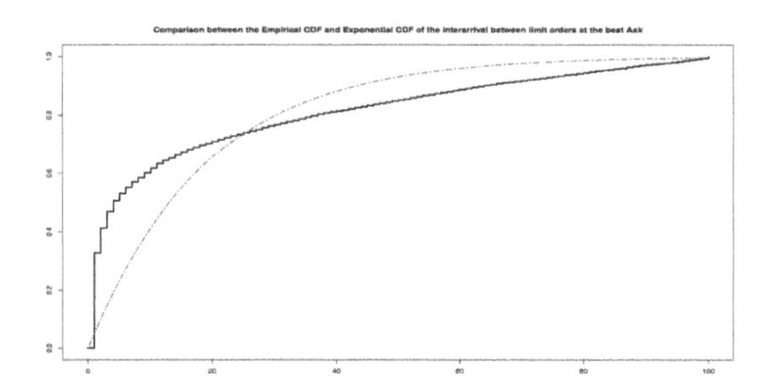

FIGURE 1.12
Comparison between Empirical and Exponential PDFs for CISCO Stock.

ii) Estimation the Parameters of our Model/Checking the Suggestion 1), Section 1.4: calibrate the empirical CDFs to some arbitrary distributions that you suggest for your inter-arrival times. Perform maximum likelihood estimation for the parameters of those distributions together with a 95% confidence interval for the parameters. Plot theoretical CDF for exponential distribution, empirical CDF and theoretical CDFs for those arbitrary distributions and compare them. The CDF that is the most close to empirical CDF is the right one.

iii) Examples of Application: To support the suggestion **i)** above, as we have seen from Table 1.3 and Table 1.4 for Microsoft and CISCO data, respectively, that the probability is not independent of the nature of the previous event.

To support suggestion **ii)** above, we calibrated the empirical CDF's $H^a(i,j,\cdot)$, $H^b(i,j,\cdot)$ to the Gamma and Weibull distributions (which are generalizations of the exponential distribution). We performed a maximum likelihood estimation of the Weibull and Gamma parameters for each one of the empirical distributions $H^a(i,j,\cdot)$, $H^b(i,j,\cdot)$ (together with a 95% confidence interval for the parameters). As we can see from Table 1.1, Section 1.4.1, the shape parameter k is always significantly different than 1 ($\sim$0.1 to 0.3), which indicates that the exponential distribution is not rich enough to fit our observations. To illustrate this, we presented the empirical CDF of $H(1,-1)$ in the case of Google Bid, Figure 1.8, and we see that Gamma and Weibull allow to fit the empirical CDF in a much better way than exponential.

Similar reasonings are valid with respect to Xetra and Frankfurt Markets stocks (Deutsche Boerse Group), on September 23, 2013, Section 1.4.2, Figures 1.9–1.11, and CISCO Data, November 3, 2014, Section 1.4.3, Figure 1.12.

1.6 Conclusion

In this introductory chapter, we presented an overview of stochastic modelling of big data in finance (including limit order books), based on price changes modelling associated with high-frequency and algorithmic trading. We introduced a big data in finance through the limit order books (LOB), and described them by Lobster data, academic data for studying LOB, and Xetra and Frankfurt Markets stocks data (Deutsche Boerse Group). Numerical results, associated with Lobster and Xetra and Frankfurt Markets stocks data, have been also presented, and explanation and justification of our method of studying of big data in finance have been considered. We also described various stochastic models for mid-price changes in the market, and explained how to use these models in practice, highlighting the methodological aspects of using the models.

Bibliography

[1] Almgren, R. and Chriss, N. (2001): *Optimal execution of portfolio transactions*, J. Risk, 3, pp. 5–40.

[2] Avellaneda, M. and Stoikov, S. (2008): *High-frequency trading in a limit order book*, Quantitative Finance, 8(3), pp. 217–224.

[3] Brémaud, P. and Massoulié, L. (1996): *Stability of nonlinear Hawkes processes.* The Annals of Probab., 24(3), 1563–1588.

[4] Cartea, A., Jaimungal, S. and Penalva, J. *Algorithmic and High-Frequency Trading*, (2015a), Cambridge: Cambridge University Press.

[5] Cartea, A. and Jaimungal, S. (2015b): *Optimal execution with limit and market orders*, Quantitative Finance, 15, pp. 1279–1291.

[6] Cont, R., Stoikov, S. and Talreja, S. (2010): *A stochastic model for order book dynamics*, Operations Research, 58, pp. 549–563.

[7] Cont, R. and de Larrard, A. (2013): *Price dynamics in a Markovian limit order market.* SIAM J. Finan. Math., 4, pp. 1–25.

[8] Fodra, P. and Pham, H. (2015a): *Semi Markov model for market microstructure*, Appl. Math. Finance, 22, pp. 261–295.

[9] Fodra, P. and Pham, H. (2015b): *High frequency trading and asymptotic for small risk aversion in a Markov renewal model*, SIAM J. Finan. Math., 6.

[10] Gould, M., Porter, M., Williams, S., McDonald, M., Fenn, D. and Howison, S. (2013): Limit order books, *Quantitative Finance*, 13(11).

[11] Hawkes, A. (1971): Spectra of some self-exciting and mutually exciting point processes. *Biometrica*, 58, pp. 83–90.

[12] Hawkes, A. and Oakes, D. (1974): A cluster process representation of a self-exciting process. *J. Applied Probab.*, 11, pp. 493–503.

[13] Deutsche Boerse Group: http://datashop.deutsche-boerse.com/1016/en) on September 23, 2013.

[14] LOBster Data: https://lobsterdata.com/info/DataSamples.php.

[15] LOBster Files: http://LOBSTER.wiwi.hu-berlin.de.

[16] Quo, Q. and Swishchuk, A. (2020): Multivariate general compound Hawkes processes and their applications in limit order books. *Wilmott*, March 2020.

[17] Swishchuk, A. and Huffman, A. (2018): General compound Hawkes processes in limit order books. *Mathematics in Science and Industry*. Submitted. Available on arXiv: https://arxiv.org/abs/1812.02298

[18] Swishchuk, A. (2018): Risk model based on general compound Hawkes process. *Wilmott*, v. 2018, Issue 94, MARCH 2018. Also available on arXiv: http://arxiv.org/abs/1706.09038.

[19] Swishchuk, A. (2017a): Risk model based on compound Hawkes process. Abstract, IME 2017, Vienna (https://fam.tuwien.ac.at/ime2017/program.php).

[20] Swishchuk, A. (2017b): General Compound Hawkes Processes in Limit Order Books. *Working Paper*, U of Calgary, 32 pages, June 2017. Available on arXiv: http://arxiv.org/abs/1706.07459.

[21] Swishchuk, A., Stochastic Modelling of Big Data in Finance. (2020): *Methodology and Computing in Applied Probability*, 22, pp. 1613–1630.

[22] Swishchuk, A., Chavez-Casillas, J., Elliott, R. and Remillard, B. (2017c): Compound Hawkes processes in limit order books. *Financial Mathematics, Volatility and Covariance Modelling*. Routledge: Taylor & Francis Group. 2018. Available also on SSRN:
https://papers.ssrn.com/sol3/papers.cfm?abstract_id=2987943

[23] Swishchuk, A. and Vadori, N. (2017): A semi-Markovian modelling of limit order markets. *SIAM J. Finan. Math.*, 8, pp. 240–273.

[24] Swishchuk, A., Cera, K., Hofmeister, T. and Schmidt, J. (2017e): General semi-Markov model for limit order books. *Intern. J. Theoret. Applied Finance*, 20, 1750019.

[25] Swishchuk, A. and Vadori, N. (2015): Strong law of large numbers and central limit theorems for functionals of inhomogeneous Semi-Markov processes. *Stochastic Analysis and Applications*, 13(2), pp. 213–243.

Part I

Semi-Markovian Modelling of Big Data in Finance

2

A Semi-Markovian Modelling of Big Data in Finance

This chapter is devoted to a semi-Markovian modelling of LOB. R. Cont and A. de Larrard [11] introduced a tractable stochastic model for the dynamics of a limit order book, computing various quantities of interest such as the probability of a price increase or the diffusion limit of the price process. As suggested by empirical observations, we extend their framework to (1) arbitrary distributions for book events inter-arrival times (possibly non-exponential) and (2) both the nature of a new book event and its corresponding inter-arrival time depend on the nature of the previous book event. We do so by resorting to Markov renewal processes to model the dynamics of the bid and ask queues. We keep analytical tractability via explicit expressions for the Laplace transforms of various quantities of interest. We justify and illustrate our approach by calibrating our model to the five stocks Amazon, Apple, Google, Intel, Microsoft on June 21, 2012, to the 15 stocks from Deutsche Börse Group (September 23, 2013) and to CISCO asset (November 3, 2014). As in [11], the bid-ask spread remains constant equal to one tick, only the bid and ask queues are modeled (they are independent from each other and get reinitialized after a price change), and all orders have the same size. We discuss possible extensions of our model for the case when the spread is not fixed, including the diffusion limit of the price dynamics in this case, and we also discuss stochastic optimal control and market making problems.

2.1 Introduction

Recently, interest in the modelling of limit order markets has increased. Some research has focused on optimal trading strategies in high-frequency environments: for example [15] studies such optimal trading strategies in a context where the stock price follows a semi-Markov process, while market orders arrive in the limit order book via a point process correlated with the stock price itself. [7] develops an optimal execution strategy for an investor seeking to execute a large order using both limit and market orders, under constraints on the volume of such orders. [29] studies optimal execution strategies for the

DOI: 10.1201/9781003265986-2

purchase of a large number of shares of a financial asset over a fixed interval of time. [8] provides an explicit closed-form startegy for an investor who executes a large order when market order-flow from all agents, including the investor's own trades, has a permanent price impact. The execution of portfolio transactions with the aim of minimizing a combination of volatility risk and transaction costs arising from permanent and temporary market impact considered in [1]. The paper [2] studies the optimal submission strategies of bid and ask orders in a limit order book. In [18], the authors proposed a framework for studying optimal market making policies in a limit order book (LOB). The bid-ask spread of the LOB is modelled by a Markov chain with finite values, multiple of the tick size, and subordinated by the Poisson process of the tick-time clock. The paper [20] provided empirical restrictions of a model of optimal order submissions in a limit order market.

On the other hand, another class of articles has aimed at modelling either the high-frequency dynamics of the stock price itself, or the various queues of outstanding limit orders appearing on the bid and the ask side of the limit order book, resulting in specific dynamics for the stock price. In [14], a semi-Markov model for the stock price is introduced: the price increments are correlated and equal to arbitrary multiples of the tick size. The correlation between these price increments occurs via their sign only, and not their (absolute) size. In [10], the whole limit order book is modeled (not only the ask and bid queues) via an integer-valued continuous-time Markov chain. Using a Laplace analysis, they compute various quantities of interest such as the probability that the mid-price increases, or the probability that an order is executed before the mid-price moves. A detailed section on parameter estimation is also presented. The paper [19] examined whether the limit order book is informative about future price changes and whether specialists use this information when trading. They used order quantities as well as option values to capture the information content of the limit order book. The article [28] presented an explicitly dynamic model of the limit order book in a one-tick market. The paper [30] considered a model of an order-driven market where fully strategic, symmetrically informed liquidity traders dynamically choose between limit and market orders, trading off execution price and waiting costs. A law of large numbers for limit order books was established in [21]. Starting from order arrival and cancelation rates for all price levels, the authors showed that the LOB dynamics can be described by a coupled PDE: ODE system when tick and order sizes tend to zero while arrival rates tend to infinity in a particular way. A key insight is that the scaling limit requires two time scales: a fast time scale for passive order arrivals and a comparably slow time scale for active order arrivals. We also mention the paper [12] where a model for the dynamics of a limit order book in a liquid market was proposed where buy and sell orders are submitted at high frequency. The authors derived a functional central limit theorem for the joint dynamics of the bid and ask queues and show that, when the frequency of order arrivals is large, the intraday dynamics of the limit order book may be approximated by a

Markovian jump-diffusion process in the positive orthant, whose characteristics are explicitly described in terms of the statistical properties of the underlying order flow. In the textbook [6], the authors developed models for algorithmic trading in contexts such as executing large orders, market making, targeting volume-weighted average price and other schedules, trading pairs or collection of assets, and executing in dark pools. The survey paper [17] highlights the insights that have emerged from the wealth of empirical and theoretical studies of limit order books. For a more thorough literature on limit order markets, we refer to the above cited articles and book, and the references thereof.

The starting point of the present manuscript is the article [11], in which a stochastic model for the dynamics of the limit order book is presented, and their approach was motivated by the observation that, if one is primarily interested in the dynamics of the price, it is sufficient to focus on the dynamics of the best bid and ask queues. And, indeed, as empirical evidence shows (see, e.g., [3]), most of the order flow is directed at the best bid and ask prices. Only the bid and ask queues are modelled in [11] (they are independent from each other and get reinitialized after a price change), the bid-ask spread remains constant equal to one tick and all orders have the same size. Their model is analytically tractable, and allows them to compute various quantities of interest such as the distribution of the duration between price changes, the distribution and autocorrelation of price changes, the probability of an upward move in the price and the diffusion limit of the price process. Among the various assumptions made in this article, we seek to challenge two of them while preserving analytical tractability:

1. the inter-arrival times between book events (limit orders, market orders, order cancellations) are assumed to be independent and exponentially distributed.
2. the arrival of a new book event at the bid or the ask is independent from the previous events.

Assumption 1 is relatively common among the existing literature ([16], [27], [9], [13], [26], [32], [10]). Nevertheless, as it will be shown later, when calibrating the empirical distributions of the inter-arrival times to the Weibull and Gamma distributions (Amazon, Apple, Google, Intel and Microsoft on June 21, 2012), we find that the shape parameter is in all cases significantly different than 1 (~ 0.1 to 0.3), which suggests that the exponential distribution is typically not rich enough to capture the behaviour of these inter-arrival times.

Regarding Assumption 2, we split the book events into 2 different types: limit orders that increase the size of the corresponding bid or ask queue, and market orders/order cancellations that decrease the size of the corresponding queue. Assimilating the former to the type "+1" and the latter to the type "−1", we find empirically that the probability to get an event of type "±1" is not independent of the nature of the previous event. Indeed, we present below the estimated transition probabilities between book events at the ask and the

TABLE 2.1
Estimated probabilities for book event arrivals. June 21, 2012.

Microsoft	Bid	Ask
$P(1,1)$	0.63	0.60
$P(-1,1)$	0.36	0.41
$P(-1,-1)$	0.64	0.59
$P(1,-1)$	0.37	0.40
$P(1)$	0.49	0.51
$P(-1)$	0.51	0.49

bid for the stock Microsoft on June 21, 2012. It is seen that the *unconditional* probabilities $P(1)$ and $P(-1)$ to obtain respectively an event of type "+1" and "−1" are relatively close to 1/2, as in [11]. Nevertheless, denoting $P(i,j)$ the *conditional* probability to obtain an event of type j given that the last event was of type i, we observe that $P(i,j)$ can significantly depend on the previous event i. For example, on the bid side, $P(1,1) = 0.63$ whereas $P(-1,1) = 0.36$.

On another front, we will show that we can obtain diffusion limit results for the stock price without resorting to the strong symmetry assumptions of [11]. In particular, the assumption that price increments are i.i.d., which is contrary to empirical observations, as shown in [14] for example.

The chapter is organized as follows: Section 2.2 introduces our semi-Markovian modelling of the limit order book, Section 2.3 presents the main probabilistic results obtained in the context of this semi-Markovian model (duration until the next price change, probability of price increase and characterization of the Markov renewal process driving the stock price process), Section 2.4 deals with diffusion limit results for the stock price process, and Section 2.5 presents some calibration results on real market data (five stocks Amazon, Apple, Google, Intel, Microsoft on June 21, 2012). Finally, we also discuss our approach for various stocks (liquid, illiquid, medium liquid) from different markets (Deutsche Börse Group (September 23, 2013) and CISCO (November 3, 2014)) in Section 2.6, as well as possible extensions of our model for the case when the spread is not fixed, including the diffusion limit of the price dynamics in this case, and stochastic optimal control and market making problems. Section 2.7 concludes the paper.

2.2 A Semi-Markovian Modelling of Limit Order Markets

2.2.1 Markov Renewal and Semi-Markov Processes

This chapter deals with models driven by so-called *semi-Markov* processes, and studies some limit theorems and financial applications in this context. At the heart of this thesis is the concept of *Markov renewal process*, together with its associated semi-Markov process. Consider a discrete-time process $(J_n, T_n)_{n \in N}$, where $J_n \in J$ represents the *state* (or *regime*) of the system after n regime transitions – taking value in the state space J (assumed to be finite) – and T_n the moment at which it entered state J_n. Such a process is called a Markov renewal process. It is characterized by the existence of a kernel Q (satisfying certain regularity conditions) such that, for $j \in J$ and $t \in R_+$:

$$P[J_{n+1} = j, T_{n+1} - T_n \leq t | J_k, T_k : k \in [|0, n|]] = Q(J_n, j, t) \text{ a.e.}$$

Denoting:

$$N(t) := \sup\{n \in N : T_n \leq t\}$$

the number of state transitions up to time t, the associated semi-Markov process X is simply defined as $X_t := J_{N(t)}$. The latter definition is valid only if we assume that $N(t) < \infty$ a.e. $\forall t$: this condition is satisfied for so-called *regular* Markov renewal processes, and we shall only work with this kind of processes. Therefore, semi-Markov processes can be seen as continuous-time extensions of Markov renewal processes, in the same way that continuous-time Markov chains are continuous-time extensions of Markov chains. In particular, a continuous-time Markov chain is a specific *Markovian* example of semi-Markov process (actually, the only one!), for which the kernel Q has the form:

$$Q(i, j, t) = P(i, j)(1 - e^{-\lambda(i)t}), \quad i, j \in J, \, t \in R_+,$$

where $\{\lambda(i)\}_{i \in J}$ are positive constants, P is a stochastic matrix (non negative entries and each row summing to one), meaning that the *sojourn times* $\{T_{n+1} - T_n\}_{n \geq 0}$ are exponentially distributed. We also point out that a semi-Markov process becomes fully Markovian if we "add" to it the process keeping track of how long the system has been in its current state: the process $(J_{N(t)}, t - T_{N(t)})_{t \geq 0}$ is Markov.

2.2.2 Semi-Markovian Modelling of Limit Order Books

Throughout this paper and to make the reading more convenient, we will use – when appropriate – the same notations as [11], as it is the starting point of the present article. In this section we introduce our model, highlighting when necessary the mains differences with the model in [11].

Let s_t, s_t^a, s_t^b be respectively the mid, the ask and the bid price processes. Denoting δ the "tick size", these quantities are assumed to be linked by the following relations:

$$\begin{aligned} s_t &= \frac{1}{2}(s_t^a + s_t^b), \\ s_t^a &= s_t^b + \delta. \end{aligned}$$

We will also assume that s_0^b is deterministic and positive. In this context, $s_0^a = s_0^b + \delta$ and $s_0 = s_0^b + \frac{\delta}{2}$ are also deterministic and positive. As shown in [11], the assumption that the bid-ask spread $s_t^a - s_t^b$ is constant and equal to one tick does not exactly match the empirical observations, but it is a reasonable assumption as [11] shows that – based on an analysis of the stocks Citigroup, General Electric, General Motors on June 26, 2008 – more than 98% of the observations have a bid-ask spread equal to 1 tick. This corresponds to a situation where the order book contains no empty levels (also called "gaps").

The price process s_t is assumed to be piecewise constant: at random times $\{T_n\}_{n \geq 0}$ (we set $T_0 := 0$), it jumps from its previous value $s_{T_n^-}$ to a new value $s_{T_n} = s_{T_n^-} \pm \delta$. By construction, the same holds for the ask and bid price processes s_t^a and s_t^b. These random times $\{T_n\}$ correspond to the times at which either the bid or the ask queue get depleted, and therefore, the distribution of these times $\{T_n\}$ will be obtained as a consequence of the dynamics that we will choose to model the bid and ask queues. Let us denote q_t^a and q_t^b the non negative integer-valued processes representing the respective sizes of the ask and bid queues at time t, namely the number of outstanding limit orders at each one of these queues. If the ask queue gets depleted before the bid queue at time T_n – i.e. $q_{T_n}^a = 0$ and $q_{T_n}^b > 0$ – then the price goes up: $s_{T_n} = s_{T_n^-} + \delta$ and both queue values $(q_{T_n}^b, q_{T_n}^a)$ are immediately reinitialized to a new value drawn according to the distribution f, independently from all other random variables. In this context, if n_b, n_a are positive integers, $f(n_b, n_a)$ represents the probability that, after a price increase, the new values of the bid and ask queues are respectively equal to n_b and n_a. On the other hand, if the bid queue gets depleted before the ask queue at time T_n – i.e. $q_{T_n}^a > 0$ and $q_{T_n}^b = 0$ – then the price goes down: $s_{T_n} = s_{T_n^-} - \delta$ and both queue values $(q_{T_n}^b, q_{T_n}^a)$ are immediately reinitialized to a new value drawn according to the distribution $\tilde{f}$, independently from all other random variables. Following the previous discussion, one can remark that the processes q_t^b, q_t^a will never effectively take the value 0, because whenever $q_{T_n}^b = 0$ or $q_{T_n}^a = 0$, we "replace" the pair $(q_{T_n}^b, q_{T_n}^a)$ by a random variable drawn from the distribution f or $\tilde{f}$. The precise construction of the processes (q_t^b, q_t^a) will be explained below.

Let $\tau_n := T_n - T_{n-1}$ the "sojourn times" between two consecutive price changes, $N_t := \sup\{n : T_n \leq t\} = \sup\{n : \tau_1 + ... + \tau_n \leq t\}$ the number of price changes up to time t, $X_n := s_{T_n} - s_{T_{n-1}}$ the consecutive price increments (which can only take the values $\pm\delta$). With these notations we have:

$$s_t = \sum_{k=1}^{N_t} X_k.$$

Let us now present the chosen model for the dynamics of the bid and ask queues. As mentioned in the introduction, we seek to extend the model [11] in the two following directions, as suggested by our calibration results:

1. inter-arrival times between book events (limit orders, market orders, order cancellations) are allowed to have an arbitrary distribution.
2. the arrival of a new book event at the bid or the ask and its corresponding inter-arrival time are allowed to depend on the nature of the previous event.

In order to do so, we will use a Markov renewal structure for the joint process of book events and corresponding inter-arrival times occurring at the ask and bid sides. Formally, for the ask side, consider a family $\{R^{n,a}\}_{n\geq 0}$ of Markov renewal processes given by:

$$R^{n,a} := \{(V_k^{n,a}, T_k^{n,a})\}_{k\geq 0}.$$

For each n, $R^{n,a}$ will "drive" the dynamics of the ask queue on the interval $[T_n, T_{n+1})$ where the stock price remains constant. $\{V_k^{n,a}\}_{k\geq 0}$ and $\{T_k^{n,a}\}_{k\geq 0}$ represent respectively the consecutive book events and the consecutive inter-arrival times between these book events at the ask side on the interval $[T_n, T_{n+1})$. At time T_{n+1} where one of the bid or ask queues gets depleted, the stock price changes and the model will be reinitialized with an independent copy $R^{n+1,a}$ of $R^{n,a}$: it will therefore be assumed that the processes $\{R^{n,a}\}_{n\geq 0}$ are independent copies of the same Markov renewal process of kernel Q^a, namely for each n:

$$\begin{aligned}
P[V_{k+1}^{n,a} = j, T_{k+1}^{n,a} \leq t | T_p^{n,a}, V_p^{n,a} : p \leq k] &= Q^a(V_k^{n,a}, j, t), \quad j \in \{-1, 1\}, \\
P[V_0^{n,a} = j] &= v_0^a(j), \quad j \in \{-1, 1\}, \\
P[T_0^{n,a} = 0] &= 1.
\end{aligned}$$

We recall that as mentioned earlier, we consider two types of book events $V_k^{n,a}$: events of type $+1$ which increase the ask queue by 1 (limit orders), and events of type -1 which decrease the ask queue by 1 (market orders and order cancellations). In particular, the latter assumptions constitute a generalization of [11] in the sense that for each n:

- $V_{k+1}^{n,a}$ depends on the previous queue change $V_k^{n,a}$: $\{V_k^{n,a}\}_{k\geq 0}$ is a Markov chain.
- the inter-arrival times $\{T_k^{n,a}\}_{k\geq 0}$ between book events can have arbitrary distributions. Further, they are not strictly independent anymore but they are independent conditionally on the Markov chain $\{V_k^{n,a}\}_{k\geq 0}$.

We use the same notations to model the bid queue – but with indexes ^{a} replaced by ^{b} – and we assume that the processes involved at the bid and at the ask are independent.

In [11], the kernel Q^a is given by (the kernel Q^b has a similar expression with indexes ^{a} replaced by ^{b}):

$$\begin{aligned} Q^a(i,1,t) &= \frac{\lambda^a}{\lambda^a+\theta^a+\mu^a}(1-e^{-(\lambda^a+\theta^a+\mu^a)t}), \qquad i\in\{-1,1\}, \\ Q^a(i,-1,t) &= \frac{\theta^a+\mu^a}{\lambda^a+\theta^a+\mu^a}(1-e^{-(\lambda^a+\theta^a+\mu^a)t}), \qquad i\in\{-1,1\}, \end{aligned}$$

where λ^a, θ^a and μ^a are rates for the limit, cancellation and market ask orders, respectively.

Given these chosen dynamics to model to the ask and bid queues between two consecutive price changes, we now specify formally the "state process":

$$\tilde{L}_t := (s_t^b, q_t^b, q_t^a)$$

which will keep track of the state of the limit order book at time t (stock price and sizes of the bid and ask queues). In the context of [11], this process $\widetilde{L}_t$ was proved to be Markovian. Here, we will need to "add" to this process the process (V_t^b, V_t^a) keeping track of the nature of the last book event at the bid and the ask to make it Markovian: in this sense we can view it as being semi-Markovian. The process:

$$L_t := (s_t^b, q_t^b, q_t^a, V_t^b, V_t^a)$$

constructed below will be proved to be Markovian.

The process L is piecewise constant and changes value whenever a book event occurs at the bid or at the ask. We will construct both the process L and the sequence of times $\{T_n\}_{n\geq 0}$ recursively on $n \geq 0$. The recursive construction starts from $n = 0$ where we have $T_0 = 0$, $s_0^b > 0$ deterministic, and $(q_0^b, q_0^a, V_0^b, V_0^a)$ is a random variable with distribution $f_0 \times v_0^b \times v_0^a$, where f_0 is a distribution on $N^* \times N^*$, and both v_0^b and v_0^a are distributions on the two-point space $\{-1, 1\}$, that is $v_0^b(1) = P[V_0^b = 1]$ is given and $v_0^b(-1) = 1 - v_0^b(1)$ (and similarly for the ask). We will need to introduce the following processes for the bid side (for the ask side, they are defined similarly):

$$\bar{T}_k^{n,b} := \sum_{p=0}^{k} T_p^{n,b}, \quad N_t^{n,b} := \sup\{k : T_n + \bar{T}_k^{n,b} \leq t\}.$$

With these notations, the book events corresponding to the interval $[T_n, T_{n+1})$ occur at times $T_n + \bar{T}_k^{n,b}$ $(k \geq 0)$ until one of the queues gets depleted, and $N_t^{n,b}$ counts the number of book events on the interval $[T_n, t]$, for $t \in [T_n, T_{n+1})$.

The joint construction of L and of the sequence of times $\{T_n\}_{n\geq 0}$ is done recursively on $n \geq 0$. The following describes the step n of the recursive construction:

- For each $T \in \{T_n + \bar{T}_k^{n,b}\}_{k \geq 1}$, the book event $v_{n,T}^b := V_{N_T^{n,b}}^{n,b}$ occurs at time T at the bid side. If $q_{T-}^b + v_{n,T}^b > 0$, there is no price change at time T and we have:

$$(s_T^b, q_T^b, q_T^a, V_T^b, V_T^a) = (s_{T-}^b, q_{T-}^b + v_{n,T}^b, q_{T-}^a, v_{n,T}^b, V_{T-}^a).$$

If on the other hand $q_{T-}^b + v_{n,T}^b = 0$, there is a price change at time T and the model gets reinitialized:

$$(s_T^b, q_T^b, q_T^a, V_T^b, V_T^a) = (s_{T-}^b - \delta, \tilde{x}_n^b, \tilde{x}_n^a, v_{0,n}^b, v_{0,n}^a),$$

where $\{(\tilde{x}_k^b, \tilde{x}_k^a)\}_{k \geq 0}$ are i.i.d. random variables, independent from all other random variables, with joint distribution $\tilde{f}$ on $N^* \times N^*$, and $\{v_{0,k}^b, v_{0,k}^a\}_{k \geq 0}$ are i.i.d. random variables, independent from all other random variables, with joint distribution $v_0^b \times v_0^a$ on the space $\{-1, 1\} \times \{-1, 1\}$. We then set $T_{n+1} = T$ and move from the step n of the recursion to the step $n + 1$.

- For each $T \in \{T_n + \bar{T}_k^{n,a}\}_{k \geq 1}$, the book event $v_{n,T}^a := V_{N_T^{n,a}}^{n,a}$ occurs at time T at the ask side. If $q_{T-}^a + v_{n,T}^a > 0$, there is no price change at time T and we have:

$$(s_T^b, q_T^b, q_T^a, V_T^b, V_T^a) = (s_{T-}^b, q_{T-}^b, q_{T-}^a + v_{n,T}^a, V_{T-}^b, v_{n,T}^a).$$

If on the other hand $q_{T-}^a + v_{n,T}^a = 0$, there is a price change at time T and the model gets reinitialized:

$$(s_T^b, q_T^b, q_T^a, V_T^b, V_T^a) = (s_{T-}^b + \delta, x_n^b, x_n^a, v_{0,n}^b, v_{0,n}^a),$$

where $\{(x_k^b, x_k^a)\}_{k \geq 0}$ are i.i.d. random variables, independent from all other random variables, with joint distribution f on $N^* \times N^*$, and $\{v_{0,k}^b, v_{0,k}^a\}_{k \geq 0}$ are the i.i.d. random variables defined above. We then set $T_{n+1} = T$ and move from the step n of the recursion to the step $n + 1$.

It results from the above construction and the Markov renewal structure of the processes $\{R^{n,a}\}_{n \geq 0}$, $\{R^{n,b}\}_{n \geq 0}$ that the process L_t is Markovian.

Since the processes $\{R^{n,a}\}_{n \geq 0}$ are independent copies of the same Markov renewal process of kernel Q^a, we will drop the index n when appropriate in order to make the notations lighter. Following this remark, we will introduce the following notations for the ask, for $i, j \in \{-1, 1\}$ (for the bid, they are defined similarly):

- $P^a(i, j) := P[V_{k+1}^a = j | V_k^a = i]$-transition probabilities for the book events V_k^a at the ask side,
- $F^a(i, t) := P[T_{k+1}^a \leq t | V_k^a = i]$-distribution function for the inter-arrival times T_{k+1}^a between book events providedl that $V_k^a = i$,

- $H^a(i,j,t) := P[T^a_{k+1} \le t | V^a_k = i, V^a_{k+1} = j]$-distribution function for the inter-arrival times T^a_{k+1} between book events provided that $V^a_k = i$ and $V^a_{k+1} = j$,
- $h^a(i,j) := \int_0^\infty tH^a(i,j,dt)$-mean value of T^a_{k+1} provided that $V^a_k = i$ and $V^a_{k+1} = j$,
- $h^a_1 := h^a(1,1) + h^a(-1,-1), \quad h^a_2 := h^a(-1,1) + h^a(1,-1)$,
- $m^a(s,i,j) := \int_0^\infty e^{-st} Q^a(i,j,dt), \quad s \in C$,-Laplace transform for $Q^a(i,j,dt)$, where $Q^a(i,j,dt)$ is defined above,
- $M^a(s,i) := m^a(s,i,-1) + m^a(s,i,1) = \int_0^\infty e^{-st} F^a(i,dt), \quad s \in C.$

Throughout this paper, we will use the following mild technical assumptions:

(A1) $0 < P^a(i,j) < 1,\ 0 < P^b(i,j) < 1, \quad i,j \in \{-1,1\}.$

(A2) $F^a(i,0) < 1,\ F^b(i,0) < 1, \quad i \in \{-1,1\}.$

(A3) $\int_0^\infty t^2 H^a(i,j,dt) < \infty,\ \int_0^\infty t^2 H^b(i,j,dt) < \infty, \quad i,j \in \{-1,1\}.$

Some brief comments on these assumptions: **(A1)** implies that each state ± 1 is accessible from each state, **(A2)** means that each inter-arrival time between book events has a positive probability to be non zero, and **(A3)** constitutes a second moment integrability assumption on the cumulative distribution functions H^a and H^b.

2.3 Main Probabilistic Results

Throughout this section and as mentioned earlier, since the processes $\{R^{n,a}\}_{n\ge 0}$ are independent copies of the same Markov renewal process of kernel Q^a, we will drop the index n when appropriate in order to make the notations lighter on the random variables $T^{n,a}_k$, $\bar{T}^{n,a}_k$, $V^{n,a}_k$ (and similarly for the bid side).

2.3.1 Duration until the next price change

Given an initial configuration of the bid and ask queues $(q^b_0, q^a_0) = (n_b, n_a)$ (n_b, n_a integers), we denote σ_b the first time at which the bid queue is depleted:

$$\sigma_b = \bar{T}^b_{k^*}, \quad k^* := \inf\{k : n_b + \sum_{m=1}^{k} V^b_m = 0\}.$$

Similarly we define σ_a the first time at which the ask queue is depleted.

The duration until the next price move is thus:

$$\tau := \sigma_a \wedge \sigma_b.$$

In order to have a realistic model in which the queues always get depleted at some point, i.e. $P[\sigma_a < \infty] = P[\sigma_b < \infty] = 1$, we impose the conditions:

$$P^a(1,1) \leq P^a(-1,-1), \quad P^b(1,1) \leq P^b(-1,-1).$$

These conditions correspond to the condition $\lambda \leq \theta + \mu$ in [11], and the proof of the proposition below shows that they are respectively equivalent to $P[\sigma_a < \infty] = 1$ and $P[\sigma_b < \infty] = 1$. Indeed, as $s \to 0$ ($s > 0$), the Laplace transform $L^a(s) := E[e^{-s\sigma_a}]$ of σ_a tends to $P[\sigma_a < \infty]$. The proposition below shows that if $P^a(1,1) > P^a(-1,-1)$, this quantity is strictly less than 1, and if $P^a(1,1) \leq P^a(-1,-1)$, this quantity is equal to 1. We have the following result which generalizes Proposition 1 in [11] (see also Remark 2.1 below):

Theorem 2.1 *The conditional law of σ_a given $q_0^a = n \geq 1$ has a regularly varying tail with:*

- *tail exponent 1 if $P^a(1,1) < P^a(-1,-1)$.*
- *tail exponent 1/2 if $P^a(1,1) = P^a(-1,-1)$.*

More precisely, we get: if $P^a(1,1) = P^a(-1,-1) = p_a$:

$$P[\sigma_a > t | q_0^a = n] \overset{t\to\infty}{\sim} \frac{\alpha^a(n)}{\sqrt{t}},$$

with:

$$\alpha^a(n) := \frac{1}{p_a\sqrt{\pi}}(n + \frac{2p_a - 1}{p_a - 1} v_0^a(1))\sqrt{p_a(1-p_a)}\sqrt{p_a h_1^a + (1-p_a)h_2^a}.$$

If $P^a(1,1) < P^a(-1,-1)$, we get:

$$P[\sigma_a > t | q_0^a = n] \overset{t\to\infty}{\sim} \frac{\beta^a(n)}{t},$$

with:

$$\begin{aligned}
\beta^a(n) &:= v_0^a(1)u_1^a + v_0^a(-1)u_2^a + (n-1)u_3^a, \\
u_1^a &:= h^a(1,-1) + \frac{P^a(1,1)}{1 - P^a(1,1)}(u_3^a + h^a(1,1)), \\
u_2^a &:= -h^a(1,1) + \frac{1 - P^a(-1,-1)}{1 - P^a(1,1)}(u_3^a + h^a(1,1)) \\
&\quad + P^a(-1,-1)h_1^a + (1 - P^a(-1,-1))h_2^a, \\
u_3^a &:= h^a(1,1) + \frac{1 - P^a(1,1)}{P^a(-1,-1) - P^a(1,1)} \\
&\quad \times (P^a(-1,-1)h_1^a + (1 - P^a(-1,-1))h_2^a).
\end{aligned}$$

Similar expressions are obtained for $P[\sigma_b > t | q_0^b = n]$, with indexes ^{a} replaced by ^{b}.

Remark 2.1 *We retrieve the results of [11]: if $P^a(1,1) = P^a(-1,-1)$, then within the context/notations of [11] we get $p_a = 1/2$ and:*

$$h^a(i,j) = \int_0^\infty 2t\lambda e^{-2\lambda t} dt = \frac{1}{2\lambda},$$

and so $\alpha^a(n) = \frac{n}{\sqrt{\pi\lambda}}$. For the case $P^a(1,1) < P^a(-1,-1)$ ($\lambda < \theta + \mu$ with their notations), we find:

$$\beta^a(n) = \frac{n}{\theta + \mu - \lambda},$$

which is different from the result of [11] that is $\beta^a(n) = \frac{n(\theta+\mu+\lambda)}{2\lambda(\theta+\mu-\lambda)}$.

Proof *Let $s > 0$ and denote $L(s,n,i) := E[e^{-s\sigma_a} | q_0^a = n, V_0^a = i]$. We have:*

$$\sigma_a = \sum_{m=1}^{k^*} T_m^a, \quad k^* := \inf\{k : n + \sum_{m=1}^{k} V_m^a = 0\}.$$

Therefore:

$$\begin{aligned}
L(s,n,i) &= E[e^{-sT_1^a} E[e^{-s(\sigma_a - T_1^a)} | q_0^a = n, V_0^a = i, V_1^a, T_1^a] | q_0^a = n, V_0^a = i] \\
&= E[e^{-sT_1^a} \underbrace{E[e^{-s(\sigma_a - T_1^a)} | q_{T_1^a}^a = n + V_1^a, V_0^a = i, V_1^a, T_1^a]}_{L(s,n+V_1^a,V_1^a)} | q_0^a = n, V_0^a = i] \\
&= E[e^{-sT_1^a} L(s, n + V_1^a, V_1^a) | q_0^a = n, V_0^a = i] \\
&= \int_0^\infty e^{-st} L(s, n+1, 1) Q^a(i, 1, dt) + \int_0^\infty e^{-st} L(s, n-1, -1) Q^a(i, -1, dt) \\
&= m^a(s, i, 1) L(s, n+1, 1) + m^a(s, i, -1) L(s, n-1, -1)
\end{aligned}$$

Denote for sake of clarity $a_n := L(s,n,1)$, $b_n := L(s,n,-1)$. These sequences therefore solve the system of coupled recurrence equations:

$$\begin{aligned}
a_{n+1} &= m^a(s,1,1)a_{n+2} + m^a(s,1,-1)b_n, \quad n \geq 0, \\
b_{n+1} &= m^a(s,-1,1)a_{n+2} + m^a(s,-1,-1)b_n, \\
a_0 = b_0 &= 1.
\end{aligned}$$

Simple algebra (computing $a_{n+1} - m^a(s,-1,-1)a_n$ on the on hand and $m^a(s,1,1)b_{n+1} - b_n$ on the other hand) gives us that both a_n and b_n solve the same following recurrence equation (but for different initial conditions):

$$m^a(s,1,1)u_{n+2} - (1 + \Delta^a(s))u_{n+1} + m^a(s,-1,-1)u_n, \quad n \geq 1,$$

with:

$$\Delta^a(s) := m^a(s,1,1)m^a(s,-1,-1) - m^a(s,-1,1)m^a(s,1,-1).$$

The parameter $\Delta^a(s)$ can be seen as a coupling coefficient and is equal to 0 when the random variable (V_k^a, T_k^a) doesn't depend on the previous state V_{k-1}^a, for example in the context of [11].

If we denote $R(X)$ the characteristic polynomial associated to the previous recurrence equation $R(X) := m^a(s,1,1)X^2 - (1+\Delta^a(s))X + m^a(s,-1,-1)$, then simple algebra gives us:

$$R(1) = \underbrace{(M^a(s,1)-1)}_{<0}\underbrace{(1-m^a(s,-1,-1))}_{>0} + \underbrace{m^a(s,1,-1)}_{>0}\underbrace{(M^a(s,-1)-1)}_{<0} < 0.$$

Note that $M^a(s,i) < 1$ for $s > 0$ because $F^a(i,0) < 1$. Since $m^a(s,1,1) > 0$, this implies that R has only one root < 1 (and an other root > 1):

$$\lambda^a(s) := \frac{1+\Delta^a(s) - \sqrt{(1+\Delta^a(s))^2 - 4m^a(s,1,1)m^a(s,-1,-1)}}{2m^a(s,1,1)}.$$

Because we have $a_n, b_n \leq 1$ for $s > 0$, then we must have for $n \geq 1$:

$$a_n = a_1\lambda^a(s)^{n-1}, \quad b_n = b_1\lambda^a(s)^{n-1}.$$

The recurrence equations on a_n, b_n give us:

$$a_1 = \frac{m^a(s,1,-1)}{1-\lambda^a(s)m^a(s,1,1)}, \quad b_1 = \frac{m^a(s,-1,1)a_1 + \Delta^a(s)}{m^a(s,1,1)}.$$

Finally, letting $L(s,n) := E[e^{-s\sigma_a}|q_0^a = n]$, we obtain:

$$L(s,n) = \sum_i L(s,n,i)v_0^a(i) = a_n v_0^a(1) + b_n v_0^a(-1).$$

The behaviour of $P[\sigma_a > t|q_0^a = n]$ as $t \to \infty$ is obtained by computing the behaviour of $L(s,n)$ as $s \to 0$, together with Karamata's Tauberian theorem. By the second moment integrability assumption on $H^a(i,j,dt)$, we note that:

$$\begin{aligned} m^a(s,i,j) &= \int_0^\infty e^{-st}Q^a(i,j,dt) = P^a(i,j)\int_0^\infty e^{-st}H^a(i,j,dt) \\ \overset{s\to 0}{\sim} P^a(i,j) &- sP^a(i,j)\int_0^\infty tH^a(i,j,dt) = P^a(i,j) - sP^a(i,j)h^a(i,j). \end{aligned}$$

Now, assume $P^a(1,1) = P^a(-1,-1) = p_a$. A straightforward but tedious Taylor expansion of $L(s,n)$ as $s \to 0$ gives us:

$$L(s,n) \overset{s\to 0}{\sim} 1 - \sqrt{\pi}\alpha^a(n)\sqrt{s}.$$

The same way, if $P^a(1,1) < P^a(-1,-1)$, a straightforward Taylor expansion of $L(s,n)$ as $s \to 0$ gives us:

$$L(s,n) \overset{s\to 0}{\sim} 1 - \beta^a(n)s.$$

We are interested in the asymptotic behaviour of the law of τ, which is, by independence of the bid/ask queues:

$$P[\tau > t|(q_0^b, q_0^a) = (n_b, n_a)] = P[\sigma_a > t|q_0^a = n_a]P[\sigma_b > t|q_0^b = n_b].$$

We get the following immediate consequence of Theorem 2.1:

Theorem 2.2 *The conditional law of τ given $(q_0^b, q_0^a) = (n_b, n_a)$ has a regularly varying tail with:*

- *tail exponent 2 if $P^a(1,1) < P^a(-1,-1)$ and $P^b(1,1) < P^b(-1,-1)$. In particular, in this case, $E[\tau|(q_0^b, q_0^a) = (n_b, n_a)] < \infty$.*
- *tail exponent 1 if $P^a(1,1) = P^a(-1,-1)$ and $P^b(1,1) = P^b(-1,-1)$. In particular, in this case, $E[\tau|(q_0^b, q_0^a) = (n_b, n_a)] = \infty$ whenever $n_b, n_a \geq 1$.*
- *tail exponent 3/2 otherwise. In particular, in this case, $E[\tau|(q_0^b, q_0^a) = (n_b, n_a)] < \infty$.*

More precisely, we get: if $P^a(1,1) = P^a(-1,-1)$ and $P^b(1,1) = P^b(-1,-1)$:

$$P[\tau > t|(q_0^b, q_0^a) = (n_b, n_a)] \overset{t\to\infty}{\sim} \frac{\alpha^a(n_a)\alpha^b(n_b)}{t};$$

if $P^a(1,1) < P^a(-1,-1)$ and $P^b(1,1) < P^b(-1,-1)$:

$$P[\tau > t|(q_0^b, q_0^a) = (n_b, n_a)] \overset{t\to\infty}{\sim} \frac{\beta^a(n_a)\beta^b(n_b)}{t^2};$$

if $P^a(1,1) = P^a(-1,-1)$ and $P^b(1,1) < P^b(-1,-1)$:

$$P[\tau > t|(q_0^b, q_0^a) = (n_b, n_a)] \overset{t\to\infty}{\sim} \frac{\alpha^a(n_a)\beta^b(n_b)}{t^{3/2}};$$

if $P^a(1,1) < P^a(-1,-1)$ and $P^b(1,1) = P^b(-1,-1)$:

$$P[\tau > t|(q_0^b, q_0^a) = (n_b, n_a)] \overset{t\to\infty}{\sim} \frac{\beta^a(n_a)\alpha^b(n_b)}{t^{3/2}}.$$

Proof *Immediate using Theorem 2.1.*

It will be needed to get the full law of τ, which is, by independence of the bid/ask queues:

$$P[\tau > t|(q_0^b, q_0^a) = (n_b, n_a)] = P[\sigma_a > t|q_0^a = n_a]P[\sigma_b > t|q_0^b = n_b].$$

We have computed explicitly the Laplace transforms of σ_a and σ_b (cf. the proof of Theorem 2.1 above). There are two possibilities: either it is possible to invert those Laplace transforms so that we can compute $P[\sigma_a > t|q_0^a = n_a]$ and $P[\sigma_b > t|q_0^b = n_b]$ in closed form and thus $P[\tau > t|(q_0^b, q_0^a) = (n_b, n_a)]$ in closed form as in [11]. If not, we will have to resort to a numerical procedure to invert the characteristic functions of σ_a and σ_b. Below we give the characteristic functions of σ_a and σ_b:

Theorem 2.3 *Let $\phi^a(t,n) := E[e^{it\sigma_a}|q_0^a = n]$ $(t \in R)$ the characteristic function of σ_a conditionally on $q_0^a = n \geq 1$. We have:*
if $m^a(-it,1,1) \neq 0$:

$$\begin{aligned}
\phi^a(t,n) &= \left(c^a(-it)v_0^a(1) + d^a(-it)v_0^a(-1)\right)\lambda^a(-it)^{n-1}, \\
c^a(z) &= \frac{m^a(z,1,-1)}{1-\lambda^a(z)m^a(z,1,1)}, \\
d^a(z) &= \frac{m^a(z,-1,1)c^a(z) + \Delta^a(z)}{m^a(z,1,1)}, \\
\Delta^a(z) &:= m^a(z,1,1)m^a(z,-1,-1) - m^a(z,-1,1)m^a(z,1,-1), \\
\lambda^a(z) &:= \frac{1+\Delta^a(z) - \sqrt{(1+\Delta^a(z))^2 - 4m^a(z,1,1)m^a(z,-1,-1)}}{2m^a(z,1,1)}.
\end{aligned}$$

and if $m^a(-it,1,1) = 0$:

$$\begin{aligned}
\phi^a(t,n) &= (m^a(-it,1,-1)v_0^a(1) + \widetilde{\lambda^a}(-it)v_0^a(-1))\widetilde{\lambda^a}(-it)^{n-1}, \\
\widetilde{\lambda^a}(z) &:= \frac{m^a(z,-1,-1)}{1-m^a(z,1,-1)m^a(z,-1,1)}.
\end{aligned}$$

The coefficient $\Delta^a(z)$ can be seen as a coupling coefficient and is equal to 0 when the random variable (V_k^a, T_k^a) doesn't depend on the previous state V_{k-1}^a, for example in the context of [11].

The characteristic function $\phi^b(t,n) := E[e^{it\sigma_b}|q_0^b = n]$ has the same expression, with indexes ^{a} replaced by ^{b}.

Proof *Similarly to the proof of Theorem 2.1, we obtain (using the same notations but denoting this time $a_n := L(-it,n,1)$, $b_n := L(-it,n,-1)$):*

$$\begin{aligned}
a_{n+1} &= m^a(-it,1,1)a_{n+2} + m^a(-it,1,-1)b_n, \qquad n \geq 0, \\
b_{n+1} &= m^a(-it,-1,1)a_{n+2} + m^a(-it,-1,-1)b_n, \\
a_0 &= b_0 = 1.
\end{aligned}$$

If $m^a(-it,1,1) = 0$, we can solve explicitly the above system to get the desired result. If $m^a(-it,1,1) \neq 0$, we get as in the proof of Prop 2.1 that both a_n and b_n solve the same following recurrence equation (but for different initial conditions):

$$m^a(-it,1,1)u_{n+2} - (1+\Delta^a(-it))u_{n+1} + m^a(-it,-1,-1)u_n, \qquad n \geq 1.$$

Because $|m^a(-it,j,-1) + m^a(-it,j,1)| = |M^a(-it,j)| = \left|\int_0^\infty e^{its}F^a(j,ds)\right| \leq 1$, tedious computations give us that $|\lambda_+^a(-it)| > 1$ whenever $t \neq 0$, where:

$$\lambda_+^a(z) := \frac{1+\Delta^a(z) + \sqrt{(1+\Delta^a(z))^2 - 4m^a(z,1,1)m^a(z,-1,-1)}}{2m^a(z,1,1)}.$$

Since both $|a_n|, |b_n| \leq 1$ for all n, it must be that:

$$a_n = a_1 \lambda^a(-it)^{n-1}, \quad b_n = b_1 \lambda^a(-it)^{n-1},$$

with a_1, b_1 being given by the recurrence equations on a_n, b_n:

$$a_1 = \frac{m^a(-it, 1, -1)}{1 - \lambda^a(-it) m^a(-it, 1, 1)}, \quad b_1 = \frac{m^a(-it, -1, 1)a_1 + \Delta^a(-it)}{m^a(-it, 1, 1)}.$$

Finally we conclude by observing that:

$$\phi^a(t, n) = a_n v_0^a(1) + b_n v_0^a(-1).$$

2.3.2 Probability of Price Increase

Starting from an initial configuration of the bid and ask queues, $(q_0^b, q_0^a) = (n_b, n_a)$, the probability that the next price change is a price increase will be denoted $p_1^{up}(n_b, n_a)$. This quantity is equal to the probability that σ_a is less than σ_b:

$$p_1^{up}(n_b, n_a) = P[\sigma_a < \sigma_b | q_0^b = n_b, q_0^a = n_a].$$

Since we know the characteristic functions of σ_a, σ_b (cf. Theorem 2.3), we can compute their individual laws up to the use of a numerical procedure. Since σ_a and σ_b are independent, the law of $\sigma_b - \sigma_a$ can be computed using the individual laws of σ_a, σ_b, and therefore $p_1^{up}(n_b, n_a)$ can be computed up to the use of numerical procedures to (1) invert the characteristic function and (2) compute an indefinite integral. Indeed, denoting $f_{n_a,a}$ the p.d.f of σ_a conditionally on $q_0^a = n_a$, and $F_{n_b,b}$ the c.d.f. of σ_b conditionally on $q_0^b = n_b$, we have:

$$p_1^{up}(n_b, n_a) = P[\sigma_a < \sigma_b | q_0^b = n_b, q_0^a = n_a] = \int_0^\infty f_{n_a,a}(t)(1 - F_{n_b,b}(t))dt,$$

where $F_{n_b,b}$ and $f_{n_a,a}$ are obtained by the following inversion formulas:

$$\begin{aligned} f_{n_a,a}(t) &= \frac{1}{2\pi} \int_R e^{-itx} \phi^a(x, n_a) dx, \\ F_{n_b,b}(t) &= \frac{1}{2} - \frac{1}{\pi} \int_0^\infty \frac{1}{x} Im\{e^{-itx} \phi^b(x, n_b)\} dx. \end{aligned}$$

2.3.3 The stock price seen as a functional of a Markov renewal process

As mentioned earlier, we can write the stock price s_t as:

$$s_t = \sum_{k=1}^{N_t} X_k,$$

where $\{X_n\}_{n\geq 0}$ are the consecutive price increments taking value $\pm\delta$, $\{\tau_n\}_{n\geq 0}$ are the consecutive durations between price changes and $\{T_n\}_{n\geq 0}$ the consecutive times at which the price changes.

In this context, the distribution of the random variable τ_{n+1} will depend on the initial configuration of the bid and ask queues at the beginning T_n of the period $[T_n, T_{n+1})$, which itself depends on the nature of the previous price change X_n: if the previous price change is a price decrease, the initial configuration will be drawn from the distribution $\tilde{f}$, and if it is an increase, the initial configuration will be drawn from the distribution f. Because for each n the random variable (X_n, τ_n) only depends on the previous increment X_{n-1}, it can be seen that the process $(X_n, \tau_n)_{n\geq 0}$ is a Markov renewal process ([25], [35]), and the stock price can therefore be seen as a functional of this Markov renewal process. We obtain the following result.

Theorem 2.4 *The process $(X_n, \tau_n)_{n\geq 0}$ is a Markov renewal process. The law of the process $\{\tau_n\}_{n\geq 0}$ is given by:*

$$F(\delta, t) := P[\tau_{n+1} \leq t | X_n = \delta] = \sum_{p=1}^{\infty}\sum_{n=1}^{\infty} f(n,p) P[\tau \leq t | (q_0^b, q_0^a) = (n,p)],$$

$$F(-\delta, t) := P[\tau_{n+1} \leq t | X_n = -\delta] = \sum_{p=1}^{\infty}\sum_{n=1}^{\infty} \tilde{f}(n,p) P[\tau \leq t | (q_0^b, q_0^a) = (n,p)].$$

The Markov chain $\{X_n\}_{n\geq 0}$ is characterized by the following transition probabilities:

$$p_{cont} := P[X_{n+1} = \delta | X_n = \delta] = \sum_{i=1}^{\infty}\sum_{j=1}^{\infty} p_1^{up}(i,j) f(i,j).$$

$$p'_{cont} := P[X_{n+1} = -\delta | X_n = -\delta] = \sum_{i=1}^{\infty}\sum_{j=1}^{\infty} (1 - p_1^{up}(i,j)) \tilde{f}(i,j).$$

The generator of this Markov chain is thus (we assimilate the state 1 to the value δ and the state 2 to the value $-\delta$):

$$P := \begin{pmatrix} p_{cont} & 1 - p_{cont} \\ 1 - p'_{cont} & p'_{cont} \end{pmatrix}.$$

Let $p_n^{up}(b,a) := P[X_n = \delta | q_0^b = b, q_0^a = a]$. We can compute this quantity explicitly:

$$\begin{aligned} p_n^{up}(b,a) &= \pi^* + (p_{cont} + p'_{cont} - 1)^{n-1} \left(p_1^{up}(b,a) - \pi^*\right), \\ \pi^* &:= \pi^*(\delta) := \frac{p'_{cont} - 1}{p_{cont} + p'_{cont} - 2}, \end{aligned}$$

where π^* is the stationary distribution of the Markov chain $\{X_n\}$:

$$\pi^* = \lim_{n\to\infty} P[X_n = \delta | X_1].$$

Further:

$$E[X_n|q_0^b = b, q_0^a = a] = \delta(2p_n^{up}(b,a) - 1),$$

and the (conditional) covariance between two consecutive price moves:

$$cov[X_{n+1}, X_n|q_0^b = b, q_0^a = a] = 4\delta^2 p_n^{up}(b,a)(1 - p_n^{up}(b,a))(p_{cont} + p'_{cont} - 1).$$

Remark 2.2 *In particular, if* $p_{cont} = p'_{cont}$, *then* $\pi^* = 1/2$ *and we retrieve the results of [11]. We also note that the sign of the (conditional) covariance between two consecutive price moves does not depend on the initial configuration of the bid and ask queues and is given by the sign of* $p_{cont} + p'_{cont} - 1$. *We also note that the quantities* p_{cont}, p'_{cont} *can be computed up to the knowledge of the quantities* $p_1^{up}(n_b, n_a)$ *which computation was discussed in the previous section. The quantities* $F(\pm\delta, t)$ *can be computed up to the knowledge of the law of* τ, *which is known up to the use of a numerical procedure to invert the characteristic functions of* σ_a *and* σ_b, *together with the results of Theorem 2.3.*

Proof *The results follow from elementary calculations in a similar way to what is done in [11]. Indeed, we have:*

$$\begin{pmatrix} p_n^{up}(b,a) & 1 - p_n^{up}(b,a) \end{pmatrix} = \begin{pmatrix} p_1^{up}(b,a) & 1 - p_1^{up}(b,a) \end{pmatrix} \begin{pmatrix} p_{cont} & 1 - p_{cont} \\ 1 - p'_{cont} & p'_{cont} \end{pmatrix}^{n-1}.$$

We also have:

$$\begin{pmatrix} p_{cont} & 1 - p_{cont} \\ 1 - p'_{cont} & p'_{cont} \end{pmatrix} = S \begin{pmatrix} 1 & 0 \\ 0 & p_{cont} + p'_{cont} - 1 \end{pmatrix} S^{-1},$$

with:

$$S = \begin{pmatrix} 1 & -\frac{1-p_{cont}}{1-p'_{cont}} \\ 1 & 1 \end{pmatrix}.$$

2.4 Diffusion Limit of the Price Process

In [11] it is assumed that $f(i,j) = \tilde{f}(i,j) = f(j,i)$ in order to make the price increments X_n independent and identically distributed. In fact, this assumption can be entirely relaxed. Indeed, as we mentioned above, $(X_n, \tau_n)_{n\geq 0}$ is in fact a Markov renewal process and therefore we can use the related theory to compute the diffusion limit of the price process. The results of this section generalize the results of Section 4 in [11].

2.4.1 Balanced Order Flow case: $P^a(1,1) = P^a(-1,-1)$ and $P^b(1,1) = P^b(-1,-1)$

Throughout this section we make the assumption:

(A4) Using the notations of Theorem 2.1, the following holds:

$$\sum_{n=1}^{\infty}\sum_{p=1}^{\infty}\alpha^b(n)\alpha^a(p)f(n,p) < \infty, \quad \sum_{n=1}^{\infty}\sum_{p=1}^{\infty}\alpha^b(n)\alpha^a(p)\tilde{f}(n,p) < \infty.$$

Using Theorem 2.2, we obtain the following result generalizing lemma 1 in [11]:

Lemma 2.1 *Under assumption* **(A4)**, *the following weak convergence holds as $n \to \infty$:*

$$\frac{1}{n\log(n)}\sum_{k=1}^{n}\tau_k \Rightarrow \tau^* := \sum_{n=1}^{\infty}\sum_{p=1}^{\infty}\alpha^b(n)\alpha^a(p)f^*(n,p),$$

where $f^*(n,p) := \pi^* f(n,p) + (1-\pi^*)\tilde{f}(n,p)$.

Proof *We have:*

$$\frac{1}{n\log(n)}\sum_{k=1}^{n}\tau_k = \sum_{i\in\{-\delta,\delta\}}\frac{N_i(n)}{n}\frac{\log(N_i(n))}{\log(n)}\frac{1}{N_i(n)\log(N_i(n))}\sum_{k=1}^{N_i(n)}\tau_{p(k,i)},$$

where for $i \in \{-\delta,\delta\}$, $N_i(n)$ represents the number of times that $X_{k-1} = i$ for $1 \le k \le n$; and $\{p(k,i) : k \ge 1\}$ the successive indexes for which $X_{k-1} = i$. By the standard theory of Markov Chains, we have for $i \in \{-\delta,\delta\}$:

$$\frac{N_i(n)}{n} \overset{a.e.}{\to} \pi^*(i),$$

and therefore we have $\frac{\log(N_i(n))}{\log(n)} \overset{a.e.}{\to} 1$. We recall that $\pi^*(\delta) := \pi^*$, and $\pi^*(-\delta) = 1-\pi^*$. For fixed $i \in \{-\delta,\delta\}$, the random variables $\{\tau_{p(k,i)} : k \ge 1\}$ are i.i.d. with distribution $F(i,\cdot)$, and with tail index equal to 1 (by Theorem 2.2). Using [11] (Lemma 1) together with Theorem 2.2, we get that:

$$\frac{1}{n\log(n)}\sum_{k=1}^{n}\tau_{p(k,\delta)} \Rightarrow \sum_{n=1}^{\infty}\sum_{p=1}^{\infty}\alpha^b(n)\alpha^a(p)f(n,p),$$

$$\frac{1}{n\log(n)}\sum_{k=1}^{n}\tau_{p(k,-\delta)} \Rightarrow \sum_{n=1}^{\infty}\sum_{p=1}^{\infty}\alpha^b(n)\alpha^a(p)\tilde{f}(n,p).$$

The latter convergence holds in probability and we finally have:

$$\frac{1}{n\log(n)}\sum_{k=1}^{n}\tau_k \overset{P}{\to} \pi^*\sum_{n=1}^{\infty}\sum_{p=1}^{\infty}\alpha^b(n)\alpha^a(p)f(n,p) + (1-\pi^*)\sum_{n=1}^{\infty}\sum_{p=1}^{\infty}\alpha^b(n)\alpha^a(p)\tilde{f}(n,p).$$

Let $s^* := \delta(2\pi^* - 1)$. Using the previous Lemma 2.1, we obtain the following diffusion limit for the re-normalized price process $s_{tn\log(n)}$:

Theorem 2.5 *Under assumption* **(A4)**, *the renormalized price process* $s_{tn\log(n)}$ *satisfies the following weak convergence in the Skorokhod topology ([31]):*

$$\left(\frac{s_{tn\log(n)}}{n}, t \geq 0\right) \overset{n\to\infty}{\Rightarrow} \left(\frac{s^* t}{\tau^*}, t \geq 0\right),$$

$$\left(\frac{s_{tn\log(n)} - N_{tn\log(n)} s^*}{\sqrt{n}}, t \geq 0\right) \overset{n\to\infty}{\Rightarrow} \frac{\sigma}{\sqrt{\tau^*}} W,$$

where W is a standard Brownian motion and σ is given by:

$$\sigma^2 = 4\delta^2 \left(\frac{1 - p'_{cont} + \pi^*(p'_{cont} - p_{cont})}{(p_{cont} + p'_{cont} - 2)^2} - \pi^*(1 - \pi^*)\right).$$

Remark 2.3 *If $p'_{cont} = p_{cont} = \pi^* = \frac{1}{2}$ as in [11], we find $s^* = 0$ and $\sigma = \delta$ as in [11]. If $p'_{cont} = p_{cont} = p$, we have $\pi^* = \frac{1}{2}$, $s^* = 0$ and:*

$$\sigma^2 = \delta^2 \frac{p}{1-p}.$$

Proof *Because $m(\pm\delta) := E[\tau_n | X_{n-1} = \pm\delta] = +\infty$ by Theorem 2.2, we cannot directly apply the well-known invariance principle results for semi-Markov processes. Denote for $t \in R_+$:*

$$R_n := \sum_{k=1}^{n} (X_k - s^*), \quad U_n(t) := n^{-1/2} \left[(1 - \lambda_{n,t}) R_{\lfloor nt \rfloor} + \lambda_{n,t} R_{\lfloor nt \rfloor + 1}\right],$$

where $\lambda_{n,t} := nt - \lfloor nt \rfloor$. We can show, following a martingale method similar to [35] (section 3), that we have the following weak convergence in the Skorokhod topology [31]:

$$(U_n(t), t \geq 0) \overset{n\to\infty}{\Rightarrow} \sigma W,$$

where W is a standard Brownian motion, and σ is given by:

$$\sigma^2 = \sum_{i \in \{-\delta, \delta\}} \pi^*(i) v(i),$$

where for $i \in \{-\delta, \delta\}$:

$$\begin{aligned} v(i) &:= b(i)^2 + p(i)(g(-i) - g(i))^2 - 2b(i)p(i)(g(-i) - g(i)), \\ b(i) &:= i - s^*, \\ p(\delta) &:= 1 - p_{cont}, \quad p(-\delta) := 1 - p'_{cont}, \end{aligned}$$

and (the vector) g is given by:

$$g = (P + \Pi^* - I)^{-1} b,$$

where Π^* is the matrix with rows equal to $(\pi^* \quad 1-\pi^*)$. After completing the calculations we get:

$$\sigma^2 = 4\delta^2 \left(\frac{1 - p'_{cont} + \pi^*(p'_{cont} - p_{cont})}{(p_{cont} + p'_{cont} - 2)^2} - \pi^*(1-\pi^*) \right).$$

For the sake of exhaustivity we also give the explicit expression for g:

$$\begin{aligned} g(\delta) &= \delta \frac{p'_{cont} - p_{cont} + 2(1-\pi^*)}{p_{cont} + p'_{cont} - 2} - s^*, \\ g(-\delta) &= \delta \frac{p'_{cont} - p_{cont} - 2\pi^*}{p_{cont} + p'_{cont} - 2} - s^*. \end{aligned}$$

Indeed, to show the above convergence of U_n, we observe that we can write R_n as the sum of a $\mathcal{F}_n$−martingale M_n and a bounded process:

$$R_n = M_n + \underbrace{g(X_n) - g(X_0) + X_n - X_0}_{unif.bounded}, M_n := \sum_{k=1}^{n} b(X_{k-1}) - g(X_k) + g(X_{k-1}),$$

where $\mathcal{F}_n := \sigma(\tau_k, X_k : k \leq n)$ and $X_0 := 0$. The process M_n is a martingale because g is the unique solution of the following Poisson equation, since $\Pi^* b = 0$:

$$[P - I]g = b.$$

The rest of the proof for the convergence of U_n follows exactly [35] (section 3).

We proved earlier (Lemma 2.1) that:

$$\frac{T_n}{n \log(n)} \Rightarrow \tau^*,$$

where $T_n := \sum_{k=1}^{n} \tau_k$. Since the Markov renewal process $(X_n, \tau_n)_{n \geq 0}$ is regular (because the state space is finite), we get $N_t \to \infty$ a.s. and therefore:

$$\frac{T_{N_t}}{N_t \log(N_t)} \Rightarrow \tau^*.$$

Observing that $T_{N_t} \leq t \leq T_{N_t+1}$ a.s., we get:

$$\frac{T_{N_t}}{N_t \log N_t} \leq \frac{t}{N_t \log N_t} \leq \frac{(N_t+1)\log(N_t+1)}{N_t \log N_t} \frac{T_{N_t+1}}{(N_t+1)\log(N_t+1)},$$

and therefore:

$$\frac{t}{N_t \log(N_t)} \Rightarrow \tau^*.$$

Let $t_n := tn\log(n)$. *We would like to show as in [11], equation* (17) *that:*

$$N_{t_n} \overset{P}{\sim} \frac{nt}{\tau^*}.$$

We have denoted by $A_n \overset{P}{\sim} B_n$ *iff* $P - \lim \frac{A_n}{B_n} = 1$. *We denote as in [11]* $\rho : (1,\infty) \to (1,\infty)$ *to be the inverse function of* $t\log(t)$, *and we note that* $\rho(t) \overset{t\to\infty}{\sim} \frac{t}{\log(t)}$. *The first equivalence in [11], equation* (17): $N_{t_n} \overset{P}{\sim} \rho\left(\frac{t_n}{\tau^*}\right)$ *is not obvious. Indeed, we have* $N_{t_n}\log(N_{t_n}) \overset{P}{\sim} \frac{t_n}{\tau^*}$, *and we would like to conclude that* $N_{t_n} = \rho(N_{t_n}\log(N_{t_n})) \overset{P}{\sim} \rho\left(\frac{t_n}{\tau^*}\right)$. *The latter implication is not true for every function* ρ, *in particular if* ρ *was exponential. Nevertheless, in our case, it is true because* $\rho(t) \overset{t\to\infty}{\sim} \frac{t}{\log(t)}$, *and therefore for any functions* f, g *going to* $+\infty$ *as* $t \to \infty$:

$$\frac{\rho(f(t))}{\rho(g(t))} \overset{t\to\infty}{\sim} \frac{f(t)}{g(t)}\frac{\log(g(t))}{\log(f(t))}.$$

Therefore we see that if $f(t) \overset{t\to\infty}{\sim} g(t)$, *then by property of the logarithm* $\log(f(t)) \overset{t\to\infty}{\sim} \log(g(t))$ *and therefore* $\rho(f(t)) \overset{t\to\infty}{\sim} \rho(g(t))$. *This allows us to conclude as in [11] that:*

$$\frac{N_{t_n}}{n} \overset{P}{\sim} \frac{t}{\tau^*}.$$

Therefore, we can make a change of time as in [35], Corollary 3.19 (see also [4], section 14), and denoting $\alpha_n(t) := \frac{N_{t_n}}{n}$, *we obtain the following weak convergence in the Skorohod topology:*

$$(U_n(\alpha_n(t)), t \geq 0) \Rightarrow (\sigma W_{\frac{t}{\tau^*}}, t \geq 0),$$

that is to say:

$$\left(\frac{s_{tn\log(n)} - N_{tn\log(n)}s^*}{\sqrt{n}}, t \geq 0\right) \Rightarrow \frac{\sigma}{\sqrt{\tau^*}}W.$$

The law of large numbers result comes from the fact that $\frac{N_{t_n}}{n} \overset{P}{\sim} \frac{t}{\tau^*}$, *together with the following fact (strong law of large numbers for Markov chains):*

$$\frac{1}{n}\sum_{k=1}^{n} X_k \to s^* \quad a.e.$$

2.4.2 Other cases: either $P^a(1,1) < P^a(-1,-1)$ or $P^b(1,1) < P^b(-1,-1)$

In this case, we know by Theorem 2.2 that the conditional expectations $E[\tau_k | q_0^b = n_b, q_0^a = n_a]$ are finite. Denoting the conditional expectations

$m(\pm\delta) := E[\tau_k | X_{k-1} = \pm\delta]$, we have:

$$m(\delta) = \sum_{p=1}^{\infty}\sum_{n=1}^{\infty} E[\tau_k | q_0^b = n, q_0^a = p] f(n,p),$$
$$m(-\delta) = \sum_{p=1}^{\infty}\sum_{n=1}^{\infty} E[\tau_k | q_0^b = n, q_0^a = p] \tilde{f}(n,p).$$

Throughout this section we will need the following assumption:

(A5) Using the previous notations, the following holds:

$$m(\pm\delta) < \infty.$$

For example, the above assumption is satisfied if the support of the distributions f and $\tilde{f}$ is compact, which is the case in practice. We obtain the following diffusion limit result as a classical consequence of invariance principle results for semi-Markov processes (see e.g. [35], section 3):

Theorem 2.6 *Under assumption* **(A5)**, *the re-normalized price process s_{nt} satisfies the following convergence in the Skorokhod topology:*

$$\left(\frac{s_{nt}}{n}, t \geq 0\right) \overset{n\to\infty}{\to} \left(\frac{s^* t}{m_\tau}, t \geq 0\right) \quad a.e.,$$
$$\left(\frac{s_{nt} - N_{nt}s^*}{\sqrt{n}}, t \geq 0\right) \overset{n\to\infty}{\Rightarrow} \frac{\sigma}{\sqrt{m_\tau}} W,$$

where W is a standard Brownian motion, σ is given in Theorem 2.5 and:

$$m_\tau := \sum_{i \in \{-\delta,\delta\}} \pi^*(i) m(i) = \pi^* m(\delta) + (1-\pi^*) m(-\delta).$$

Proof *This is an immediate consequence of strong law of large numbers and invariance principle results for Markov renewal processes satisfying $m(\pm\delta) < \infty$ (see e.g. [35] section 3). In the previous article [35], the proof of the invariance principle is carried on using a martingale method similar to the one of the proof of proposition 2.5.*

2.5 Numerical Results

In this section, we present calibration results which illustrate and justify our approach.

In [11], it is assumed that the queue changes V_k^b, V_k^a do not depend on their previous values V_{k-1}^b, V_{k-1}^a. Empirically, it is found that $P[V_k^b = 1] \approx P[V_k^b =$

TABLE 2.2
Average time between orders (ms) & Average number of stocks per order. June 21, 2012.

	Amazon		Apple		Google		Intel		Microsoft	
	Bid	Ask	Bid	Ask	Bid	Ask	Bid	Ask	Bid	Ask
Avg time btw. orders (ms)	910	873	464	425	1123	1126	116	133	130	113 —
Avg nb. of stocks per order	100	82	90	82	84	71	502	463	587	565 —

TABLE 2.3
Estimated transition probabilities of the Markov Chains V_k^b, V_k^a. June 21, 2012.

	Amazon		Apple		Google		Intel		Microsoft	
	Bid	Ask	Bid	Ask	Bid	Ask	Bid	Ask	Bid	Ask
$P(1,1)$	0.48	0.57	0.50	0.55	0.48	0.53	0.55	0.61	0.63	0.60
$P(-1,1)$	0.46	0.42	0.40	0.42	0.46	0.49	0.44	0.40	0.36	0.41
$P(-1,-1)$	0.54	0.58	0.60	0.58	0.54	0.51	0.56	0.60	0.64	0.59
$P(1,-1)$	0.52	0.43	0.50	0.45	0.52	0.47	0.45	0.39	0.37	0.40
$P(1)$	0.47	0.497	0.44	0.48	0.47	0.51	0.495	0.505	0.49	0.508
$P(-1)$	0.53	0.503	0.56	0.52	0.53	0.49	0.505	0.495	0.51	0.492

$-1] \approx 1/2$ (and similarly for the ask side). Here, we challenge this assumption by estimating and comparing the probabilities $P(-1,1)$ Vs. $P(1,1)$ on the one side and $P(-1,-1)$ Vs. $P(1,-1)$ on the other side to check whether or not they are approximately equal to each other, for both the ask and the bid. We also give – for both the bid and ask – the estimated probabilities $P[V_k = 1]$, $P[V_k = -1]$ that we call respectively $P(1)$, $P(-1)$, to check whether or not they are approximately equal to $1/2$ as in [11].

The results below correspond to the 5 stocks Amazon, Apple, Google, Intel, Microsoft on June 21, 2012[1]. The probabilities are estimated using the strong law of large numbers. We also give for indicative purposes the average time between order arrivals (in milliseconds (ms)) as well as the average number of stocks per order.

Findings: First of all, we find as in [11] that for all stocks, $P[V_k = 1] \approx P[V_k = -1] \approx 1/2$, except maybe in the case of Apple Bid. It is worth mentioning that we always have $P(1) < P(-1)$ except in 3 cases: Google Ask, Intel Ask and Microsoft Ask. Nevertheless, in these cases, $P(1)$ and $P(-1)$ are very close to each other and so they could be considered to fall into the case $P(1) = P(-1)$ of [11]. These 3 cases also correspond to the only 3 cases where $P(1,1) > P(-1,-1)$, which is contrary to our assumption $P(1,1) \leq P(-1,-1)$. Nevertheless, in these 3 cases, $P(1,1)$ and $P(-1,-1)$

[1]The data was taken from the webpage *https://lobster.wiwi.hu-berlin.de/info/DataSamples.php*

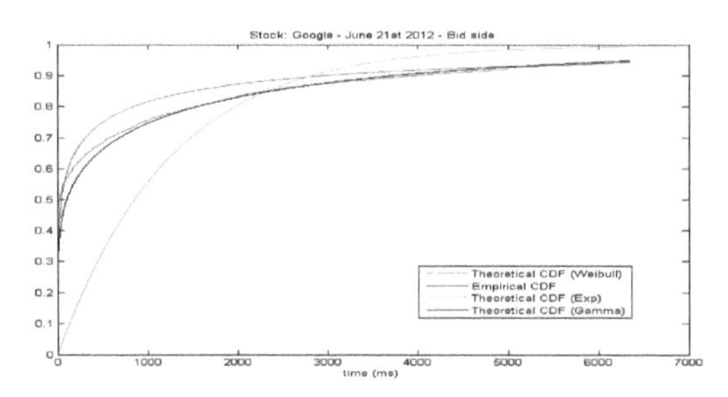

FIGURE 2.1
H(1,-1) – Google Bid – June 21, 2012.

are very close to each other so we can consider them to fall into the case $P(1,1) = P(-1,-1)$.

More importantly, we notice that the probabilities $P(-1,1)$, $P(1,1)$ can be significantly different from each other – and similarly for the probabilities $P(-1,-1)$, $P(1,-1)$ – which justifies the use of a Markov Chain structure for the random variables $\{V_k^b\}, \{V_k^a\}$. This phenomenon is particularly visible for example on Microsoft (Bid+Ask), Intel (Bid+Ask), Apple (Bid+Ask) or Amazon Ask. Further, regarding the comparison of $P(1,1)$ and $P(-1,-1)$, it turns out that they are often very smilar, except in the cases Amazon Bid, Apple Bid, Google Bid.

The second assumption of [11] that we would like to challenge is the assumed exponential distribution of the order arrival times T_k^a, T_k^b. To this end, on the same data set as used to estimate the transition probabilities $P^a(i,j)$, $P^b(i,j)$, we calibrate the empirical c.d.f.'s $H^a(i,j,\cdot)$, $H^b(i,j,\cdot)$ to the Gamma and Weibull distributions (which are generalizations of the exponential distribution). We recall that the p.d.f.'s of these distributions are given by:

$$\begin{aligned} f_{Gamma}(x) &= \frac{1}{\Gamma(k)\theta^k} x^{k-1} e^{-\frac{x}{\theta}} 1_{x>0}, \\ f_{Weibull}(x) &= \frac{k}{\theta}\left(\frac{x}{\theta}\right)^{k-1} e^{-\left(\frac{x}{\theta}\right)^k} 1_{x>0}. \end{aligned}$$

Here, $k > 0$ and $\theta > 0$ represent respectively the shape and the scale parameter. The variable k is dimensionless, whereas θ will be expressed in ms^{-1}. We perform a maximum likelihood estimation of the Weibull and Gamma parameters for each one of the empirical distributions $H^a(i,j,\cdot)$, $H^b(i,j,\cdot)$ (together with a 95% confidence interval for the parameters). As we can see on the Table 2.4, the shape parameter k is always significantly different than 1 ($\sim$ 0.1 to 0.3), which indicates that the exponential distribution is not rich enough to fit our observations. To illustrate this, we present below the empirical c.d.f. of $H(1,-1)$ in the case of Google Bid, and we see that Gamma and Weibull allow to fit the empirical c.d.f. in a much better way than Exponential.

We summarize our calibration results in the Tables 2.5–2.13.

TABLE 2.4
Amazon Bid: Fitted Weibull and Gamma parameters. 95% confidence intervals in brackets. June 21, 2012.

Amazon Bid	$H(1,1)$	$H(1,-1)$	$H(-1,-1)$	$H(-1,1)$
Weibull θ	99.1	185.5	87.7	87.0
	(90.2-109.0)	(171.3-200.8)	(80.1-96.0)	(78.7-96.1)
Weibull k	0.279	0.323	0.285	0.258
	(0.274-0.285)	(0.317-0.329)	(0.280-0.290)	(0.253-0.263)
Gamma θ	4927	4321	4712	5965
	(4618-5257)	(4075-4582)	(4423-5019)	(5589-6366)
Gamma k	0.179	0.215	0.179	0.165
	(0.174-0.184)	(0.209-0.220)	(0.175-0.184)	(0.161-0.169)

TABLE 2.5
Amazon Ask: Fitted Weibull and Gamma parameters. 95% confidence intervals in brackets. June 21, 2012.

Amazon Ask	$H(1,1)$	$H(1,-1)$	$H(-1,-1)$	$H(-1,1)$
Weibull θ	80.8	197.8	57.9	137.0
	(74.4-87.7)	(181.9-215.1)	(52.8-63.4)	(124.2-151.2)
Weibull k	0.274	0.324	0.279	0.276
	(0.269-0.278)	(0.317-0.330)	(0.274-0.285)	(0.270-0.281)
Gamma θ	4732	4623	3845	5879
	(4475-5004)	(4345-4919)	(3609-4095)	(5502-6283)
Gamma k	0.174	0.215	0.173	0.181
	(0.170-0.178)	(0.209-0.221)	(0.168-0.177)	(0.176-0.186)

TABLE 2.6
Apple Bid: Fitted Weibull and Gamma parameters. 95% confidence intervals in brackets. June 21, 2012.

Apple Bid	$H(1,1)$	$H(1,-1)$	$H(-1,-1)$	$H(-1,1)$
Weibull θ	75.9	180.9	31.5	78.2
	(71.6-80.5)	(172.6-189.7)	(29.5-33.6)	(73.4-83.3)
Weibull k	0.317	0.400	0.271	0.300
	(0.313-0.321)	(0.394-0.405)	(0.267-0.274)	(0.296-0.304)
Gamma θ	2187	1860	2254	2711
	(2094-2284)	(1787-1935)	(2157-2355)	(2592-2835)
Gamma k	0.206	0.276	0.168	0.196
	(0.202-0.210)	(0.271-0.282)	(0.165-0.171)	(0.192-0.199)

TABLE 2.7
Apple Ask: Fitted Weibull and Gamma parameters. 95% confidence intervals in brackets. June 21, 2012.

Apple Ask	$H(1,1)$	$H(1,-1)$	$H(-1,-1)$	$H(-1,1)$
Weibull θ	46.6	152.5	27.7	95.5
	(44.1-49.2)	(145.5-159.8)	(26.0-29.6)	(90.0-101.5)
Weibull k	0.298	0.394	0.271	0.308
	(0.294-0.301)	(0.388-0.399)	(0.267-0.275)	(0.303-0.312)
Gamma θ	2019	1666	1995	2740
	(1942-2099)	(1603-1732)	(1907-2087)	(2624-2861)
Gamma k	0.189	0.271	0.168	0.204
	(0.186-0.192)	(0.266-0.277)	(0.165-0.171)	(0.200-0.208)

TABLE 2.8
Google Bid: Fitted Weibull and Gamma parameters. 95% confidence intervals in brackets. June 21, 2012.

Google Bid	$H(1,1)$	$H(1,-1)$	$H(-1,-1)$	$H(-1,1)$
Weibull θ	113.9	158.5	67.9	56.8
	(102.8-126.2)	(143.4-175.3)	(60.6-76.0)	(50.5-63.8)
Weibull k	0.276	0.284	0.261	0.246
	(0.270-0.282)	(0.278-0.290)	(0.255-0.266)	(0.241-0.251)
Gamma θ	6720	6647	6381	7025
	(6263-7210)	(6204-7122)	(5913-6886)	(6517-7571)
Gamma k	0.174	0.185	0.160	0.151
	(0.169-0.179)	(0.180-0.191)	(0.155-0.165)	(0.147-0.156)

TABLE 2.9
Google Ask: Fitted Weibull and Gamma parameters. 95% confidence intervals in brackets. June 21, 2012.

Google Ask	$H(1,1)$	$H(1,-1)$	$H(-1,-1)$	$H(-1,1)$
Weibull θ	196.7	271.6	38.1	57.0
	(180.6-214.2)	(248.5-296.8)	(33.8-43.0)	(51.3-63.3)
Weibull k	0.290	0.310	0.258	0.263
	(0.285-0.295)	(0.303-0.316)	(0.253-0.264)	(0.258-0.268)
Gamma θ	6081	6571	4304	4698
	(5734-6450)	(6165-7003)	(3971-4664)	(4380-5040)
Gamma k	0.195	0.209	0.156	0.164
	(0.190-0.200)	(0.203-0.215)	(0.151-0.161)	(0.159-0.168)

TABLE 2.10
Intel Bid: Fitted Weibull and Gamma parameters. 95% confidence intervals in brackets. June 21, 2012.

Intel Bid	$H(1,1)$	$H(1,-1)$	$H(-1,-1)$	$H(-1,1)$
Weibull θ	2.76	2.56	3.33	2.01
	(2.66-2.86)	(2.45-2.67)	(3.21-3.45)	(1.92-2.10)
Weibull k	0.227	0.226	0.267	0.209
	(0.226-0.229)	(0.225-0.228)	(0.265-0.269)	(0.208-0.211)
Gamma θ	1016	912	543	1093
	(991-1040)	(888-937)	(530-557)	(1063-1124)
Gamma k	0.129	0.130	0.151	0.120
	(0.128-0.130)	(0.129-0.131)	(0.150-0.152)	(0.119-0.121)

TABLE 2.11
Intel Ask: Fitted Weibull and Gamma parameters. 95% confidence intervals in brackets. June 21, 2012.

Intel Ask	$H(1,1)$	$H(1,-1)$	$H(-1,-1)$	$H(-1,1)$
Weibull θ	1.33	5.46	4.63	5.15
	(1.28-1.38)	(5.21-5.73)	(4.45-4.80)	(4.90-5.41)
Weibull k	0.235	0.231	0.256	0.225
	(0.234-0.237)	(0.230-0.233)	(0.254-0.257)	(0.224-0.227)
Gamma θ	705	1219	884	1305
	(688-723)	(1183-1256)	(862-907)	(1266-1345)
Gamma k	0.126	0.137	0.146	0.133
	(0.125-0.127)	(0.136-0.139)	(0.144-0.147)	(0.132-0.135)

TABLE 2.12
Microsoft Bid: Fitted Weibull and Gamma parameters. 95% confidence intervals in brackets. June 21, 2012.

Microsoft Bid	$H(1,1)$	$H(1,-1)$	$H(-1,-1)$	$H(-1,1)$
Weibull θ	0.79	2.98	2.68	2.64
	(0.76-0.82)	(2.83-3.13)	(2.59-2.78)	(2.50-2.78)
Weibull k	0.215	0.221	0.259	0.211
	(0.214-0.217)	(0.219-0.223)	(0.257-0.260)	(0.209-0.213)
Gamma θ	1012	1315	664	1488
	(987-1039)	(1274-1358)	(648-681)	(1440-1537)
Gamma k	0.112	0.125	0.142	0.120
	(0.111-0.113)	(0.124-0.127)	(0.141-0.143)	(0.118-0.121)

TABLE 2.13
Microsoft Ask: Fitted Weibull and Gamma parameters. 95% confidence intervals in brackets. June 21, 2012.

Microsoft Ask	$H(1,1)$	$H(1,-1)$	$H(-1,-1)$	$H(-1,1)$
Weibull θ	0.85	1.57	2.07	1.43
	(0.82-0.89)	(1.50-1.64)	(2.00-2.15)	(1.36-1.50)
Weibull k	0.218	0.223	0.259	0.210
	(0.217-0.219)	(0.222-0.225)	(0.258-0.261)	(0.208-0.211)
Gamma θ	1004	1081	574	1138
	(980-1028)	(1051-1112)	(560-588)	(1105-1171)
Gamma k	0.113	0.121	0.140	0.116
	(0.112-0.114)	(0.120-0.122)	(0.139-0.141)	(0.115-0.117)

2.6 More Big Data

In this section, we discuss our approach for various assets (different from 5 stocks considered in Section 2.5) (liquid, illiquid, medium liquid) from different markets, namely for 15 stocks from Deutsche Börse Group (September 23, 2013) and for CISCO (November 3, 2014), and motivation for some of our assumptions. We also discuss a possible extension of our model for the case when the spread is not fixed, including the diffusion limit of the price dynamics in this case, and about the optimal liquidation/acquisition and market making problems.

2.6.1 More Data

Of course, the five stocks we have chosen are perhaps the most active (at least in NASDAQ) and our numerical results might be misleading when considering a more typical stocks. Here, we would like to point out that our assumptions about the non-Markovian behaviour of the LOB and non-exponential distribution of inter-arrival events are valid not only for those five stocks but also for bunch of many others.

We use the financial instruments traded on Xetra and Frankfurt markets (Deutsche Börse Group, web: http://datashop.deutsche-boerse.com/1016/en) on September 23, 2013. The description of all instruments is presented in Figure 2.2: the first column gives German security identification number, the second-international security identification number, the third-security name, and the last one-common name. We divided 15 assets, presented in the Figure 2.2, by three groups: (1) Liquid assets (every 372–542 milliseconds (ms) in average an order arrives), (2) Medium liquid assets (every 1392–1415 ms in average an order arrives), and (3) Illiquid assets (every 8392–8467 ms in average an order arrives). The thresholds for these three groups of assets are

WKN	ISIN	INSTRUMENT NAME	COMMON NAME
		LIQUID ASSETS:	
A1JEAN	LU0665646815	UBS-ETF-MSCI EU.IN.2035 I	UBS-ETF MSCI Europe Infrastructure I
A1JVYM	IE00B7KMTJ66	UBS(I)ETF-SOL.G.P.GD IDDL	Solactive Global Pure Gold Miners UCITS ETF
A1JEAJ	LU0665646229	UBS-ETF-MSCI JA.IN.2035 I	UBS-ETF MSCI Japan Infrastructure I
A1JVYN	IE00B7KYPQ18	UBS(I)ETF-SOL.G.O.EQ.IDDL	Solactive Global Oil Equities UCITS ETF I
A1JVCB	IE00B7KL1H59	UBS(I)ETF-MSCI WORLD IDDL	MSCI World UCITS ETF I
		MEDIUM LIQUID ASSETS	
ETC057	DE000ETC0571	COMMERZBANK ETC UNL.	Coba ETC -3x WTI Oil Daily Short Index
ETC015	DE000ETC0159	COMMERZBANK ETC UNL.	Coba ETC -1x Gold Daily Short Index
ETC030	DE000ETC0308	COMMERZBANK ETC UNL.	Coba ETC 4x Brent Oil Daily Long Index
A0X8SE	IE00B3VWMM18	ISHSVII-MSCI EMU SC U.ETF	iShares MSCI EMU Small Cap UCITS ETF
A0JMFG	FR0010296061	LYXOR ETF MSCI USA D-EO	Lyxor UCITS ETF MSCI USA D-EUR
		ILLIQUID ASSESTS	
A1JB4P	DE000A1JB4P2	I.II-IS. D.J.G.S.S.UTS DZ	iShares Dow Jones Global Sustainability Screened UCITS ETF
630500	DE0006305006	DEUTZ AG O.N.	DEUTZ AG O.N.
A1T8GD	IE00B9CQXS71	SPDR S+P GL.DIV.ARIST.ETF	SPDR® S&P® Global Dividend Aristocrats UCITS ETF
851144	US3696041033	GENL EL. CO. DL -,06	General Electric STK
113541	DE0001135416	BUNDANL.V. 10/20	Bundesrepublik Deutschland 2,250% 9/2020 BOND

FIGURE 2.2
15 Stocks from Deutsche Börse Group.

arbitrary as long as we captured by this division the whole range of assets in this set of data (for measure liquidity we refer to [6], Section 2.1.3). Comparisons of ask PDF for the 5 Liquid assets presented in Figures 2.3–2.7, for the 5 Illiquid assets presented in Figures 2.13–2.17, and for the 5 Medium Liquid assets in Figures 2.8–2.12. As we can see, the best fit for these set of assets gives the Burr type XII distribution $F(x) = 1 - (1 + x^c)^{-k}$, ($x > 0, c > 0, k > 0$, both c and k are shape parameters (see [34])), not exponential. We note, that all graphs contains comparison for empirical, exponential, Gamma, Weibull, Pareto, Power law and Burr distributions (7 in total).

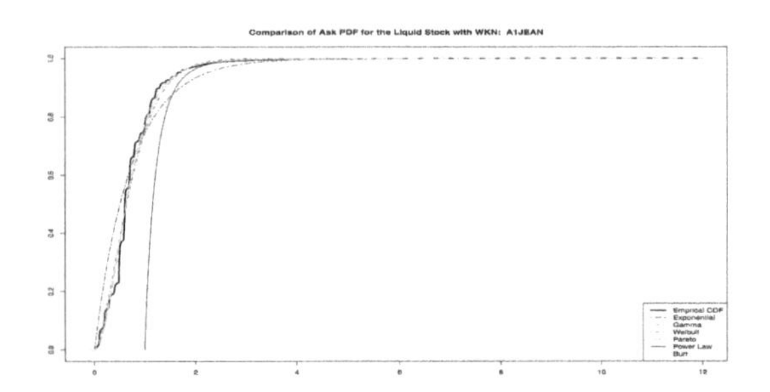

FIGURE 2.3
Comparison of Ask PDF for Liquid Stock with WKN: A1JEAN.

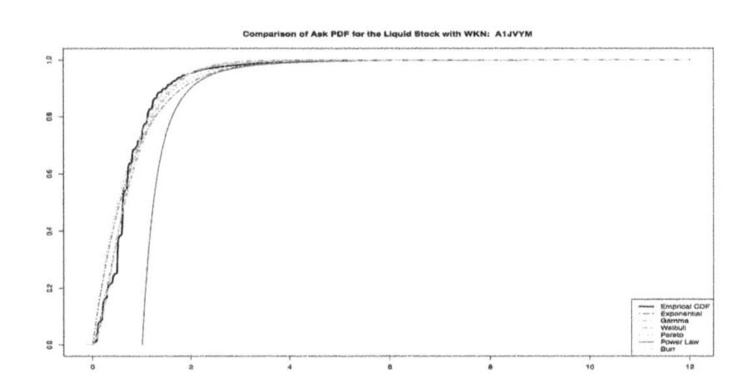

FIGURE 2.4
Comparison of Ask PDF for Liquid Stock with WKN: A1JVYM.

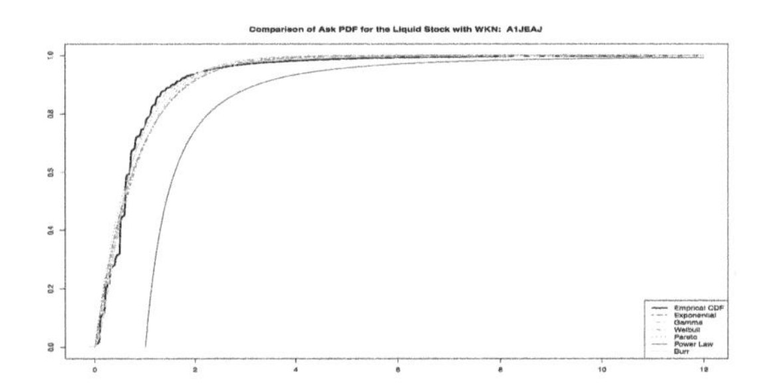

FIGURE 2.5
Comparison of Ask PDF for Liquid Stock with WKN: A1JEAN.

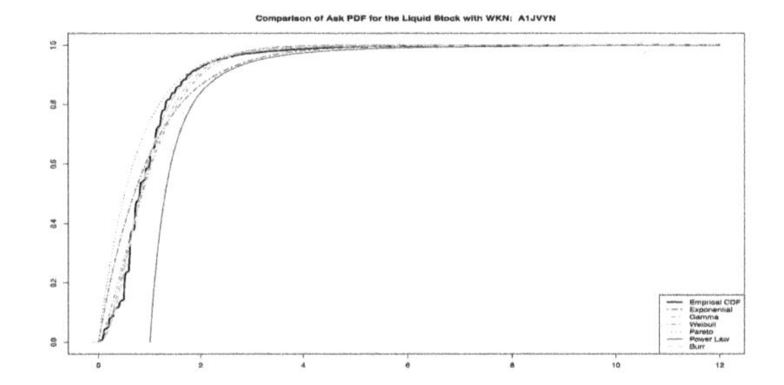

FIGURE 2.6
Comparison of Ask PDF for Liquid Stock with WKN: A1JVYN.

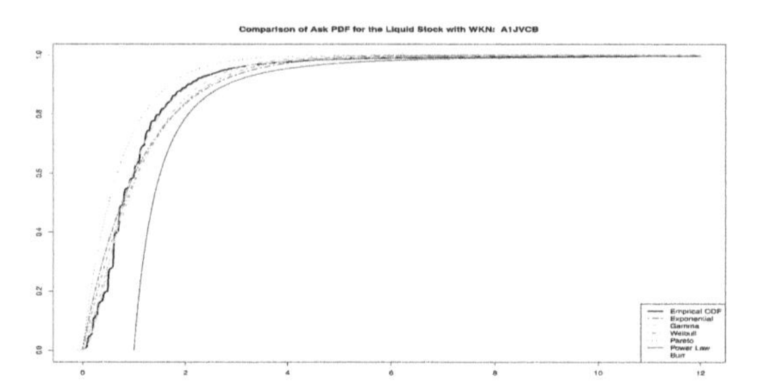

FIGURE 2.7
Comparison of Ask PDF for Liquid Stock with WKN: A1JVCB.

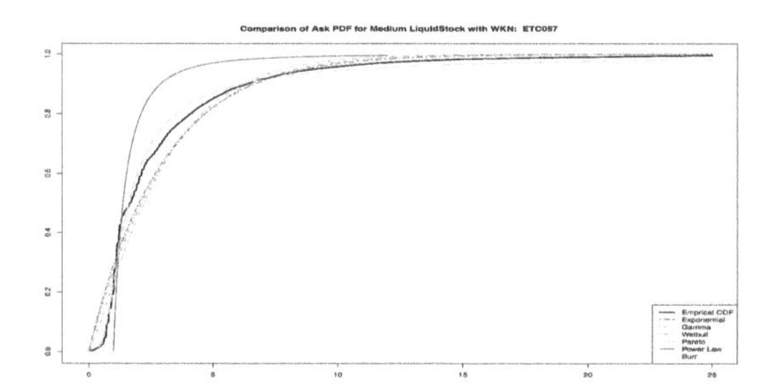

FIGURE 2.8
Comparison of Ask PDF for Medium Stock with WKN: ETC057.

Moreover, we used one more set of data, namely, CISCO on November 3, 2014 (courtesy: [6]), to show that inter-arrival times between limit orders at the best ask does not follow exponential distribution: see Figure 2.20. Plots of the frequencies of inter-arrival times in absolute and log scales for CISCO on November 3, 2014, are also presented in Figures 2.18–2.19 for completeness.

We also mention that non-exponential property of the inter-arrival times was indicated, for example, in [6] for AAPL on July 30, 2013: their graphs (page 49, Figures 3.4–3.5) indicates that the inter-arrival times have a power-law distribution with heavy tails. Also, non-Markovian nature of price changes was indicated in [6], Section 3.5, as well.

2.6.2 Estimated Probabilities

Concerning the estimated probabilities for another set of data, we took CISCO data on November 3, 2014 (courtesy: [6]) and estimated the transition probabilities between book events at the best ask, using tsimilar approach we have used for Microsoft on June 21, 2012 (see Table in the Introduction). We get the following entries:

As we can see, for the CISCO data the probabilities $P(i, j)$ also significally depend on the previous event i $(i, j = 1, 2)$.

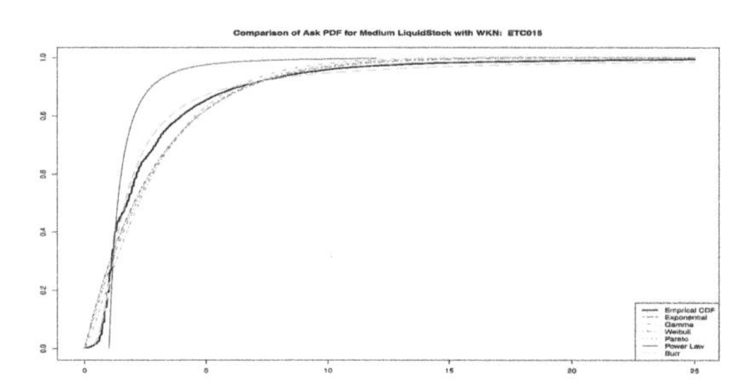

FIGURE 2.9
Comparison of Ask PDF for Medium Stock with WKN: ETC015.

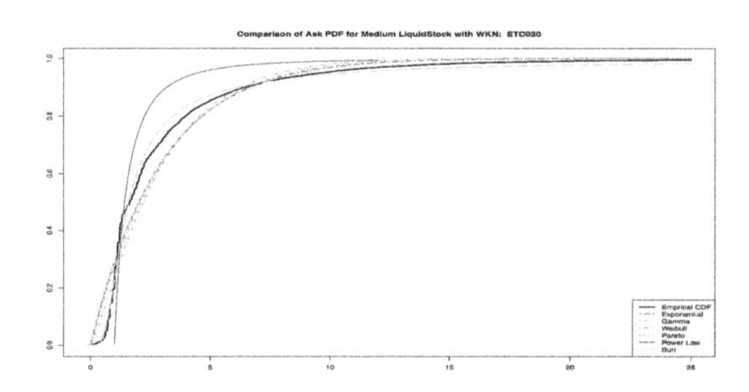

FIGURE 2.10
Comparison of Ask PDF for Medium Stock with WKN: ETC030.

The dependency goes even further. As an example, below are the probabilities for the best ask (level I, as always) for CISCO (the same day as before) for the events $[k|i, j]$, where $k, i, j \in \{1, -1\}$:

$P[1|1, 1] = 0.1993023,$
$P[1|1, -1] = 0.09849214,$
$P[1|-1, 1] = 0.1033831,$
$P[1|-1, -1] = 0.09559463,$
$P[-1|1, 1] = 0.09849214,$
$P[-1|1, -1] = 0.1004856,$
$P[-1|-1, 1] = 0.09559463,$

TABLE 2.14
Estimated probabilities for book event arrivals for CISCO, November 3, 2014.

CISCO	Ask
$P(1, 1)$	0.5994541
$P(-1, 1)$	03954211
$P(-1, -1)$	0.6045789
$P(1, -1)$	0.4005226

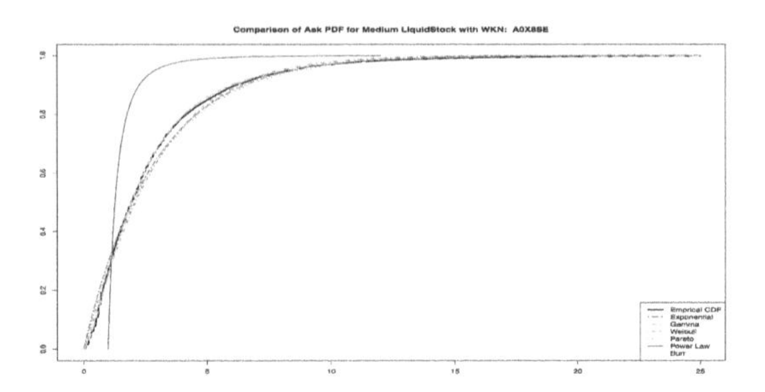

FIGURE 2.11
Comparison of Ask PDF for Medium Stock with WKN: AOX8SE.

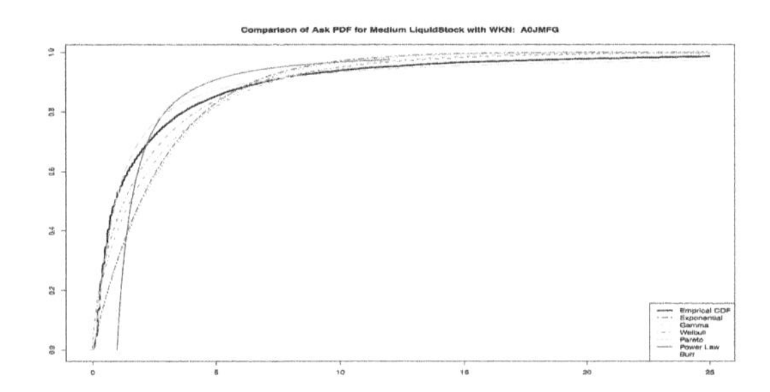

FIGURE 2.12
Comparison of Ask PDF for Medium Stock with WKN: AOJMFG.

$P[-1|-1,-1] = 0.2086323.$

They are not symmetric, and clearly the events $[1|1,1[$ and $[-1|-1,-1]$ have larger probability than the rest. The cases where the probabilities are mixed seem to have close probabilities.

Moreover, we would also like to point out that some limit order markets have long memory. For example, in the paper [23], the authors made use of a data set from the London Stock Exchange (LSE), which contains a full record of individual orders and cancellations, to show that stocks in the LSE display a remarkable long-memory property. They labeled each event as either a buy or a sell order, and assign it ± 1 accordingly. The autocorrelation of the resulting time series shows a power law autocorrelation function, with exponents α that are typically about 0.6, in the range $0.36 < \alpha < 0.77$. Positive autocorrelation coefficients are seen at statistically significant levels over lags of many thousand events, spanning many days. In the paper [5], the authors also independently discovered the same long-memory effect (that the previous paper [23] reported) for market order flow in the Paris Stock Exchange. Thus the memory of some markets is remarkably long.

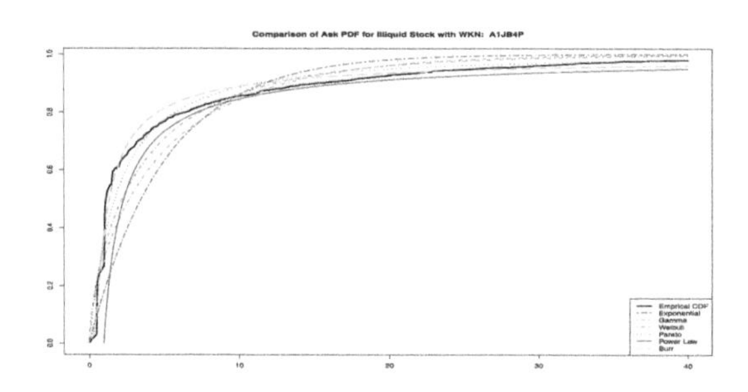

FIGURE 2.13
Comparison of Ask PDF for Illiquid Stock with WKN: A1JB4P.

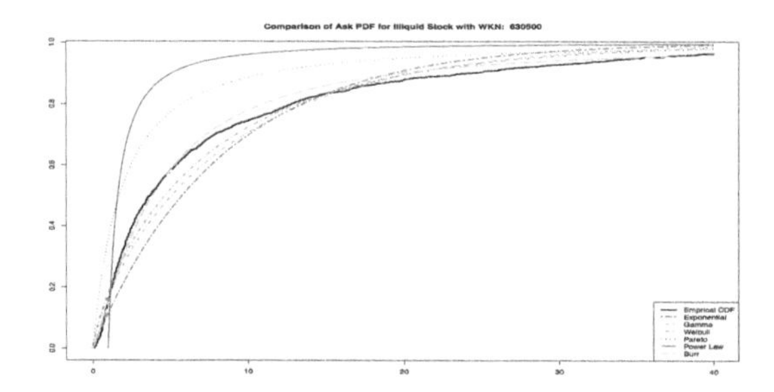

FIGURE 2.14
Comparison of Ask PDF for Illiquid Stock with WKN: 630500.

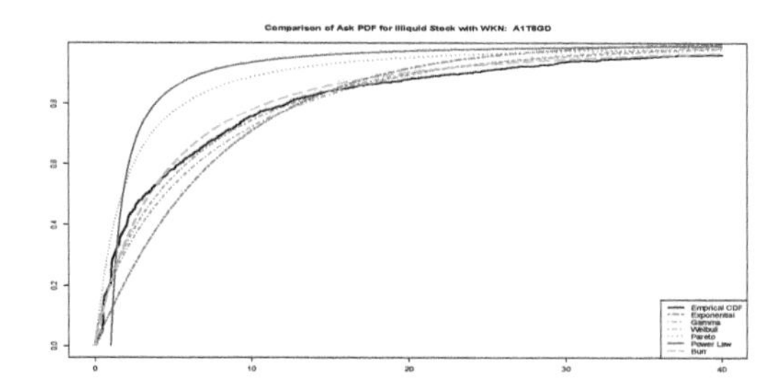

FIGURE 2.15
Comparison of Ask PDF for Illiquid Stock with WKN: A1T8GD.

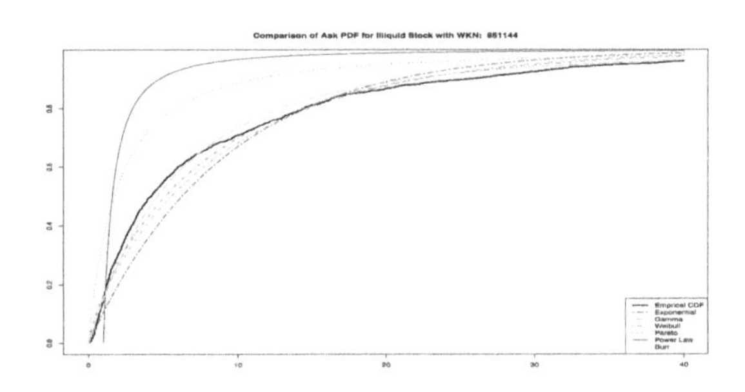

FIGURE 2.16
Comparison of Ask PDF for Illiquid Stock with WKN: 851144.

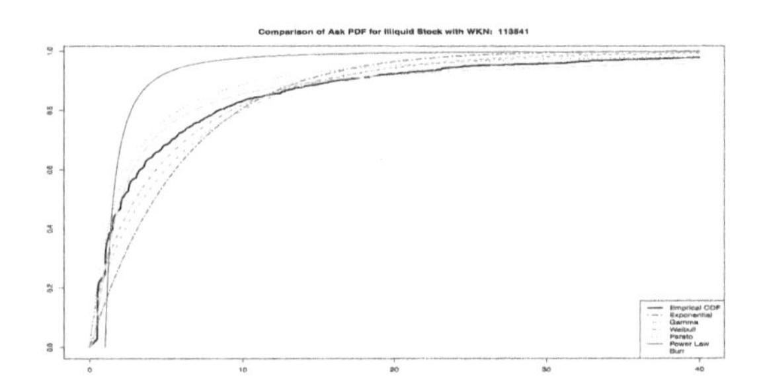

FIGURE 2.17
Comparison of Ask PDF for Illiquid Stock with WKN: 113541.

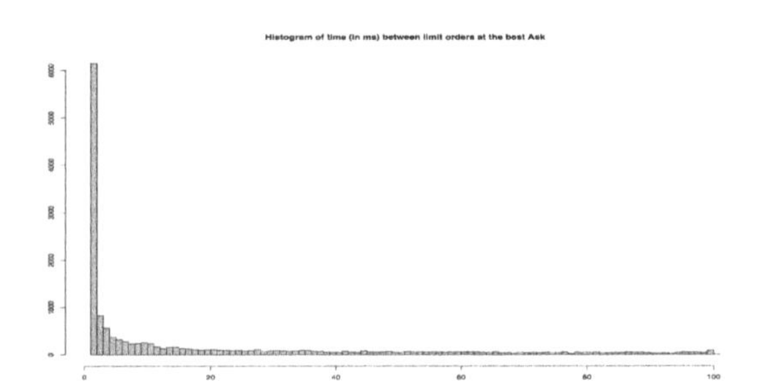

FIGURE 2.18
Histogram of time (ms) between limit orders at the best ask for CISCO on November 3, 2014.

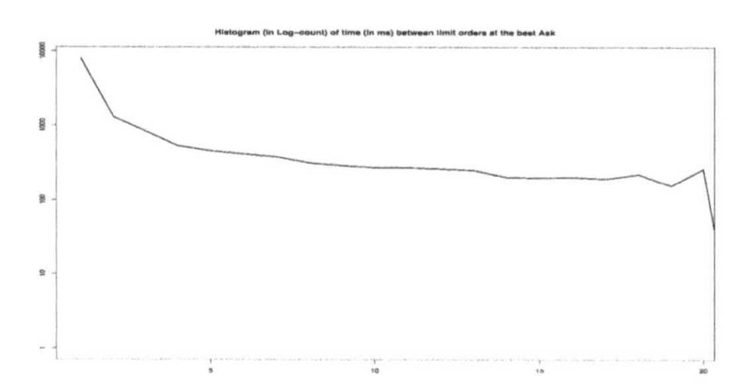

FIGURE 2.19
Histogram (log-count) of time (ms) between limit orders at the best ask for CISCO on November 3, 2014.

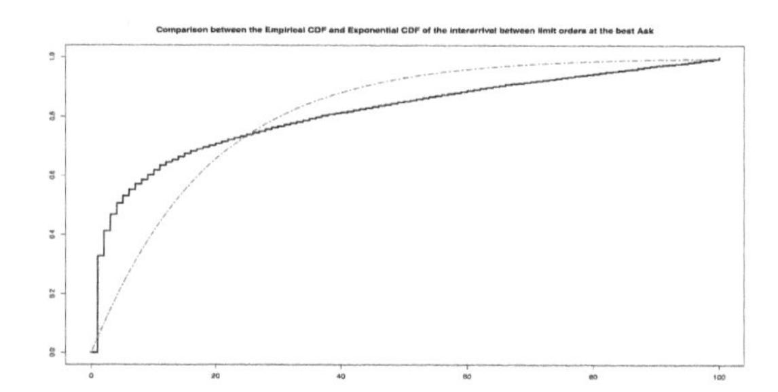

FIGURE 2.20
Comparison between the Empirical CDF and Exponential CDF of the inter-arrival between limit orders at the best Ask for CISCO on November 3, 2014.

2.6.3 Assumption on Distributions f and $\tilde{f}$

In [11] it is assumed that $f(i,j) = \tilde{f}(i,j) = f(j,i)$ to make the price increments X_n independent and identically distributed. Namely: if $q^a_{t-} = 0$, then (q^b_t, q^a_t) is a random variable with distribution f independent from $\mathcal{F}_{t-}$; if $q_t{-}^b = 0$, then (q^b_t, q^a_t) is a random variable with distribution $\tilde{f}$ independent from $\mathcal{F}_{t-}$. Then the process (q^b_t, q^a_t) is Markovian, otherwise it becomes non-Markovian. This is exactly our case: we entirely relax this assumption, $f(i,j) \neq \tilde{f}(i,j) \neq f(j,i)$, and that is why we can use the related theory of semi-Markov processes to compute the diffusion limit of the price process s_t, because (X_n, τ_n) is a Markov renewal process. From the construction of the processes q^b_t and q^a_t in our case it follows that they never take simultaneously the value 0, as whenever $q^b_t = 0$ or $q^a_t = 0$, the pair $(q^b_{T_n}, q^a_{T_n})$ is replaced by a random variable drawn from the distribution f or $\tilde{f}$, respectively. This assumption is motivated and justified by using data. For example, we used CISCO on November 3, 2014 to check this out. The total number of entries is 304, 386. According to the data the simultaneous depletion happened only 6 times, but all of those times were within less than 17.5 seconds after the market opened (9:30 am), and that period is very transitional, because the outstanding orders from the previous day are being posted, and this is the case when things are getting started for the trading day. After 18 seconds elapsed (at 9:30:18 am) it never happened again. Thus, the probability of this event is very small, $6/304,386 \approx 0.00002$.

2.6.4 Diffusion Limit (Not-Fixed Spread)

In this section, we discuss the case when the spread is not fixed at one tick δ. In this case, we can consider the following model for the price process:

$$s_t = \sum_{k=1}^{N(t)} a(X_k),$$

where $N(t)$ is the counting process defined in Section 2.2, X_k is a Markov chain with two states $1, 2$, and function $a(x)$ is uniformly bounded. Then, in the case of fixed tick, we can take $a(1) = \delta$ and $a(2) = -\delta$, which means that this model is more general than the model considered in Section 2. We present here only the result on the diffusion limit of the price dynamics for the balanced case (so-called the "heavy traffic" case). Suppose that assumption (**A4**) from Subsection 4.1 is satisfied and $(\pi^*(1), \pi^*(2))$ are stationary probabilities of the Markov chain X_k. Then, in the balanced order flow case (proof follows from general CLT for inhomogeneous SMP, see [35] for more details):

$$\left(\frac{s_{tn\log(n)} - N_{tn\log(n)}a^*}{\sqrt{n}}, t \geq 0\right) \overset{n\to\infty}{\Rightarrow} \frac{\sigma^*}{\sqrt{\tau^*}} W_t, \tag{2.1}$$

where W_t is a standard Brownian motion, $a^* := \pi^*(1)a(1) + \pi^*(2)a(2)$, $(\sigma^*)^2 := \sum_{i\in\{1,2\}} \pi^*(i)v^*(i)$, $v^*(i) := (a(i) - a^*)^2 + p(i)(g(-i) - g(i))^2 - 2b(i)p(i)(g(-i) - g(i))$, $g(i) := (P + \Pi^* - I)(a(i) - a^*)$, Π^* is a matrix with rows equal to $(\pi^*(1), \pi^*(2))$. P, τ^* are defined in Subsection 3.3 and 4.1, respectively. We mention that this result is even more general than in Theorem 4.1: if we take $a(1) = \delta$ and $a(2) = -\delta$, then $a^* = s^*$ (here, $\pi^*(1) = \pi^*$ and $\pi^*(2) = 1 - \pi^*$) and $\sigma^* = \sigma$ (because $a(i) = i \in \{\delta, -\delta\}$ and, hence, $v^*(i) = v(i)$) (see proof of Theorem 4.1), then the result of Theorem 4.1 follows.

Similar, in the other non-balanced cases, the price process s_{tn} satisfies the following convergence in the Skorokhod topology:

$$\left(\frac{s_{tn} - N_{tn}a^*}{\sqrt{n}}, t \geq 0\right) \overset{n\to\infty}{\Rightarrow} \frac{\sigma^*}{\sqrt{m_\tau}} W_t, \tag{2.2}$$

where a^* and σ^* are defined above in (1), $m_\tau := \pi^*(1)m(1) + \pi^*(2)m(2)$, and $m(i), i = 1, 2$, are defined in Subsection 4.2.

We could present other results, considered in this chapter, associated with this model for the price process, but it is beyond the scope of the chapter and the book.

2.6.5 The Optimal Liquidation/Acquisition Problems

Here, we would like to discuss some ideas of how we can use our approach in the typical algo problems such as optimal liquidation or acquisition strategies that imploy limit orders. We have two situation here: long time scale and short time scale.

The first situation (long time scale) is associated with trading decisions that affect only an agent's wealth process, but not the dynamics of the asset which is trading. In many applications such as order executions (e.g., an agent seeks to maximize expected wealth by trading in a risky asset and the risk-free bank account), the metric of success is the VWAP, so someone (the agent) may be interested in the dynamics of order flow over a large time scale, typically tens of seconds or minutes. In this way, the agent's trading decisions affect only his wealth process, but not the dynamics of the asset which he is trading. On long time scales this is reasonable assumption, as it was mentioned in [6]. In this case, we can use the diffusion limit model for the price process on long time scale, i.e., for the balanced case, we could use result in (1) above, and for other non-balanced cases,-result in (2) above, respectively. In both cases, the price process is diffusion process on long scale with specific drifts and diffusion coefficients that can be recovered from (1) and (2). Thus, the optimal liquidation/aquisition problem in this cases reduces to the control for diffusion processes, which is described in [6], Section 5. And for solving this problem, we could proceed in this direction.

The second situation (short time scale) is associated with the trading decision that effects both, an agent's wealth and the dynamics of the price itself. For example, the situation when the agent is attempting to acquire (or sell) a large number of assets in a short period of time. In this case, we should consider the price process in the form of

$$s_t^u = \sum_{k=1}^{N(t)} a(X_k, u_k), \tag{2.3}$$

where $u = (u_k)$ is the control process (e.g., Markov chain), $N(t)$ is counting process, $a(x, u)$ is uniformly bounded function. The process s_t^u in (3) is controlled jump process with jump sizes $a(x, u)$. Hence, the optimal liquidation/aquisition problem in this case reduces to the control for jump process in (3), which is described in [6] as well, Section 5.4.3. We could consider even the case when the counting process $N(t)$ also depends on control u, $N = N^u(t)$. (See also [24] for more information about the control of semi-Markov random evolutions, in particular, functionals of a Markov renewal processes, and their applications).

We hope to pursue further some of these ramifications in future work.

2.6.6 Market Making

Let us discuss how would one use our approach in the typical algo problems such as market making. We consider the following model for the mid price of the stock:

$$s_t = \sum_{k=1}^{N(t)} a(X_k),$$

where $N(t)$ is a counting process, X_k is a Markov chain with two states $\{1,2\}$, and function $a(x)$ is uniformly bounded. Let p_t^a and p_t^b be the quote limit orders around this mid-price. Also, let the number of stocks bought N_t^b is a Poisson process with intensity $\lambda^b(p^b - s)$, and the number of stocks sold N_t^a is a Poisson process with intensity $\lambda^a(p^a - s)$. The wealth in cash will be $dX_t = p^a dN_t^a - p^b dN_t^b$ with the inventory process $q_t = N_t^b - N_t^a$. With this set up we could consider the following market maker's objectives (among others): (1) maximizing exponential utility

$$u(s,x,q,t) = \max_{p_t^a, p_t^b, 0 \le t \le T} E_t[-e^{-\gamma(X_T + q_T s_T)}]$$

or (2) the mean/variance objective

$$v(s,x,q,t) = \max_{p_t^a, p_t^b, 0 \le t \le T} \{E_t[(X_T + q_T s_T)] - \frac{\gamma}{2} Var[(X_T + q_T s_T)]\}.$$

For solving these optimization problems, we need to find the Hamilton-Jacobi-Bellman (HJB) equation in this case. Taking into account the fact that $(s_t, t - T_{N(t)})$ is a Markov process with infinitesimal generator

$$\frac{\partial}{\partial t} f(i,s,t) + \frac{dF(i,t)/dt}{1 - F(i,t)} \sum_{j \in \{1,2\}} [P(i,j) f(i, s + a(j), t) - f(i,s,t)]$$

(see, e.g., [22]), where $T_n, N(t)$ and $F(i,t)$ are defined in Section 2.2, the HJB equation can be written straightforward. The details we leave for our future work.

2.7 Conclusion

In this chapter, we introduced a semi-Markovian modelling of limit order books in order to match empirical observations. We extended the model of [11] in the following ways:

1. inter-arrival times between book events (limit orders, market orders, order cancellations) are allowed to have an arbitrary distribution.

2. the arrival of a new book event at the bid or the ask and its corresponding inter-arrival time are allowed to depend on the nature of the previous event.

In order to do so, both the bid and ask queues are driven by Markov renewal processes. It results from these chosen dynamics that the price process can be expressed as a functional of another Markov renewal process, which we characterized explicitly. In this context, we obtained probabilistic results such as the duration until the next price change, the probability of price increase and the characterization of the Markov renewal process driving the stock price process (Section 2.3). In Section 2.4, we obtained diffusion limit results for the stock price process generalizing those of [11]. We presented in Section 2.5 calibration results on real market data (five stocks Amazon, Google, Apple, Intel and Microsoft) in order to illustrate and justify our approach. Finally, we discussed our approach for various stocks (liquid, illiquid, medium liquid) from different markets in Section 2.6, as well as possible extensions of our model for the case when the spread is not fixed, including the diffusion limit of the price dynamics in this case, and also stochastic optimal control and market making problems.

Bibliography

[1] R. Almgren and N. Chriss, *Optimal execution of portfolio transactions*, J. Risk, (Winter 2001), 3, pp. 5–40.

[2] M. Avellaneda and S. Stoikov, *High-frequency trading in a limit order book*, Quantitative Finance, (2008), 8(3), pp. 217–224.

[3] B. Biais and P. Hillion, *An empirical analysis of the order flow and order book in the Paris bourse*, J. Finance, (1995), 50 (5), pp. 1655–1689.

[4] P. Billingsley, *Convergence of Probability Measures* (1999), Wiley & Sons Inc.

[5] J.-P. Bouchaud, Y. Gefen, M. Potters and M. Wyart, *Fluctuations and response in financial markets: the subtle nature of 'random' price changes*, Quantitative Finance, (2004), 4, pp. 176–190.

[6] A. Cartea, S. Jaimungal and J. Penalva *Algorithmic and High-Frequency Trading*, (2015), Cambridge: Cambridge University Pres.

[7] A. Cartea, and S. Jaimungal, *Optimal Execution with Limit and Market Orders*, Quantitative Finance, forthcoming.

[8] A. Cartea, and S. Jaimungal, *Incorporating order-flow into optimal execution*, Mathematics and Financial Economics, (2016), pp. 1–26.

[9] K. J. Cohen, R. M. Conroy and S. F. Maier, *Order flow and the quality of the market*, in Market Making and the Changing Structure of the Securities Industry, (1985).

[10] R. Cont, S. Stoikov and R. Talreja, *A stochastic model for order book dynamics*, Operations Research, (2010), 58, pp. 549–563.

[11] R. Cont and A. de Larrard, *Price dynamics in a Markovian limit order book market*, SIAM Journal for Financial Mathematics, (2013), 4, No 1, pp. 1–25.

[12] R. Cont and A. De Larrard, *Order book dynamics in liquid markets: Limit theorems and diffusion approximations*, (2012), Available at SSRN: http://ssrn.com/abstract=1757861.

[13] I. Domowitz and J. Wang, *Auctions as algorithms*, J. Economic Dynamics and Control, (1994), 18, pp. 29–60.

[14] P. Fodra and H. Pham, *Semi Markov model for market microstructure*, ArXiv:1305v1 [q-fin.TR], 1 May (2013).

[15] P. Fodra and H. Pham, *High frequency trading and asymptotics for small risk aversion in a Markov renewal model*, ArXiv:1310.1765v2 [q-fin.TR], 4 Jan (2015).

[16] M. Garman, *Market microstructure*, J. Financial Economics, (1976), 3, pp. 257–275.

[17] M. Gould, M. Porter, S. Williams, M. McDonald, D. Fenn and S. Howison, *Limit order books*, Quantitative Finance, (2013), 13(11).

[18] F. Guilbaud and H. Pham *Optimal high-frequency trading with limit and market orders*, Quantitative Finance, (2013), 13, pp. 79–94.

[19] L. Harris and V. Panchapagesan, *The information content of the limit order book: evidence from NYSE specialist trading decision*, J. Financial Markets, (2005), 8(1), pp. 25–67.

[20] B. Hollifield, R. Miller and P. Sandas, *Empirical analysis of limit order markets*, Review of Economic Studies, (2004), 71(4), pp. 1027–1063.

[21] U. Horst and M. Paulsen, *A Law of Large Numbers for limit order books*, (2015), ArXiv: 1501.00843v1[q-fin.MF] 5 Jan 2015.

[22] V. S. Korolyuk and A. V. Swishchuk, *Semi-Markov Random Evolutions*, Kluwer Academic Publishers, Dordrecht, The Netherlands, (1995).

[23] F. Lillo and J. D. Farmer, *The long memory of the efficient market*, (2004), 8(3).

[24] N. LIMNIOS AND A. SWISHCHUK, *Discrete-time semi-Markov random evotuions and their applications*, Adv. Appl. Prob. (2013), 45, pp. 213–240.

[25] N. LIMNIOS AND G. OPRISAN, *Semi-Markov Processes and Reliability*, Birkhauser, (2001).

[26] H. LUCKOCK, *A steady-state model of a continuous double auction*, Quant. Finance, (2003), 3, pp. 385–404.

[27] H. MENDELSON, *Market behaviour in a clearing house*, Econometrica, (1982), 50, pp. 1505–1524.

[28] C. PARLOUR, *Price dynamics in limit order markets*, Review of Financial Studies, (1998), 11(4), pp. 789–816.

[29] S. PREDOIU, G. SHAIKHET AND S. SHREVE, *Optimal execution in a general one-sided limit order book*, SIAM J. Financial Math., (2011), 2(1), pp. 183–212.

[30] I. ROSU A DYNAMIC MODEL OF THE LIMIT ORDER BOOK, Review of Financial Studies, (2009), 22, pp. 4601–4641.

[31] A. SKOROKHOD, *Studies in the Theory of Random Processes*, Addison-Wesley, Reading, Mass., (1965) (Reprinted by Dover Publications, NY).

[32] E. SMITH, J. D. FARMER, L. GILLEMOT AND S. KRISHNAMURTHY (2003): *Statistical theory of the continuous double auction*, Quant. Finance, 3, 481–514.

[33] SWISHCHUK, A. AND VADORI, N. (2017): *A semi-Markovian modelling of limit order markets*. SIAM J. Finan. Math., 8, pp. 240–273.

[34] P. TADIKAMALLA *A look at the Burr and related distributions*, Intern. Statis. Review, (1980), 48, pp. 337–344.

[35] N. VADORI AND A. SWISHCHUK, *Strong Law of Large Numbers and Central Limit Theorems for Functionals of Inhomogeneous Semi-Markov Processes*, Stochastic Analysis and Applications, 33(2) (2015), pp. 213–243.

3

General Semi-Markovian Modelling of Big Data in Finance

The chapter considers a general semi-Markov model for limit order books with two states, which incorporates price changes that are not fixed to one tick. Furthermore, we introduce an even more general case of the semi-Markov model for limit order books that incorporates an arbitrary number of states for the price changes. For both cases the justifications, diffusion limits, implementations and numerical results are presented for different limit order book data: Apple, Amazon, Google, Microsoft, Intel on 2012/06/21 and Cisco, Facebook, Intel, Liberty Global, Liberty Interactive, Microsoft, Vodafone from 2014/11/03 to 2014/11/07. To illustrate an application in risk management, the first model extension is used to set up a statistical process control model.

3.1 Introduction

One of the main approaches of modelling limit order books is the zero intelligence approach (see [4]), which assumes all quantities of interest in the limit order book are governed by stochastic processes. Of the zero-intelligence models developed so far, the approach of [2] is an attractive starting point for modelling limit order flow in continuous time due to the tractability of the model and it's reduced dimensionality. They calculate various quantities of interest such as the probability of a price increase or the diffusion limit of the price process.

Having found evidence in empirical observations, in Chapter 2 we extended the framework of [2]. We incorporated an arbitrary distribution for the inter-arrival times of the book events as well as a dependency of both, the type of a book event and its corresponding inter-arrival times, on the type of the previous book event. Therefore we used a Markov renewal process to model the dynamics of the bid and ask queues, which are assumed to be independent of each other. After a price change they are reinitialized. The model remains analytically tractable. As in [2] the bid/ask spread remains equal to one tick and all orders have the same size.

DOI: 10.1201/9781003265986-3

As a reminder, we briefly recap the model used in [9] (see also Chapter 2) highlighting the notations and definitions we will use in this chapter. As the model from [2], price changes are assumed to happen at each time T_n at which the ask or the bid queue is depleted. The sojourn times are notated as $\tau_n := T_n - T_{n-1}$. The changes in the queue sizes are modelled by a Markov chain V^a for the ask side and V^b for the bid side. Their state space is $\{-1, 1\}$. When a limit order appears at time t, the queue size increases by one unit and $V_t^a = 1$, when a market order or cancellation appears, it decreases by one and $V_t^a = -1$. The notations for the bid side are defined accordingly.

The paper defines a balanced and a unbalanced case. The classification is dependent on the transition probabilities of the Markov chain modelling the queue sizes:

$$P^a(i,j) := P[V_{k+1}^a = j | V_k^a = i], \ i, j \in \{-1, 1\}.$$

P_t^b is defined accordingly.

The balanced case is defined in the following way: $P^a(1,1) = P^a(-1,-1)$ and $P^b(1,1) = P^b(-1,-1)$. The unbalanced case is on hand if $P^a(1,1) < P^a(-1,-1)$ or $P^b(1,1) < P^b(-1,-1)$.

Built on this model, [9] proposes the following jump model for the stock price s_t based on a counting process $N(t)$ and a Markov Chain X_t:

$$s_t = \sum_{k=1}^{N(t)} X_k,$$

where $N(t)$ counts how often the price changes and X_t keeps track of which direction the price changed at each time of T_n, meaning $X_t \in \{-\delta, \delta\}$. A price change is assumed to happen at every time the bid or the ask queue of the limit order book is depleted.

Remark 3.1 *We note that results of [12] were fist announced at IPAM FMWSI, UCLA, March 23-27, 2015 (see [8]). Also available at SRRN (see [9]) and arXiv (see [11]).*

3.1.1 Motivation for Generalizing the Model

As in [2], [9] assume that all price changes occurring in the price process are of magnitude δ, a single tick. Table 3.1 demonstrates the average price change for each stock in our data set. For Apple midprice data we observe 53,654 price changes, from which only 9007 are of magnitude δ. Meaning, 83.21% (44,647) of the price changes were different from one tick.

Possible extensions of the proposed model were discussed in [12] for the case when the size of price changes is not fixed, including the diffusion limit of the price dynamics in this case. We will illustrate their model with the corresponding proofs and another model extension generalizing the model to allowing for more than two price changes.

TABLE 3.1
Mid-price changes in ticks

	Apple	Amazon	Google
Avg. up movements	1.7	1.3	3.1
Avg. down movements	−1.7	−1.3	−3.0
Min price change	−18.5	−11.5	−30.5
Max price change	15.0	16.5	30.5

For both model extensions we show results for the diffusion coefficients and verify them using the same approach as in [2].

3.1.2 Data

To test empirical validity of our more general model we use the following freely available data:

- Level 1 LOB data provided by [5]: Apple, Amazon, Google, Microsoft and Intel on 2012/06/21
- LOB data provided in [1]: Cisco, Facebook, Intel, Liberty Global, Liberty Interactive, Microsoft, Vodafone from 2014/11/03 to 2014/11/07

3.2 Reviewing the Assumptions with Our New Data Sets

The objective of this section is to test the validity of the assumptions of [9] and [2] with respect to new data.

3.2.1 Liquidity of Our Data

Table 3.2 demonstrates the liquidity of the new data. Note that in calculating the average number of orders occurring in 10 seconds, we limited the orders to only those occurring at the best bid or ask price, i.e. at level 1.

TABLE 3.2
Stock liquidity of AAPL, AMZN, GOOG, INTC, and MSFT on 2012/06/21.

	Average no. of orders in 10s	Price changes in 1 day
AAPL	51	64,350
AMZN	25	27,557
GOOG	21	24,084
INTC	173	3,217
MSFT	176	4,060

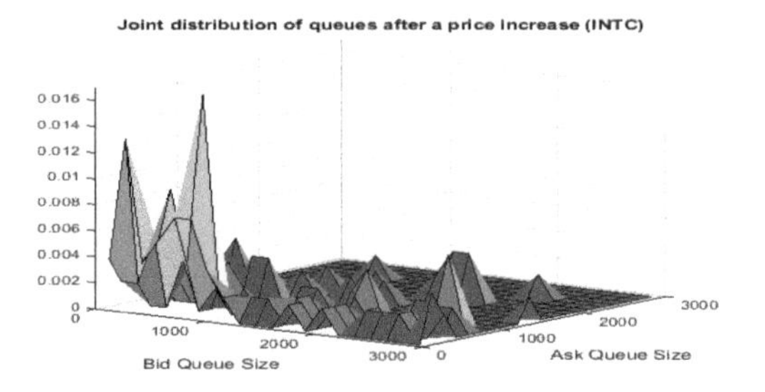

FIGURE 3.1
Empirical Joint Density after a price increase, $f(q_t^b, q_t^a)$.

While the average number of orders in 10 seconds is significantly less than that reported in [2], we note that most of these equities undergo more price changes in one day. The high number of daily price changes implies that we can use asymptotic analysis in order to approximate long-run volatility using order flow by finding the diffusion limit of the price process.

3.2.2 Empirical Distributions of Initial Queue Sizes and Calculated Conditional Probabilities

Like [2] and [9], the generalized model uses empirical distribution functions $f(q_t^b, q_t^a)$ and $\tilde{f}(q_t^b, q_t^a)$ to initialize the bid and ask queues after either a price increase, or decrease. In order to generalize their models, we trim our data to only points where the spread is a single tick. The resulting empirical distribution $f(q_t^b, q_t^a)$ for INTC from [5] is displayed in Figure 3.1.

One of the accomplishments of [2] is the formula for the conditional probability of a price increase conditional on the state of the order book,

$$p_1^{up}(n,p) = \frac{1}{\pi}\int_0^{\pi}\left(2-\cos(t)-\sqrt{(2-cos(t))^2-1}\right)^p \frac{\sin(nt)\cos\left(\frac{t}{2}\right)}{\sin\left(\frac{t}{2}\right)}dt.$$

We can compare this to the quantity calculated in [9] as

$$p_1^{up}(n,p) = \int_0^{\infty} f_{p,a}(t)(1-F_{n,b}(t))dt$$

where

$$f_{p,a}(t) = \frac{1}{2\pi}\int_{\mathbb{R}} e^{itx}\varphi^a(x,p)dx$$
$$F_{n,b}(t) = \frac{1}{2} - \frac{1}{\pi}\int_0^{\infty}\frac{1}{x}Im\{e^{-itx}\varphi^b(x,n)\}dx$$

and $\varphi^a(t), \varphi^b(t)$ are the characteristic functions of σ_a and σ_b, respectively.

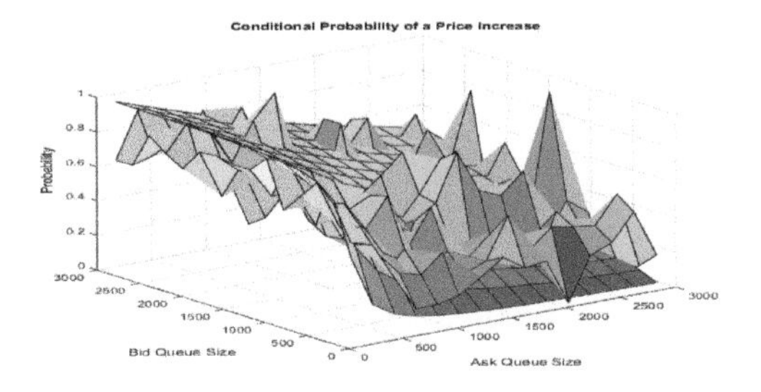

FIGURE 3.2
Conditional probability of a price increase conditional on size of bid and ask queues.

Figure 3.2 shows, for INTC, the conditional probability of a price increase conditional on the size of bid and ask queues, as calculated using the formula from [2].

While the empirical data does not fit $p_1^{up}(n,p)$ as closely as in [2], we can see that the calculated value from the model still matches the empirical frequencies to some extent.

3.2.3 Inter-arrival Times of Book Events

While [2] assume the inter-arrival times between book events follow independent exponential distributions, [9] challenges this assumption. We have calculated the empirical distribution functions of the inter-arrival times between book events for our data and get the same result as [9]: The exponential distribution does not fit the data as well as alternative distributions do. Figures 3.3 and 3.4 illustrates the example for Amazon ask and bid, rspectively. We have similar figures for Apple, Google, Intel and Microsoft illustrating the same finding.

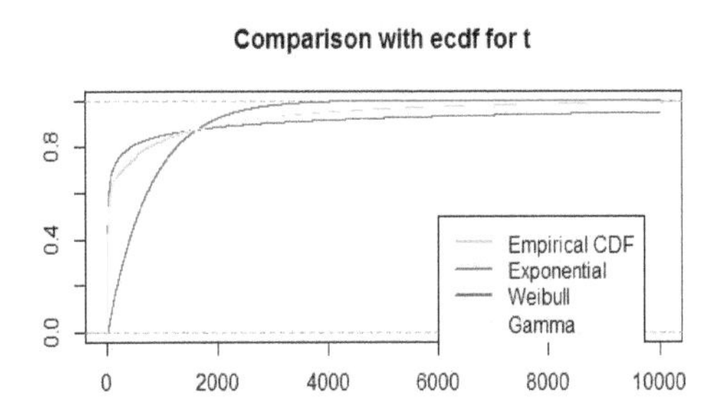

FIGURE 3.3
Distribution of inter-arrival times Amazon ask.

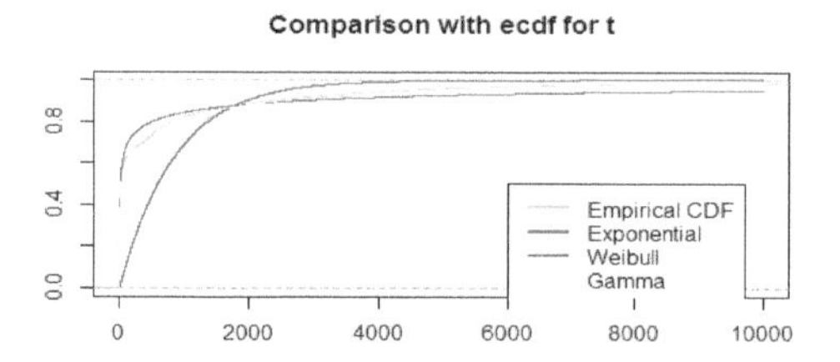

FIGURE 3.4
Distribution of inter-arrival times Amazon bid.

3.2.4 Asymptotic Analysis

The asymptotic analysis presented in [2] is another strength of their paper. Using the relevant formula when the rate of incoming limit orders is assumed to equal the combined rate of incoming market orders and cancellations, we compute the diffusion coefficients for our new data. [2] demonstrates the linear relationship between $\sqrt{\frac{\lambda}{D(f)}}$ and the 10 minute standard deviation of various equities, where λ is the intensity of incoming orders and $D(f)$ is the square of the average depth of the bid and ask queues after a price change. Figure 3.5 contains the same plot generated for the new data. The linear relationship between $\sqrt{\frac{\lambda}{D(f)}}$ and the 10 minute standard deviation is not as significant for the new data set. We will see later on a huge improvement for our considered model extensions.

3.3 General Semi-Markov Model for the Limit Order Book with Two States

We develop the model proposed in [12] (see Section 3.6: Discussion), incorporating price changes which are not fixed to one tick. The authors already

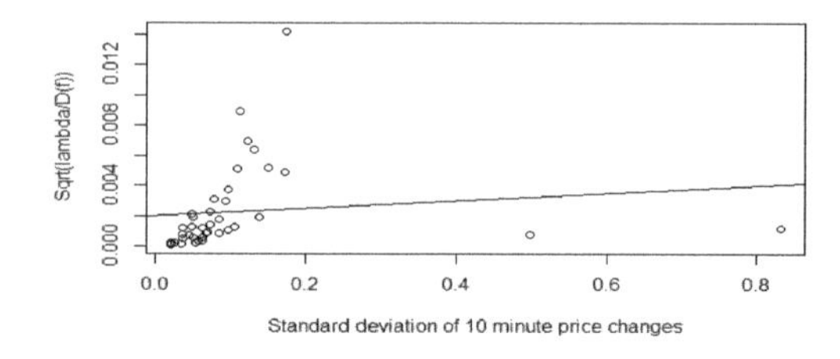

FIGURE 3.5
$\sqrt{\lambda/D(f)}$ compared to 10 minute standard deviation.

introduced the following model for the price process:

$$s_t = \sum_{k=1}^{N(t)} a(X_k),$$

where $N(t)$ is the counting process for the price changes, X_k is a two state Markov chain with state space $\mathcal{S} = \{1, 2\}$ and $a(x)$ is an uniformly bounded function. To consider one tick spreads we set $a(1) = \delta$ and $a(2) = -\delta$.

3.3.1 Diffusion Limits

For the calculation of the diffusion limits we will use the following two assumptions from [9].

Assumption 3.1 *We assume the following inequalities to be true:*

$$\sum_{n=1}^{\infty} \alpha^b(n)\alpha^a(p)f(n,p) < \infty,$$

$$\sum_{n=1}^{\infty} \alpha^b(n)\alpha^a(p)\tilde{f}(n,p) < \infty,$$

with $\alpha^a(n) := \frac{1}{p_a\sqrt{\pi}}(n + \frac{2p_a-1}{p_a-1}v_0^a(1))\sqrt{p_a(1-p_a)}\sqrt{p_a h_1^a + (1-p_a)h_2^a}$, h_1^a *is defined on page 7 in [9],* $p_a = P^a(1,1) = P^a(-1,-1)$, $\alpha^b(n)$ *is defined accordingly.*

Assumption 3.2 *In this section we assume* $m(1) < \infty$ *and* $m(2) < \infty$*, where* $m(i) = E[\tau_k | X_{k-1} = i], i \in \{1, 2\}$.

Theorem 3.1 *Given that assumption 3.1 is satisfied for the balanced case and assumption 3.2 for the unbalanced case, we can proof the following weak convergences in the Skorokhod topology (see [7]):*

$$\left(\frac{s_{tnlog(n)} - N_{tnlog(n)}a^*}{\sqrt{n}}, t \geq 0\right) \overset{n\to\infty}{\Rightarrow} \frac{\sigma^*}{\sqrt{\tau^*}}W_t, \text{ for the balanced case and}$$

$$\left(\frac{s_{tn} - N_{tn}a^*}{\sqrt{n}}, t \geq 0\right) \overset{n\to\infty}{\Rightarrow} \frac{\sigma^*}{\sqrt{m_\tau}}W_t, \text{ for the unbalanced case,}$$

where W_t *is a standard Brownian motion,* $a_i := a(i)$, $a^* = \pi_1^* a_1 + \pi_2^* a_2$ *and* (π_1^*, π_2^*) *is the stationary distribution of the Markov chain* $a(X)$. τ^*, m_τ *and* $(\sigma^*)^2$ *are given by:*

$$\tau^* = \lim_{t\to+\infty} \frac{t}{N_t log(N_t)} \quad \text{(see [9], p.19)}$$

$$m_\tau = \pi_1^* m(1) + \pi_2^* m(2)$$

$$
\begin{aligned}
(\sigma^*)^2 = {} & \pi_1^* a_1^2 + \pi_2^* a_2^2 + (\pi_1^* a_1 + \pi_2^* a_2) \\
& \times [-2a_1\pi_1^* - 2a_2\pi_2^* + (\pi_1^* a_1 + \pi_2^* a_2)(\pi_1^* + \pi_2^*)] \\
& + \frac{(\pi_1^*(1 - p_{cont}) + \pi_2^*(1 - p'_{cont}))(a_1 - a_2)^2}{(p_{cont} + p'_{cont} - 2)^2} \\
& + 2(a_2 - a_1) \cdot \left[\frac{\pi_2^* a_2(1 - p'_{cont}) - \pi_1^* a_1(1 - p_{cont})}{p_{cont} + p'_{cont} - 2} \right. \\
& \left. + \frac{(\pi_1^* a_1 + \pi_2^* a_2)(\pi_1^* - p_{cont}\pi_1^* - \pi_2^* + p'_{cont}\pi_2^*)}{p_{cont} + p'_{cont} - 2} \right]
\end{aligned}
$$

p_{cont} is the probability of two subsequent increases of the stock price, p'_{cont} the probability of two subsequent decreases of the stock price.

Proof *We get the following law of large numbers:*

$$\frac{s_{tnlog(n)}}{n} \overset{n\to\infty}{\Rightarrow} \frac{a^* t}{\tau^*}.$$

The proof follows that of Proposition 8 in [9], except of the calculation for σ. For $t \in R_+$ we first of all consider the following processes:

$$R_n := \sum_{k=1}^{n} (a(X_k) - a^*),$$

$$U_n(t) := n^{-1/2}[(1 - \lambda_{n,t})R_{\lfloor nt \rfloor} + \lambda_{n,t} R_{\lfloor nt \rfloor + 1}],$$

where $\lambda_{n,t} := nt - \lfloor nt \rfloor$. We can show the following weak convergence in the Skorokhod topology (see [7]) similar to the approach of [10]:

$$(U_n(t), t \geq 0) \overset{n\to\infty}{\Rightarrow} \sigma^* W,$$

where W is a standard Brownian motion, and σ is given by:

$$(\sigma^*)^2 := \sum_{i \in \{1,2\}} \pi_i^* v^*(i)$$

$$v^*(1) := (a_1 - a^*)^2 + p(1)(g_2 - g_1)^2 - 2(a_1 - a^*)p(1)(g_2 - g_1),$$

$$v^*(2) := (a_2 - a^*)^2 + p(2)(g_1 - g_2)^2 - 2(a_2 - a^*)p(2)(g_1 - g_2),$$

$$\begin{pmatrix} g_1 \\ g_2 \end{pmatrix} = (P + \Pi^* - I)^{-1} \begin{pmatrix} a_1 - a^* \\ a_2 - a^* \end{pmatrix},$$

$$p(1) := 1 - p_{cont}, p(2) = 1 - p'_{cont},$$

Π^* is the matrix of the stationary distribution consisting of rows equal to $(\pi_1^* \ \pi_2^*)$.

We get the following calculation for g:

$$
\begin{aligned}
g :=& (P + \Pi^* - I)^{-1}(a - a^*) \\
=& \left[\begin{pmatrix} p_{cont} & 1 - p_{cont} \\ 1 - \pi' & p'_{cont} \end{pmatrix} + \begin{pmatrix} \pi_1^* & \pi_2^* \\ \pi_1^* & \pi_2^* \end{pmatrix} - \begin{pmatrix} 1 & 0 \\ 0 & 1 \end{pmatrix}\right]^{-1} \begin{pmatrix} a_1 - a^* \\ a_2 - a^* \end{pmatrix} \\
=& \frac{1}{(p_{cont} + \pi_1^* - 1)(p'_{cont} + \pi_2^* - 1) - (1 - p_{cont} + \pi_2^*)(1 - p'_{cont} + \pi_1^*)} \\
& \cdot \begin{pmatrix} p_{cont} + \pi_2^* - 1 & p_{cont} - \pi_2^* - 1 \\ p'_{cont} - \pi_1^* - 1 & p_{cont} + \pi_1^* - 1 \end{pmatrix} \begin{pmatrix} a_1 - a^* \\ a_2 - a^* \end{pmatrix} \\
=& \begin{pmatrix} \dfrac{a_1(p'_{cont} + \pi_2^* - 1) + a_2(p_{cont} - \pi_2^* - 1)}{(\pi_1^* + \pi_2^*)(p_{cont} + p'_{cont} - 2)} - \dfrac{a^*}{\pi_1^* + \pi_2^*} \\[2ex] \dfrac{a_1(p'_{cont} - \pi_1^* - 1) + a_2(p_{cont} + \pi_1^* - 1)}{(\pi_1^* + \pi_2^*)(p_{cont} + p'_{cont} - 2)} - \dfrac{a^*}{\pi_1^* + \pi_2^*} \end{pmatrix}
\end{aligned}
$$

In order to get a nice form for $(\sigma^*)^2$ we firstly calculate the summand for $i = 1$:

$$
\begin{aligned}
\pi_1^* v(1) =& \pi_1^* \Bigg[(a_1 - a^*)^2 + (1 - p_{cont}) \Bigg(\frac{a_1(p'_{cont} - \pi_1^* - 1) + a_2(p_{cont} + \pi_1^* - 1)}{(p_{cont} + p'_{cont} - 2)(\pi_1^* + \pi_2^*)} \\
& - \frac{a^*}{\pi_1^* + \pi_2^*} - \frac{a_1(p'_{cont} + \pi_2^* - 1) + a_2(p_{cont} - \pi_2^* - 1)}{(p_{cont} + p'_{cont} - 2)(\pi_1^* + \pi_2^*)} + \frac{a^*}{\pi_1^* + \pi_2^*} \Bigg)^2 \\
& -2(a_1 - a^*)(1 - p_{cont}) \Bigg(\frac{a_1(p'_{cont} - \pi_1^* - 1) + a_2(p_{cont} + \pi_1^* - 1)}{(p_{cont} + p'_{cont} - 2)(\pi_1^* + \pi_2^*} \\
& - \frac{a^*}{\pi_1^* + \pi_2^*} - \frac{a_1(p'_{cont} + \pi_2^* - 1) + a_2(p_{cont} - \pi_2^* - 1)}{(p_{cont} + p'_{cont} - 2)(\pi_1^* + \pi_2^*)} + \frac{a^*}{\pi_1^* + \pi_2^*} \Bigg) \Bigg] \\
=& \pi_1^* \Bigg[(a_1 - a^*)^2 + (1 - p_{cont}) \left(\frac{a_2 - a_1}{p_{cont} + p'_{cont} - 2} \right)^2 \\
& -2(a_1 - a^*)(1 - p_{cont}) \left(\frac{a_2 - a_1}{p_{cont} + p'_{cont} - 2} \right) \Bigg] \\
=& \pi_1^* \Bigg[(a_1 - (\pi_1^* a_1 + \pi_2^* a_2))^2 + (1 - p_{cont}) \frac{(a_2 - a_1)^2}{(p_{cont} + p'_{cont} - 2)^2} \\
& -2(a_1 - (\pi_1^* a_1 + \pi_2^* a_2))(1 - p_{cont}) \frac{a_2 - a_1}{p_{cont} + p'_{cont} - 2} \Bigg]
\end{aligned}
$$

Similarly we calculate the summand for $i = 2$:

$$\pi_2^* v(2) = \pi_2^* \Bigg[(a_2 - a^*)^2 + (1 - p'_{cont}) \Bigg(\frac{a_1(p'_{cont} + \pi_2^* - 1) + a_2(p_{cont} - \pi_2^* - 1)}{(p_{cont} + p'_{cont} - 2)(\pi_1^* + \pi_2^*)}$$
$$- \frac{a^*}{\pi_1^* + \pi_2^*} - \frac{a_1(p'_{cont} - \pi_1^* - 1) + a_2(p_{cont} + \pi_1^* - 1)}{(p_{cont} + p'_{cont} - 2)(\pi_1^* + \pi_2^*)} + \frac{a^*}{\pi_1^* + \pi_2^*} \Bigg)^2$$
$$-2(a_2 - a^*)(1 - p'_{cont}) \Bigg(\frac{a_1(p'_{cont} + \pi_2^* - 1) + a_2(p_{cont} - \pi_2^* - 1)}{(p_{cont} + p'_{cont} - 2)(\pi_1^* + \pi_2^*)}$$
$$- \frac{a^*}{\pi_1^* + \pi_2^*} - \frac{a_1(p'_{cont} - \pi_1^* - 1) + a_2(p_{cont} + \pi_1^* - 1)}{(p_{cont} + p'_{cont} - 2)(\pi_1^* + \pi_2^*)} + \frac{a^*}{\pi_1^* + \pi_2^*} \Bigg) \Bigg]$$
$$= \pi_2^* \Bigg[(a_2 - a^*)^2 + (1 - p'_{cont}) \left(\frac{a_1 - a_2}{p_{cont} + p'_{cont} - 2} \right)^2$$
$$-2(a_2 - a^*)(1 - p'_{cont}) \left(\frac{a_1 - a_2}{p_{cont} + p'_{cont} - 2} \right) \Bigg]$$
$$= \pi_2^* \Bigg[(a_2 - (\pi_1^* a_1 + \pi_2^* a_2))^2 + (1 - p'_{cont}) \frac{(a_1 - a_2)^2}{(p_{cont} + p'_{cont} - 2)^2}$$
$$-2(a_2 - (\pi_1^* a_1 + \pi_2^* a_2))(1 - p'_{cont}) \frac{a_1 - a_2}{p_{cont} + p'_{cont} - 2} \Bigg]$$

From this it follows:

$$\sigma^2 = \pi_1^* v(a_1) + \pi_2^* v(a_2)$$
$$= \pi_1^* \left(a_1^2 - 2a_1(\pi_1^* a_1 + \pi_2^* a_2) + (\pi_1^* a_1 + \pi_2^* a_2)^2 \right)$$
$$+ \pi_2^* \left(a_2^2 - 2a_2(\pi_1^* a_1 + \pi_2^* a_2) + (\pi_1^* a_1 + \pi_2^* a_2)^2 \right)$$
$$+ \frac{(\pi_1^*(1 - p_{cont}) + \pi_2^*(1 - p'_{cont}))(a_1 - a_2)^2}{(p_{cont} + p'_{cont} - 2)}$$
$$+ 2(a_2 - a_1) \Bigg[\frac{\pi_2^* a_2(1 - p'_{cont}) - \pi_1^* a_1(1 - p_{cont})}{p_{cont} + p'_{cont} - 2}$$
$$+ \frac{(\pi_1^* a_1 + \pi_2^* a_2)(\pi_1^* - p_{cont}\pi_1^* - \pi_2^* + p'_{cont}\pi_2^*)}{p_{cont} + p'_{cont} - 2} \Bigg]$$
$$= \pi_1^* a_1^2 + \pi_2^* a_2^2 + (\pi_1^* a_1 + \pi_2^* a_2)[-2a_1\pi_1^* - 2a_2\pi_2^* + (\pi_1^* a_1 + \pi_2^* a_2)(\pi_1^* + \pi_2^*)]$$
$$+ \frac{(\pi_1^*(1 - p_{cont}) + \pi_2^*(1 - p'_{cont}))(a_1 - a_2)^2}{(p_{cont} + p'_{cont} - 2)^2}$$
$$+ 2(a_2 - a_1) \Bigg[\frac{\pi_2^* a_2(1 - p'_{cont}) - \pi_1^* a_1(1 - p_{cont})}{p_{cont} + p'_{cont} - 2}$$
$$+ \frac{(\pi_1^* a_1 + \pi_2^* a_2)(\pi_1^* - p_{cont}\pi_1^* - \pi_2^* + p'_{cont}\pi_2^*)}{p_{cont} + p'_{cont} - 2} \Bigg]$$

The continuing proof directly follows the proof of Proposition 8 and Proposition 10 in [9].

Remark 3.3.1 *When inserting $a_1 = \delta, a_2 = -\delta$, we get $a^* = s^*, \pi_1^* = \pi^*$ and $\pi_2^* = 1 - \pi^*$ and therewith*

$$\sigma^2 = 4\delta^2 \left(\frac{1 - p'_{cont} + \pi^*(p'_{cont} - p_{cont})}{(p_{cont} + p'_{cont} - 2)^2} - \pi^*(1 - \pi^*) \right).$$

This is the same result as in Proposition 8 in [9], which shows that our model is a generalization of their model.

Proof *When inserting $a_1 = \delta, a_2 = -\delta$ in σ^2 we get $a^* = s^*, \pi_1^* = \pi^*$ and $\pi_2^* = 1 - \pi^*$ and therewith*

$$\begin{aligned}
\sigma^2 &= \pi^*\delta^2 + (1 - \pi^*)(-\delta)^2 + (\pi^*\delta + (1 - \pi^*)(-\delta))(-2\pi^*\delta - 2(1 - \pi^*)(-\delta) \\
&\quad + (\pi^*\delta + (1 - \pi^*)(-\delta))(\pi^* + (1 - \pi^*))) \\
&\quad + \frac{(\pi^*(1 - p_{cont}) + (1 - \pi^*)(1 - p'_{cont}))(\delta - (-\delta))^2}{(p_{cont} + p'_{cont} - 2)^2} \\
&\quad + 2(-\delta - \delta) \left[\frac{(1 - \pi^*)(-\delta)(1 - p'_{cont}) - \pi^*\delta(1 - p_{cont})}{p_{cont} + p'_{cont} - 2} \right. \\
&\quad \left. + \frac{(\pi^*\delta + (1 - \pi^*)(-\delta))(\pi^* - p_{cont}\pi^* - (1 - \pi^*) + (1 - \pi^*)p'_{cont})}{p_{cont} + p'_{cont} - 2} \right] \\
&= \delta^2 + (2\pi^*\delta - \delta)(-2\pi^*\delta + \delta) + 4\delta^2 \frac{1 - p'_{cont} + \pi^*p'_{cont} - \pi^*p_{cont}}{(p_{cont} + p'_{cont} - 2)^2} \\
&\quad - 4\delta^2 \frac{4(\pi^*)^2 - 4\pi^* - 2(\pi^*)^2 p_{cont} - 2(\pi^*)^2 p'_{cont} + 2\pi^* p_{cont} + 2\pi^* p'_{cont}}{p_{cont} + p'_{cont} - 2} \\
&= 4\pi^*\delta^2 - 4(\pi^*)^2\delta^2 + 4\delta^2 \left(\frac{1 - p'_{cont} + \pi^*(p'_{cont} - p_{cont})}{(p_{cont} + p'_{cont} - 2)^2} \right. \\
&\quad \left. - \frac{2\pi^*(2\pi^* - 2 - \pi^*p_{cont} - \pi^*p'_{cont} + p_{cont} + p'_{cont})}{p_{cont} + p'_{cont} - 2} \right) \\
&= 4\delta^2 \left(\pi^* - (\pi^*)^2 + \frac{1 - p'_{cont} + \pi^*(p'_{cont} - p_{cont})}{(p_{cont} + p'_{cont} - 2)^2} \right. \\
&\quad \left. - \frac{2\pi^*(1 - \pi^*)(p_{cont} + p'_{cont} - 2)}{p_{cont} + p'_{cont} - 2} \right) \\
&= 4\delta^2 \left(\frac{1 - p'_{cont} + \pi^*(p'_{cont} - p_{cont})}{(p_{cont} + p'_{cont} - 2)^2} + \pi^* - (\pi^*)^2 - 2\pi^* + 2(\pi^*)^2 \right) \\
&= 4\delta^2 \left(\frac{1 - p'_{cont} + \pi^*(p'_{cont} - p_{cont})}{(p_{cont} + p'_{cont} - 2)^2} - \pi^*(1 - \pi^*) \right).
\end{aligned}$$

3.3.2 Implementation

This subsection explains how we realized the implementation of the diffusion limits given in Subsection 3.3.1.

It is clear that assumptions 3.1 and 3.2 are fulfilled, as we are considering finite data sets. Therefore the calculations are straightforward.

The question arises how to define the function $a(X_k)$. It is enough to define the values of a_1 and a_2, in lieu of the whole function a. We set a_1 to the average of upward movements and a_2 to the one of downwards movements respectively.

The matrix P, which is defined as

$$P = \begin{pmatrix} p_{cont} & 1 - p_{cont} \\ 1 - p'_{cont} & p'_{cont} \end{pmatrix},$$

contains the transition probabilities of the Markov chain X_k. We assign each positive stock price change to the state 1 and each negative one to 2. We count their absolute frequencies and calculate the relative frequencies to get p_{cont} and p'_{cont}.

The stationary distribution π^* of the transition matrix P satisfies $\pi^* = \pi^* P$ and the exclusively positive entries have to sum up to 1 since it is a probability distribution. This is equivalent to solving the problem

$$\begin{pmatrix} p_{cont} - 1 & 1 - p'_{cont} \\ 1 & 1 \end{pmatrix} \begin{pmatrix} \pi_1^* \\ \pi_2^* \end{pmatrix} = \begin{pmatrix} 0 \\ 1 \end{pmatrix}.$$

We need to remark at that point why the stationary distribution exists. As stated in several books about the theory of Markov chains, as e.g. in [6], a unique stationary distribution exists for a Markov chain with a finite space set as long as the Markov chain is irreducile. The state space is finite for every set of our data and looking at the transition matrices, we can see that there are no closed sets and that the Markov chain is irreducible.

We calculate τ^* by using the following result from Lemma 7 in [9]:

$$\frac{1}{nlog(n)} \sum_{k=1}^{n} \tau_k \to \tau^*,$$

where n is the total number of price changes.

For the computation of m_τ, σ^* and the diffusion coefficients we use the formulas given in Subsection 3.3.1.

3.3.3 Numerical Results

The following Table 3.3 depict the results gained from our computations for the [5] data for Apple, Amazon, Google, Intel and Microsoft on 06/21/2012. We do not consider the first and last 15 minutes after opening and before closing, because we do not want to include opening and closing auctions.

In Subsections 3.2 and 3.3 in [2] the authors state that their diffusion coefficients are linearly related to the standard deviation of the ten minute price changes. Therefore we calculated the standard deviations and diffusion

TABLE 3.3
Numerical results for two states

	p_{cont}	p'_{cont}	a_1	a_2	σ^2	τ^*	m_τ	$\frac{\sigma}{\sqrt{\tau^*}}$	$\frac{\sigma}{\sqrt{m_\tau}}$
Apple	0.4932	0.4956	0.0170	−0.0172	0.0003	0.0370	0.4026	0.0881	0.0267
Amazon	0.4576	0.4635	0.0133	−0.0134	0.0002	0.0892	0.9001	0.0412	0.0130
Google	0.4461	0.4769	0.0308	−0.0302	0.0008	0.1145	1.1291	0.0834	0.0266
Intel	0.5588	0.6106	0.0050	−0.0050	0.00004	1.3151	10.0897	0.0052	0.0019
Microsoft	0.5827	0.6269	0.0050	−0.0050	0.00004	0.8944	7.1657	0.0065	0.0023

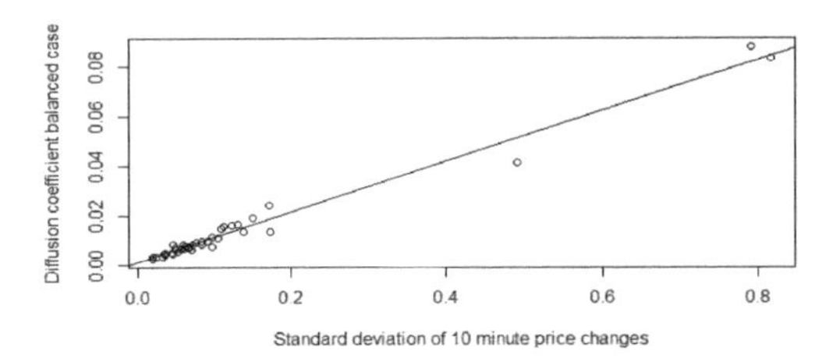

FIGURE 3.6
Linear relationship of coefficients and standard deviation of 10 minutes mid-price changes for two states balanced case.

coefficients for all our data, including the data provided by [5] and [1], resulting in 40 data points for the regression. We get an adjusted R^2 of 0.9788 for the linear regression of the diffusion coefficients on the standard deviation for the balanced case and one of 0.9821 for the unbalanced case. Figures 3.6 and 3.7 depict the regression. It follows, that the standard deviation of the ten minutes price changes is linearly dependent on the diffusion coefficients. In both cases the regression coefficients are highly significant.

As a comparison and justification for our model, we also calculated the regression for the diffusion coefficients resulting from the basic model of [9] allowing only for price changes of one tick. We used exactly the same data as for the regression above. The regression is plotted in Figures 3.8 and 3.9. The adjusted R^2 for the balanced case is 0.3916, the one for the unbalanced case is 0.3813. It can clearly be seen that the linear relationship is better captured by our extended model.

3.3.4 Application of the Model

In this subsection we illustrate how our model could be used for applications in risk management. We implemented a statistical process control model which gives a signal to warn about abnormal processes in the limit order book. We tested our model for a constructed sample day.

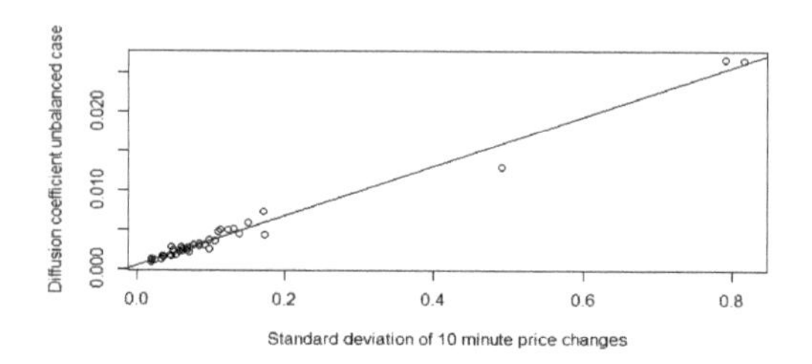

FIGURE 3.7
Linear relationship of coefficients and standard deviation of 10 minutes mid-price changes for two states unbalanced case.

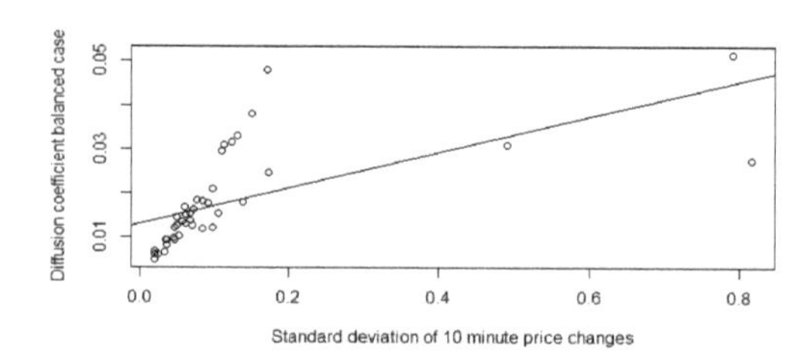

FIGURE 3.8
Linear relationship of coefficients and standard deviation of 10 minutes mid-price changes for one tick jump sizes balanced case.

3.3.4.1 Examination of the Data

Before implementing our model, we wanted to explore how our calculated quantities and parameters behave for different regimes with different magnitudes of volatility.

We had a closer look at the mid-price of Cisco on 2014/11/07. It is plotted in Figure 3.10. The first part of the day seems a lot more volatile than the second. Therefore we decided to use this data to have a look at how our calculated quantities and parameters change from part 1 to part 2.

We divided the data into two parts: One up to time 40,946.07 seconds after midnight, meaning around 11:22 am. The other part includes the data starting from 11:22 am. The standard deviation of the first part is 25,636.77, of the second part 730.757. We calculated the mean inter-arrival time between book events mean_t and estimated the distribution parameters of the inter-arrival times between book events for the exponential and for the gamma distribution.

The following table summarizes the results.

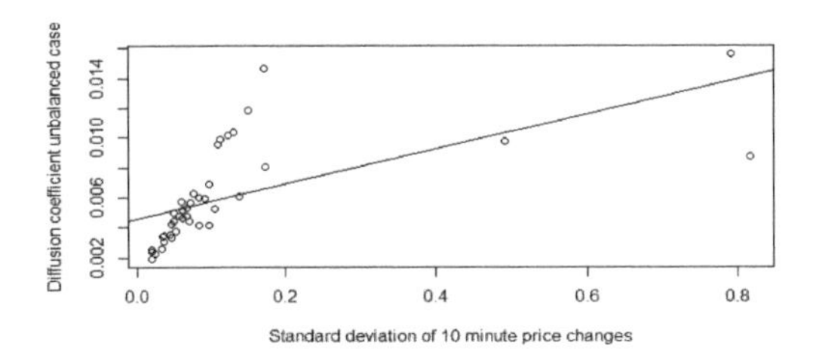

FIGURE 3.9
Linear relationship of coefficients and standard deviation of 10 minutes mid-price changes for one tick jump sizes unbalanced case.

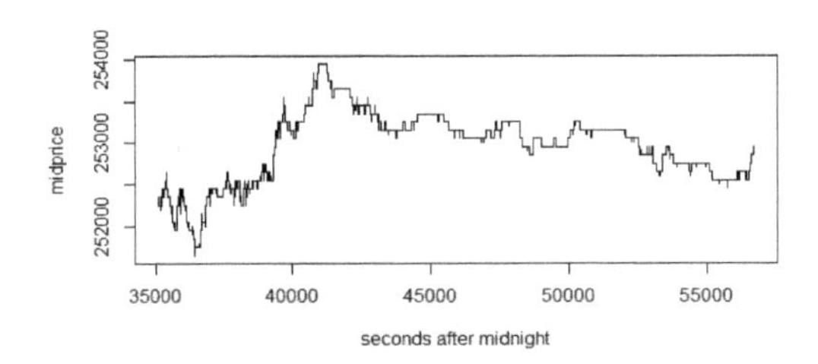

FIGURE 3.10
Midprice of Cisco on 2014/11/07.

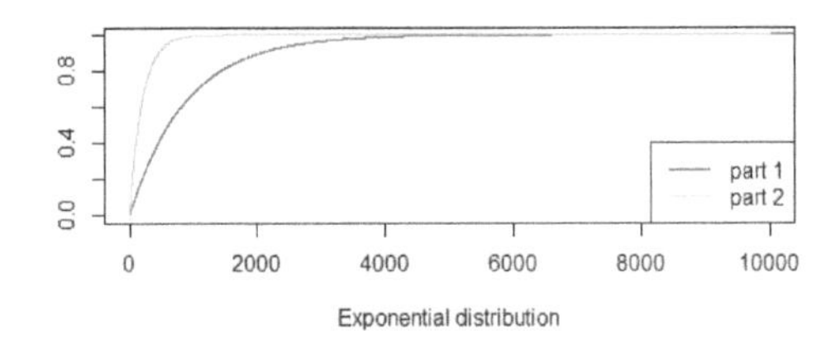

FIGURE 3.11
Distributions of inter-arrival times between book events, Cisco ask data on 2014/11/07.

The change in the distribution of the inter-arrival times between book events is illustrated in Figures 3.11 and 3.12 for the ask side data. The distribution is shifted to the left for the more volatile part.

We calculated the diffusion coefficients for the mid prices of the two parts of the data using the model extension we presented in this section.

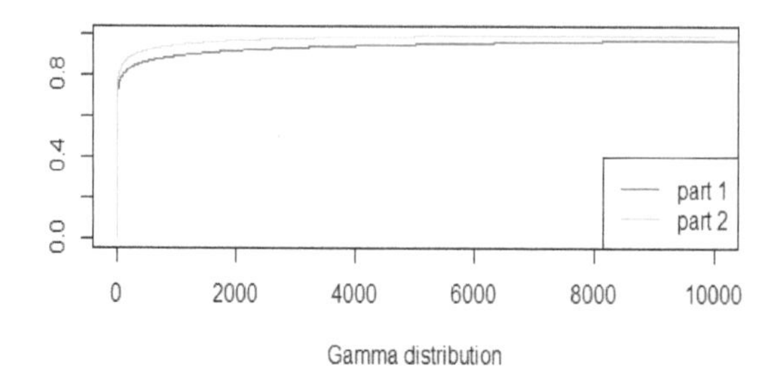

FIGURE 3.12
Distributions of inter-arrival times between book events, Cisco ask data on 2014/11/07.

TABLE 3.4
mean_t and estimated distribution parameters for book events inter-arrival times

	Ask		Bid	
	Part 1	Part 2	Part 1	Part 2
mean_t	89.52	201.59	103.02	233.73
Exp distr rate	0.011	0.005	0.010	0.004
Gam distr shape	0.045	0.046	0.046	0.050
Gam distr scale	1995	4328	2257	4685

TABLE 3.5
Diffusion coefficients

	Part 1	Part 2
Balanced diff coeff	0.00368	0.00170
Unbalanced diff coeff	0.00146	0.00069

The statistics behave as we would have intuitively expected: As it can be seen in Table 3.4 mean_t was smaller during the more volatile part meaning more book events happened. Table 3.5 shows that the diffusion coefficients were higher during the more volatile part of the trading day. These results made us confident that a statistical process control model could bring the desired results.

3.3.4.2 Model Implementation

We decided to use the unbalanced diffusion coefficient as order statistic in our model as it yields the highest adjusted R^2 in the regression against the volatility of the coarse-grained midprice changes.

Our approach was to calibrate our model using all available days except the last one and to test the model for the last day. We calculated the unbalanced

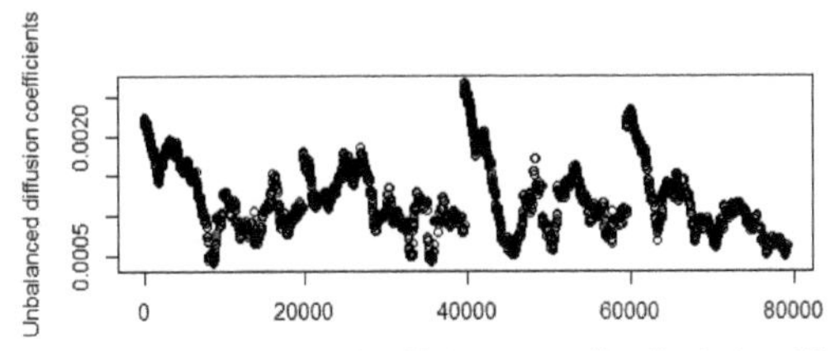

FIGURE 3.13
Unbalanced diffusion coefficients on the calibration days of Cisco.

diffusion coefficients for 30 minutes rolling time windows that we shifted bit by bit by one second.

There is empirical evidence that the frequency of trading follows patterns over the course of the day. This can be seen in Figure 3.13, which shows the unbalanced diffusion coefficients during the first four days of Cisco. The well-known patterns of intra-day activity are also described in the fourth chapter of [1]. We implemented two models, one including seasonality and one without taking seasonality into account.

For the model without seasonality we simply computed the mean and the standard deviation over all considered time windows of the calibration days. For the test day we had a look whether the calculated diffusion coefficients exceed the mean by more than three standard deviations. We define any deviation from these boundaries as a signal.

To include the seasonality we calculated the mean and the standard deviation not over all time windows of the calibration days, but only per time window on the calibration days. Therefore the diffusion coefficient of a time window on the test day is only compared to the diffusion coefficients of the same time windows on the calibration days and not to all diffusion coefficients.

3.3.4.3 Results for Constructed Sample Day

We ran the model for the Cisco data, meaning we used the first four days for calibration and the fifth day to test the model. The model including seasonality produced a signal to warn for low volatility on the fifth day of Cisco (see Figures 3.14 and 3.15), the model not including seasonality produced no signal. We decided to test our model on a constructed day as well.

We added an artificial jump to the original data of Cisco on 11/07/2014. Therefore we started at the 85th minute after opening to increase the midprice by 1 cent for 500 consecutive book events of the original data, meaning a price increase of 5$ within 5.6 seconds. Further we halved the inter-arrival times in this section. After this artificial jump we added 5$ to each midprice and adapted the time stamp. Since we cut off the last 15 minutes, there is

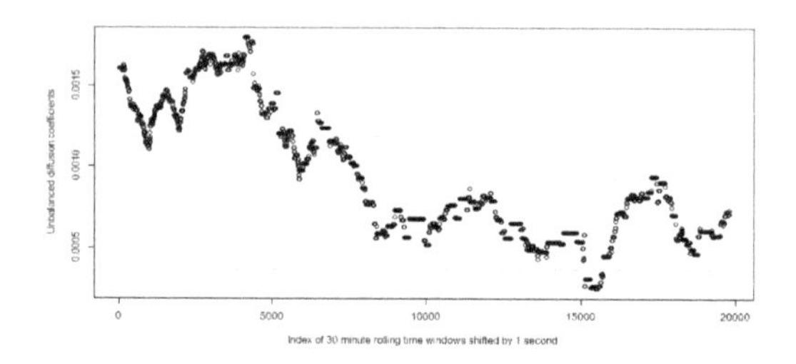

FIGURE 3.14
Unbalanced coefficients for Cisco on 11/07/2014, no signal, without seasonality.

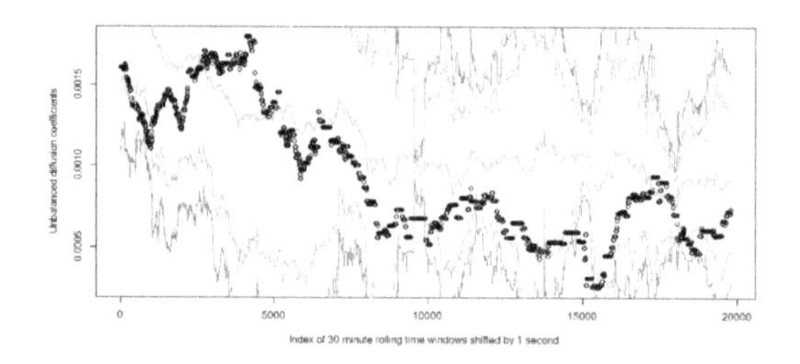

FIGURE 3.15
Unbalanced coefficients for Cisco on 11/07/2014, no signal, with seasonality.

no problem with halving the time in the specific jump points. Otherwise we would have a trading day shorter than the others. The midprice of the sample day can be seen in Figure 3.16.

We calculated the unbalanced diffusion coefficients for 30 minutes rolling time windows shifted by one second. The result can be seen in Figure 3.17. The model detects the artificial jump within 1.8 seconds after the begin of the

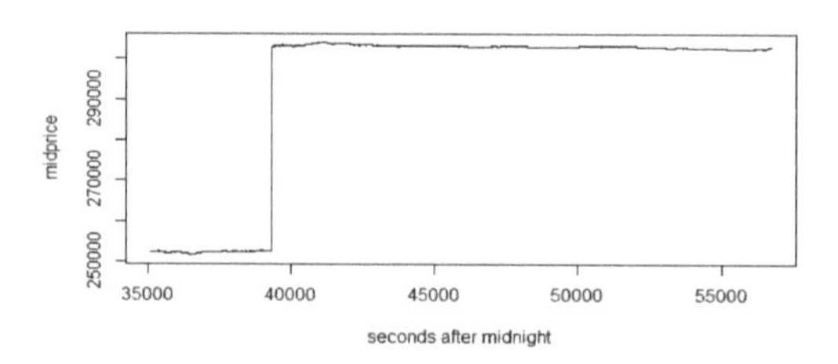

FIGURE 3.16
Midprice of the constructed sample day.

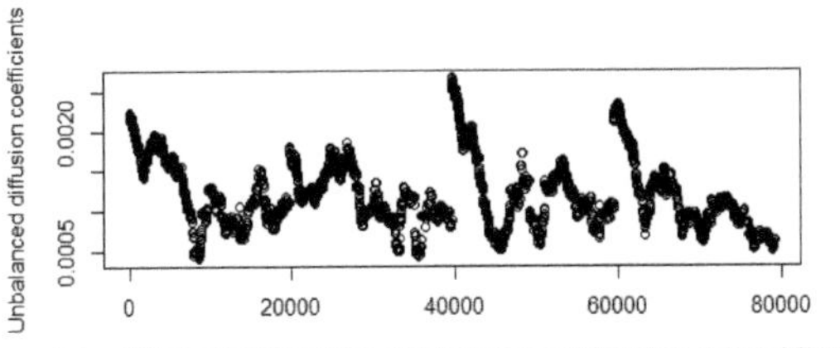

FIGURE 3.17
Unbalanced coefficients for constructed sample day of Cisco, without seasonality.

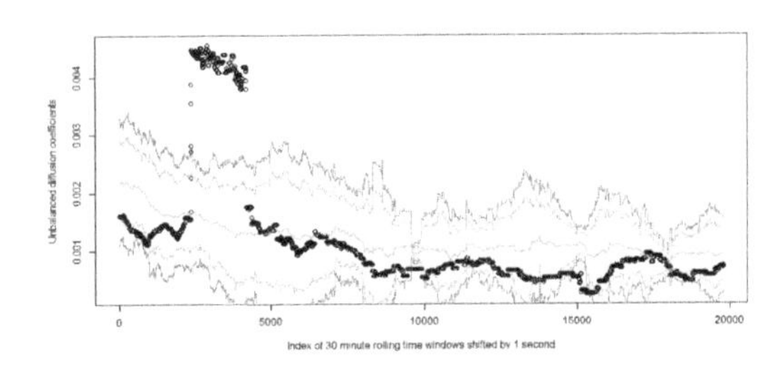

FIGURE 3.18
Unbalanced coefficients for constructed sample day of Cisco, with seasonality.

price increase, after the price has increased by \$0.78. Thus we are confident that it will deliver very good results for even smaller shifts. We can see again that our model detects jumps in volatility.

For the model including seasonality we get the result shown in Figure 3.18. The shortfall of this approach is that we only use four data points to calculate the mean and the standard deviation. We believe that practitioners should have access to more than only four calibration days.

3.4 General Semi-Markov Model for the Limit Order Book with arbitrary number of states

3.4.1 Justification

Our next goal is to further generalize the modelling of the stock price. In the last section we assumed that the jump sizes of the stock prices can only take

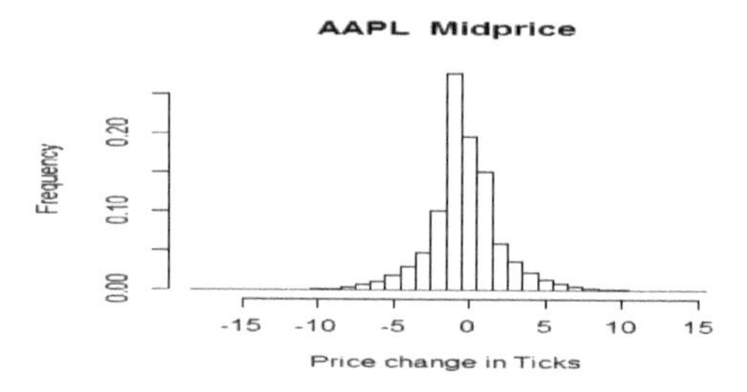

FIGURE 3.19
Jump sizes Apple Midprices.

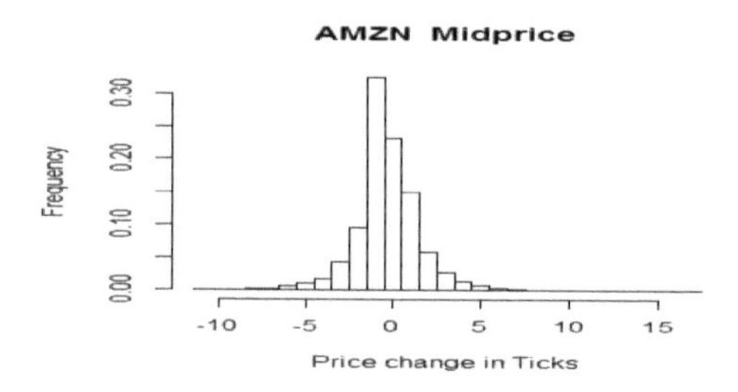

FIGURE 3.20
Jump sizes Amazon Midprices.

two values $a(1)$ and $a(2)$. Some of the available data give evidence that the price changes can take more than two values. This can be seen in Figures 3.19–3.21.The exact numbers of different price changes for Apple, Amazon and Google are stated in Table 3.6, which justifies the modelling of more than two states.

We will consider the following model:

$$s_t = \sum_{k=1}^{N(t)} a(X_k),$$

where X_k is a Markov chain with n states, meaning that the state space is extended to $\mathcal{S} = \{1, ..., n\}$.

3.4.2 Diffusion Limits

For the extended model, balanced and unbalanced case are defined as in Subsection 3.3.1. Assumption 3.1 does not change as well.

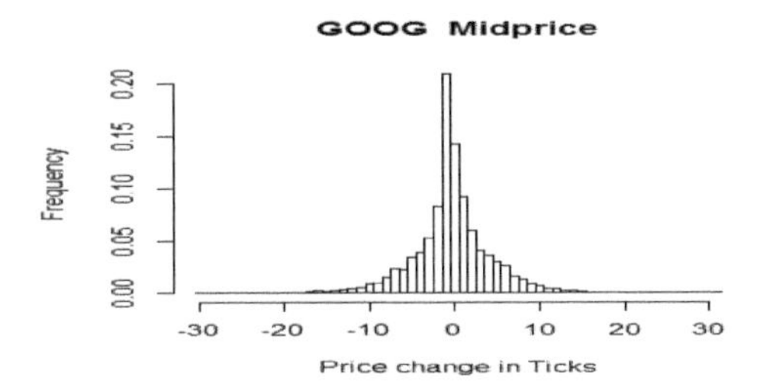

FIGURE 3.21
Jump sizes Google Midprices.

Assumption 3.3 *We assume $m(i) < \infty$ for all $i = 1, 2, \ldots n$, where $m(i)$ is defined like in Subsection 3.3.1.*

TABLE 3.6
Number of different price changes of mid prices

	Apple	Amazon	Google
States	60	46	87

We get the following results for the diffusion limits in the new model:

Theorem 3.2 *Given that assumption 3.1 is satisfied for the balanced case and assumption 3.3 is satisfied for the unbalanced case, we can proof the following weak convergences in the Skorokhod topology (see [7]):*

$$\left(\frac{s_{tnlog(n)} - N_{tnlog(n)}a^*}{\sqrt{n}}, t \geq 0\right) \overset{n\to\infty}{\Rightarrow} \frac{\sigma^*}{\sqrt{\tau^*}} W_t, \text{ for the balanced case and}$$

$$\left(\frac{s_{tn} - N_{tn}a^*}{\sqrt{n}}, t \geq 0\right) \overset{n\to\infty}{\Rightarrow} \frac{\sigma^*}{\sqrt{m_\tau}} W_t, \text{ for the unbalanced case,}$$

where W_t is a standard Brownian motion, $a^* = \sum_{i\in\mathcal{S}} \pi_i^* a(i)$ and $m_\tau = \sum_{i\in\mathcal{S}} \pi_i^* m(i)$. τ^* and $(\sigma^*)^2$ are given by:

$$\tau^* = \lim_{t\to+\infty} \frac{t}{N_t log(N_t)} \text{ (see [9], p.19)}$$

$$(\sigma^*)^2 = \sum_{i\in\mathcal{S}} \pi_i v(i)$$

$$v(i) = b(i)^2 + \sum_{j\in\mathcal{S}} (g(j) - g(i))^2 P(i,j) - 2b(i) \sum_{j\in\mathcal{S}} (g(j) - g(i)) P(i,j),$$

where

$$b = (b(1), b(2), ..., b(n))',$$
$$b(i) := a(X_i) - a^* := a(i) - a^* \text{ and}$$
$$g := (P + \Pi^* - I)^{-1} b.$$

P is a transition probability matrix, where $P(i,j) = P(X_{k+1} = j | X_k = i)$. Π^* denotes the stationary distribution of P and $g(j)$ is the jth entry of g.

Proof *We get the following law of large numbers:*

$$\frac{s_{tnlog(n)}}{n} \overset{n\to\infty}{\Rightarrow} \frac{a^* t}{\tau^*}.$$

The proof follows the one in [9], in which the calculation for σ has to be adapted. Therefore we consider the more general result given on page 28 in [10]. For simplification we mainly use our notations instead of the ones they used.

Denote for $t \in R_+$ (as in [9]):

$$R_n := \sum_{k=1}^{n} (a(X_k) - a^*),$$

$$U_n(t) := n^{-1/2}[(1 - \lambda_{n,t}) R_{\lfloor nt \rfloor} + \lambda_{n,t} R_{\lfloor nt \rfloor + 1}],$$

where $\lambda_{n,t} := nt - \lfloor nt \rfloor$. In [10] (page 28) it is shown for a more general case that we have the following weak convergence in the Skorokhod topology:

$$(U_n(t), t \geq 0) \overset{n\to\infty}{\Rightarrow} \sigma W,$$

where W is a standard Brownian motion, $\sigma^2 := \sum_i \pi(i) v(i)$, and for $i \in \mathcal{S}$:

$$v(i) = \sum_{j\in\mathcal{S}} \int_0^\infty (f(i,j,u) - \alpha_f(u))^2 Q(i,j,du) + \sum_{j\in\mathcal{S}} (g(j) - g(i))^2 P(i,j)$$
$$-2 \sum_{j\in\mathcal{S}} (g(j) - g(i)) P(i,j) \int_0^\infty (f(i,j,u) - \alpha_f(u)) H(i,j,du).$$

Q is the kernel of the Markov renewal process $Q(X_n, j, t) = \mathbb{P}[X_{n+1} = j, \tau_{n+1} \leq t | X_n]$ and can also be written as $Q(i,j,t) = P(i,j)H(i,j,t)$. P is the transition matrix of the Markov chain and $H(i,j,t) := \mathbb{P}[\tau_{n+1} \leq t | X_n = i, X_{n+1} = j]$. Further the result in [10] contains time inhomogeneity. As we consider time homogeneity, α_f is a constant equal to a^* of Section 3.3. Since the bounded function $a(\cdot)$ is only dependent on one variable, we simply have a

look at the case where $f(i) = a(i)$. Inserting the known simplifications we get

$$\begin{aligned}
v(i) &= \sum_{j\in\mathcal{S}} \int_0^\infty (a(i)-a^*)^2 Q(i,j,du) + \sum_{j\in\mathcal{S}} (g(j)-g(i))^2 P(i,j) \\
&\quad - 2\sum_{j\in\mathcal{S}} (g(j)-g(i))P(i,j) \int_0^\infty (a(i)-a^*)H(i,j,du) \\
&= \sum_{j\in\mathcal{S}} \int_0^\infty b(i)^2 P(i,j)H(i,j,du) + \sum_{j\in\mathcal{S}} (g(j)-g(i))^2 P(i,j) \\
&\quad - 2\sum_{j\in\mathcal{S}} (g(j)-g(i))P(i,j) \int_0^\infty b(i)H(i,j,du) \\
&= b(i)^2 \sum_{j\in\mathcal{S}} P(i,j) \int_0^\infty H(i,j,du) + \sum_{j\in\mathcal{S}} (g(j)-g(i))^2 P(i,j) \\
&\quad - 2\sum_{j\in\mathcal{S}} (g(j)-g(i))P(i,j)b(i) \int_0^\infty H(i,j,du) \\
&= b(i)^2 \sum_{j\in\mathcal{S}} P(i,j) + \sum_{j\in\mathcal{S}} (g(j)-g(i))^2 P(i,j) - 2\sum_{j\in\mathcal{S}} (g(j)-g(i))P(i,j)b(i) \\
&= b(i)^2 + \sum_{j\in\mathcal{S}} (g(j)-g(i))^2 P(i,j) - 2b(i)\sum_{j\in\mathcal{S}} (g(j)-g(i))P(i,j).
\end{aligned}$$

The continuing proof directly follows the proofs of Proposition 8 and Proposition 10 in [9].

Remark 3.4.1 *If we have a look at a state space containing only two states $\mathcal{S} = \{1, 2\}$, we get*

$$\begin{aligned}
v(1) &= b(1)^2 + P(1,2)(g(2)-g(1))^2 - 2b(1)P(1,2)(g(2)-g(1)) \\
v(2) &= b(2)^2 + P(2,1)(g(1)-g(2))^2 - 2b(2)P(2,1)(g(1)-g(2))
\end{aligned}$$

where $b(i) = a(i) - a^*$. This is the same result as the one derived in Section 3.3, with $P(1,2) = p(1)$, $P(2,1) = p(2)$, $g(1) = g_1$ and $g(2) = g_2$.

3.4.3 Implementation

The approaches used to calculate the diffusion coefficients for the model described above are similar to the ones explained in Subsection 3.3.2.

The question arises how to choose the values of $a(i)$ for $i \in \mathcal{S}$. Our quantile-based approach is illustrated in this section.

Having calculated the price jumps, we split the data into two parts: One containing all the negative price changes and one containing the positive price jumps. Then evenly distributed quantiles are calculated for both sets of data. Depending on the data there might occur equal values for the quantiles. In this

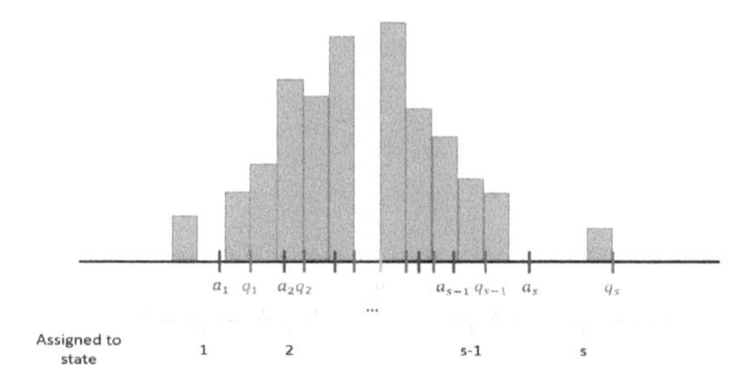

FIGURE 3.22
Illustration.

case we decrease the number of states as long as necessary. The state values $a(i)$ are set in the following way: We calculate the average of price changes located in between two quantiles or respectively below the first quantile and above the last one.

The single price changes are assigned in the following way to the states:

- price changes smaller than the smallest quantile are assigned to state 1
- price changes between the ith and j^{th} quantile are assigned to state j

The approach is illustrated in Figure 3.22.

Based on these assignments, we follow the same approach as in Subsection 3.3.2 to calculate the transition matrix and do the remaining calculations.

3.4.4 Numerical Results

Table 3.7 depicts the results gained from our computations for the [5] data of Apple, Amazon, Google, Intel and Microsoft on 2012/06/21. The number of states we used to get the data, the matrix P and the states $a(i)$ can be seen in Figures 3.23 and 3.24. These two stocks serve here as examples and we can provide the same figures for Apple, Intel and Microsoft, if requested. As in Subsection 3.3.3 we do not consider the first and last 15 minutes of trading.

TABLE 3.7
Numerical Results for many states, Mid Prices

	σ^2	τ^*	m_τ	$\frac{\sigma}{\sqrt{\tau^*}}$	$\frac{\sigma}{\sqrt{m_\tau}}$
Apple	0.00031	0.0370	0.4026	0.0915	0.0277
Amazon	0.00017	0.0892	0.9001	0.0433	0.0136
Google	0.00090	0.1145	1.1291	0.0885	0.0282
Intel	0.00004	1.3151	10.0839	0.0052	0.0019
Microsoft	0.00004	0.8944	7.1647	0.0065	0.0023

P(ij)	1	2	3	4	5	6	a(i)
1	0.464	0.278	0.259				-0.013
2	0.506	0.338	0.157				0.006
3	0.594	0.259	0.146				0.024
4							
5							
6							

FIGURE 3.23
Amazon.

P(ij)	1	2	3	4	5	6	a(i)
1	0.187	0.214	0.099	0.178	0.321		-0.052
2	0.202	0.353	0.239	0.130	0.077		-0.008
3	0.144	0.306	0.318	0.115	0.118		0.006
4	0.397	0.228	0.145	0.119	0.112		0.013
5	0.506	0.115	0.160	0.107	0.113		0.068
6							

FIGURE 3.24
Google.

We can show the linear relationship between the standard deviation of the ten minute price changes and the calculated diffusion coefficients, like we did in Subsection 3.3.3. We calculate these standard deviations for our Apple, Amazon, Google, Intel and Microsoft data (provided in [5]) and the data which is provided with [1]. We get an adjusted R^2 of 0.9814 for the linear regression of the diffusion coefficients on the standard deviation for the balanced case and one of 0.9839 for the unbalanced case. See Figures 3.25 and 3.26. The difference to the fit in 3.3.3 is very small, which can be explained by having 25 of 40 stocks for which the algorithm sets only two states. In both cases the regression coefficients are highly significant.

3.5 Discussion on Price Spreads

A simplifying assumption used in [2] and [9] is that the spread, $p_t^a - p_t^b$ is fixed at a single tick, δ. They justify this assumption by observing that over 98% of all data points in their sample have a spread of δ. Furthermore, they observe that the average lifetime, in ms, of a spread larger than a single tick

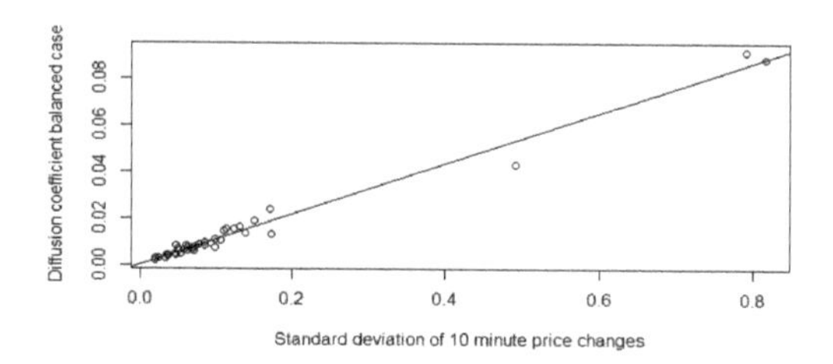

FIGURE 3.25
Linear relationship of coefficients and standard deviation of 10 minutes mid-price changes for many states, Balanced case.

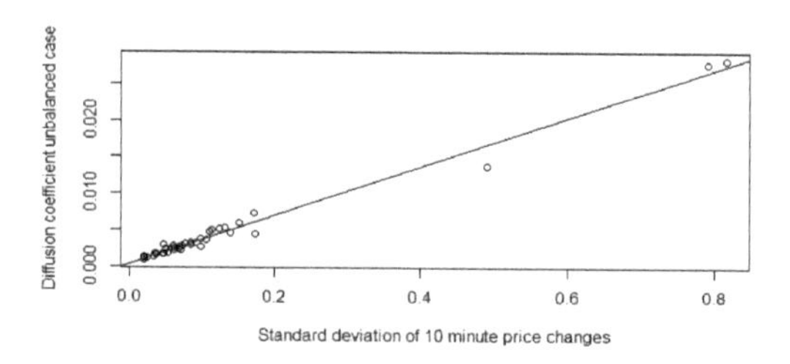

FIGURE 3.26
Linear relationship of coefficients and standard deviation of 10 minutes mid-price changes for many states, Unbalanced case.

is extremely small on average, often less than a single ms.
In reproducing their figures, using our data, we found that over 90% of all observations for AAPL, AMZN, and GOOG on 2012/06/21 has a spread strictly greater than 5δ. Table 3.8 displays the percentage of observations in some of our data sets at various tick sizes.

Even for our MSFT and INTC data the assumption that $p_t^a - p_t^b = \delta$ seems less warranted, given that a significant proportion of our observed data includes spreads greater than δ. Figure 3.27 displays histograms, for MSFT only, of the lifetimes of a spread greater than one tick and the lifetimes of a spread equal to one tick.

Further calculations demonstrate that, for MSFT and INTC, roughly 85% of spreads greater than one tick had a lifetime less than 5 ms. While the assumption that spreads greater than δ are instantly filled do not hold as well for our data set, given the longer time scales used for asymptotic analysis we can justify this assumption for a subset of our sample.

We believe that the larger spreads observed in our empirical data has consequences for estimated intensities of order flow. If we follow the method of [3] to estimate the rates of incoming Limit orders (λ), Market orders (μ)

TABLE 3.8
Percentage of Observations at various tick sizes for MSFT and INTC 2012/06/21

Spread	1 Tick	2 Ticks	3 Ticks	4 Ticks	5 Ticks	≥ 5 Ticks	Avrg. Spread
AAPL	0.79	1.79	2.10	2.44	2.81	90.09	15.50
AMZN	1.31	1.52	1.74	2.23	2.78	90.43	13.59
GOOG	0.24	0.37	0.35	0.45	0.63	97.96	31.11
INTC	66.82	33.14	0.04	0.00	0.00	0.00	1.33
MSFT	65.63	34.31	0.05	0.00	0.00	0.00	1.34

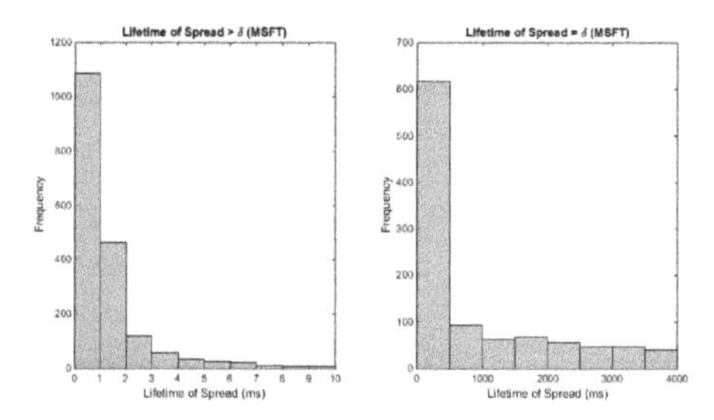

FIGURE 3.27
Histograms of (left) lifetime of spread greater than one tick and (right) lifetime of spread equal to one tick for MSFT, 2012/06/21.

and Cancellations (θ), we only use the data points where $p_t^a - p_t^b = \delta$. The resulting estimations are included in Table 3.9.

For AAPL, AMZN, and GOOG, this means we use a small sample size to estimate these parameters; decreasing the accuracy of our estimations. Furthermore, since the vast majority of incoming orders occur at the best bid and ask (see [2]), and the best bid and ask are rarely δ apart, this means that we will have a much lower flow for these stocks empirically than observed in data where $p_t^a - p_t^b = \delta$ more often. For a subset of our sample, where the spread is much more frequently δ, we find higher observed order intensities. However, even with the increased intensity of incoming order flow, we see that $\lambda > \mu + \theta$, which violates an important assumption of [2]. We believe that the increased instances of $p_t^a - p_t^b > \delta$ are also causing this violation. MSFT and INTC both have roughly 30% of observed data points with spread 2δ. When the spread is greater than δ, there is an incentive for traders to post limit orders within the spread so that their orders are executed first. We believe the resulting limit orders account for the increased estimated limit order intensity$\hat{\lambda}$.

Given that our model focuses explicitly on the midprice, which does not include information about the observed spread, we assume $p_t^a - p_t^b = \delta$ for our model even though it is difficult to justify empirically using our data sets. As

TABLE 3.9
Estimates for intensity of limit orders, market orders + cancellations in number of shares per second on 2012/06/21.

	$\hat{\lambda}$	$\hat{\mu}+\hat{\theta}$
MSFT	3000.78	3000.72
INTC	2483.13	2427.63
GOOG	0.61	0.20
AMZN	3.76	1.15
AAPL	3.09	1.59

demonstrated, the diffusion limit calculated using the midprice is still a valid approximation for long run volatility of the midprice.

3.6 Conclusion

After reviewing some of the assumptions from [9] and [2] for our sets of data, we showed how to develop the model presented in [12] (see Section 3.6: Discussion). In Section 3.3, we illustrated a model that considers two possible price changes different from one tick, as well as our numerical results. Further we illustrated a possible application in risk management for our model. In Section 3.4, we further generalized the model by now considering an arbitrary number of possible price changes.

As proposed in [2], we compared the diffusion coefficients to the standard deviation of the ten minute price changes. Applying a linear regression using our available data, we tested how good our model extensions describe the linear relationship. We showed a large improvement in the adjusted R^2 with the first extension, where we allowed sizes of price jumps to have a magnitude different to one tick. The second model extension included a higher number of possible sizes of price changes. The adjusted R^2 increased again.

Bibliography

[1] A. Cartera, S. Jaimungal & J. Penalva (2015) *Algorithmic and High-Frequency Trading.* Cambridge University Press.

[2] R. Cont & A. de Larrard (2013) Price dynamics in a Markovian Limit Order Book market, *SIAM Journal for Financial Mathematics* **4** (1): 1–25.

[3] R. Cont, S. Stoikov & R. Talreja (2010) A stochastic model for order book dynamics, *Operations Research* **58** (3), 549–563.

[4] M.D. Gould, M.A. Porter, S. Williams, M. McDonald, D.J. Fenn, S.D. Howison (2013) Limit Order Books, *Quantitative Finance* **13** (10), 1709–1742.

[5] Humbold Universität zu Berlin, Germany (2013) LOBSTER: Limit Order Book System – The Efficient Reconstructor. http://LOBSTER.wiwi.hu-berlin.de. Accessed: 2016-07-20

[6] J.R. Norris (1997) *Markov Chains.* Cambridge Series in Statistical and Probabilistic Mathematics.

[7] A. Skorokhod (1965) *Studies in the Theory of Random Processes.* Reading: Addison Wesley. Reprinted by Dover Publications, NY.

[8] A. Swishchuk & N. Vadori (2015a) Semi-Markov Model for the price dynamics in limit order markets. IPAM FMWSI, UCLA, March 23-27, 2015 (people.ucalgary.ca/ aswish/IPAM_UCLA_2015.pdf).

[9] A. Swishchuk & N. Vadori (2015b) Semi-Markov Model for the price dynamics in limit order markets. SSRN: papers.ssrn.com/sol3/papers.cfm?abstract_id=2579865.

[10] A. Swishchuk & N. Vadori (2015c) Strong Law of Large Numbers and Central Limit Theorems for Functionals of Inhomogeneous Semi-Markov Processes.

[11] A. Swishchuk & N. Vadori (2016a) A semi-Markovian modelling of limit order markets. arXiv: 1601.01710.

[12] A. Swishchuk & N. Vadori (2016b) A semi-Markovian modelilng of limit order markets. *SIAM J. Finan. Math.*, **8**, 240–273.

[13] A. Swishchuk, K. Cera, T. Hofmeister and J. Schmidt (2017) General semi-Markov model for limit order books. *Intern. J. Theoret. Applied Finance*, **20**, 1750019.

Part II

Modelling of Big Data in Finance with Hawkes Processes

4

A Brief Introduction to Hawkes Processes

This introductory chapter contains some basics definitions of HPs, their stylized properties, and the descriptions of CHP and GCHP. The stylized properties include non-exponential distribution of inter-arrival times, clustering affect of trades, and non-independency of the mid-price changes. We use the set of real data, namely EBAY, FB, MU, PCAR, SMH, CSCO, (provided by Cartea et al. (2015), [9], book), to check, justify and illustrate our approach in using Hawkes process.

4.1 Introduction

Trading activity, which is not completely memoryless process, leads to time series of irregularly spaced points that show a clustering behaviour. This stylized property suggests the use of the Hawkes process (HP), which is an extension of the classical Poisson process that is not suitable for modelling trade arrival times. Consequently, the Poisson process is not suitable for modelling trade arrival times. (See, e.g., QQ-plot of inter-arrival times of trades against an exponential distribution for Eurostoxx futures on March 03, 2011 (Da Fonseca et al. (2013))). Trades tend to cluster. (See, e.g., a his-togram of the number of trades occurring every minute during a trading day for the Eurostoxx futures, March 03, 2011. The clustering is graphically clear (Da Fonseca et al. (2013)).

The Hawkes process (HP) is named after its creator Alan Hawkes (1971, 1974), [27], [28]. The HP is a simple point process equipped with a self-exciting property, clustering effect and long run memory. Through its dependence on the history of the process, the HP captures the temporal and cross sectional dependence of the event arrival process as well as the "self-exciting" property observed in our empirical data on limit order books. Self-exciting point processes have recently been applied to high frequency data for price changes [54] or order arrival times [55]. HPs have seen their application in many areas, like genetics (2010) [11], occurrence of crime (2010) [50], bank defaults [51] and earthquakes [52].

Point processes gained a significant amount of attention in statistics during the 1950s and 1960s. Cox (1955) [16] introduced the notion of a doubly stochastic Poisson process (called the Cox process now) and Bartlett (1963)

DOI: 10.1201/9781003265986-4

[8] investigated statistical methods for point processes based on their power spectral densities. Lewis (1964) [34] formulated a point process model (for computer power failure patterns) which was a step in the direction of the HP. A nice introduction to the theory of point processes can be found in Daley et al. (1988) [17]. The first type of point process in the context of market microstructure is the autoregressive conditional duration (ACD) model introduced by Engel et al. (1998) [19].

A recent application of HP is in financial analysis, in particular limit order books. In this chapter, we introduce various new Hawkes processes, namely general compound Hawkes processes to model the price process in limit order books. The general compound Hawkes process was first introduced in [40] to model the risk process in insurance and studied in detail in [41]. In the paper [43] we obtained functional CLTs and LLNs for general compound Hawkes processes with dependent orders and regime-switching compound Hawkes processes.

Bowsher (2007) [6] was the first one who applied the HP to financial data modelling. Cartea et al. (2011) [9] applied HP to model market order arrivals. Filimonov and Sornette (2012) [25] and Filimonov et al. (2013) [26] applied the HPs to estimate the percentage of price changes caused by endogenous self-generated activity rather than by the exogenous impact of news or novel information. Bauwens and Hautsch (2009) [7] used a five dimensional HP to estimate multivariate volatility between five stocks, based on price intensities. Hewlett (2006) [29] used the instantaneous jump in the intensity caused by the occurrence of an event to qualify the market impact of that event, taking into account the cascading effect of secondary events causing further events. Hewlett (2006) [29] also used the Hawkes model to derive optimal pricing strategies for market makers and optimal trading strategies for investors given that the rational market makers have the historic trading data. Large (2007) [32] applied a Hawkes model for the purpose of investigating market impact, with a specific interest in order book resiliency. Specifically, he considered limit orders, market orders and cancellations on both the buy and sell side, and further categorizes these events based on their level of aggression, resulting in a ten dimensional Hawkes process. Other econometric models based on marked point processes with stochastic intensity include autoregressive conditional intensity (ACI) models with the intensity depending on its history. Hasbrouck (1999) [30] introduced a multivariate point process to model the different events of an order book but did not parametrize the intensity. We note that Brémaud et al. (1996) [4] generalized the HP to its nonlinear form. Also, a functional central limit theorem for nonlinear Hawkes processes was obtained in Zhu (2013) [49]. The "Hawkes diffusion model" introduced in Ait-Sahalia et al. (2013) [1] attempted to extend previous models of stock prices to include financial contagion. Chavez-Demoulin et al. (2012) [12] used Hawkes processes to model high-frequency financial data. An application of affine point processes to portfolio credit risk may be found in Errais et al. (2010) [24]. Some applications of Hawkes processes to financial data are also given in Embrechts et al. (2011) [23].

Cohen et al. (2014) [14] derived an explicit filter for Markov modulated Hawkes processes. Vinkovskaya (2014) [47] considered a regime-switching Hawkes process to model its dependency on the bid-ask spread in limit order book. Regime-switching models for pricing of European and American options were considered in Buffington et al. (2000) [2] and Buffington et al. (2002) [3], respectively. Semi-Markov processes were applied to limit order books in [44] to model the mid-price. We also note that level-1 limit order books with time dependent arrival rates $\lambda(t)$ were studied in [13], including the asymptotic distribution of the price process.

The paper by Bacry et al. (2015) [5] proposes an overview of the recent academic literature devoted to the applications of Hawkes processes in finance. It is a nice survey of applications of Hawkes processes in finance. In general, the main models in high-frequency finance can be divided into univariate models, price models, impact models, order-book models and some systemic risk models, models accounting for news, high-dimensional models and clustering with graph models. The book by Cartea et al. (2015) [10] developed models for algorithmic trading such as methods for executing large orders, market making, trading pairs of collections of assets, and executing in the dark pool. This book also contains a link from which several datasets can be downloaded, along with **MATLAB** code to assist in experimentation with the data.

A detailed description of the mathematical theory of Hawkes processes is given in Liniger (2009) [33]. The paper by Laub et al. (2015) [35] provides background, introduces the field and historical development, and touches upon all major aspects of Hawkes processes.

4.2 Definition of Hawkes Processes (HPs)

In this section we give various definitions and some properties of Hawkes processes which can be found in the existing literature (see, e.g., [27], [28], [23] and [48], to name a few). They include in particular one dimensional and non-linear Hawkes processes.

Definition 4.1 (Counting Process) *A counting process is a stochastic process $N(t)$ with $t \geq 0$, where $N(t)$ takes positive integer values and satisfies $N(0) = 0$. It is almost surely finite and a right-continuous step function with increments of size $+1$.*

Denote by $\mathcal{F}^N(t)$, $t \geq 0$, the history of the arrival up to time t, that is, $\mathcal{F}^N(t)$, $t \geq 0$, is a filtration (an increasing sequence of σ-algebras).

A counting process $N(t)$ can be interpreted as a cumulative count of the number of arrivals into a system up to the current time t. The counting process can also be characterized by the sequence of random arrival times $(T_1, T_2, \dots)$ at which the counting process $N(t)$ has jumped. The process defined by these arrival times is called a point process (see [17]).

Definition 4.2 (Point Process) *If a sequence of random variables $(T_1, T_2, \dots)$, taking values in $[0, \infty)$ has $P(0 \leq T_1 \leq T_2 \leq \dots) = 1$, and the number of points in a bounded region is almost surely finite, then $(T_1, T_2, \dots)$ is called a point process.*

Definition 4.3 (Conditional Intensity Function) *Consider a counting process $N(t)$ with associated histories $\mathcal{F}^N(t)$, $t \geq 0$. If a non-negative function $\lambda(t)$ exists such that*

$$\lambda(t) = \lim_{h \to 0} \frac{E[N(t+h) - N(t) \mid \mathcal{F}^N(t)]}{h} \tag{4.1}$$

then it is called the conditional intensity function of $N(t)$ (see [35]). We note that originally this function was called the hazard function (see [16]).

Definition 4.4 (One-dimensional Hawkes Process) *The one-dimensional Hawkes process (see [35], [28]) is a point process $N(t)$ which is characterized by its intensity $\lambda(t)$ with respect to its natural filtration:*

$$\lambda(t) = \lambda + \int_0^t \mu(t-s) dN(s) \tag{4.2}$$

where $\lambda > 0$, and the response function $\mu(t)$ is a positive function that satisfies $\int_0^\infty \mu(s)ds < 1$.

The constant λ is called the background intensity and the function $\mu(t)$ is sometimes called the excitation function. To avoid the trivial case of a homogeneous Poisson process, we assume $\mu(t) \neq 0$. Thus, the Hawkes process is a non-Markovian extension of the Poisson process.

With respect to the Definitions of $\lambda(t)$ in 4.3 and $N(t)$ in 4.4, it follows that

$$P(N(t+h) - N(t) = m \mid \mathcal{F}^N(t)) = \begin{cases} \lambda(t)h + o(h) & m = 1 \\ o(h) & m > 1 \\ 1 - \lambda(t)h + o(h) & m = 0 \end{cases}$$

The interpretation of Equation (4.2) is that the events occur according to an intensity with a background intensity λ which increases by $\mu(0)$ at each new event, eventually decaying back to the background intensity value according to the evolution of the function $\mu(t)$. Choosing $\mu(0) > 0$ leads to a jolt in the intensity at each new event, and this feature is often called the self-exciting feature. In other words, if an arrival causes the conditional intensity function $\lambda(t)$ in Equations (4.1)–(4.2) to increase then the process is called self-exciting.

We would like to mention that the conditional intensity function $\lambda(t)$ in Equations (4.1)–(4.2) can be associated with the compensator $\Lambda(t)$ of the counting process $N(t)$, that is

$$\Lambda(t) = \int_0^t \lambda(s)ds \tag{4.3}$$

We note that $\Lambda(t)$ is the unique non-decreasing, $\mathcal{F}^N(t)$, $t \geq 0$, predictable function, with $\Lambda(0) = 0$ such that

$$N(t) = M(t) + \Lambda(t) \text{ a.s.}$$

where $M(t)$ is an $\mathcal{F}^N(t)$, $t \geq 0$, local martingale (existence of which is guaranteed by the Doob-Meyer decomposition).

A common choice for the function $\mu(t)$ in Equation (4.2) is the one of exponential decay (see [27])

$$\mu(t) = \alpha e^{-\beta t} \tag{4.4}$$

with parameters α, $\beta > 0$. In this case, the Hawkes process is called the Hawkes process with exponentially decaying intensity.

In the case of Equation (4.4), Equation (4.2) becomes

$$\lambda(t) = \lambda + \int_0^t \alpha e^{-\beta(t-s)} dN(s) \tag{4.5}$$

We note that in the case of Equation (4.4), the process $(N(t), \lambda(t))$ is a continuous-time Markov process, which is not the case for a general choice of excitation function in Equation (4.1).

With some initial condition $\lambda(0) = \lambda_0$, the conditional intensity $\lambda(t)$ in Equation (4.5) with exponential decay in Equation (4.4) satisfies the SDE

$$d\lambda(t) = \beta(\lambda - \lambda(t))dt + \alpha dN(t), t \geq 0 \tag{4.6}$$

which can be solved using stochastic calculus as

$$\lambda(t) = e^{-\beta t}(\lambda_0 - \lambda) + \lambda + \int_0^t \alpha e^{-\beta(t-s)} dN(s), \tag{4.7}$$

which is an extension of Equation (4.5).

Another choice for $\mu(t)$ is a power law function

$$\lambda(t) = \lambda + \int_0^t \frac{k}{(c + (t-s))^p} dN(s) \tag{4.8}$$

with positive parameters (c, k, p). This power law form for $\lambda(t)$ in Equation (4.8) was applied in the geological model called Omori's law, and used to predict the rate of aftershocks caused by an earthquake.

Definition 4.5 (D-dimensional Hawkes Process) *The D-dimensional Hawkes process (see [23]) is a point process $\vec{N}(t) = (N^i(t))_{i=1}^D$ which is characterized by its intensity vector $\vec{\lambda}(t) = (\lambda^i(t))_{i=1}^D$ such that:*

$$\lambda^i(t) = \lambda^i + \int_0^t \mu^{ij}(t-s)dN^j(s) \tag{4.9}$$

where $\lambda^i > 0$, and $M(t) = (\mu^{ij}(t))$ is a matrix-valued kernel such that:

1. *it is component-wise non-negative: $(\mu^{ij}(t)) \geq 0$ for each $1 \leq i,j \leq D$*
2. *it is component-wise L^1-integrable*

In matrix-convolution form, Equation (4.9) can be written as

$$\vec{\lambda}(t) = \vec{\lambda} + M * d\vec{N}(t) \tag{4.10}$$

where $\vec{\lambda}(t) = (\lambda^i)_{i=1}^D$.

Definition 4.6 (Non-linear Hawkes Process) *The non-linear Hawkes process (see, e.g., [48]) is defined by the intensity function in the following form:*

$$\lambda(t) = h\left(\lambda + \int_0^t \mu(t-s)dN(s)\right) \tag{4.11}$$

where $h(\cdot)$ is a non-linear function with support in R^+. Typical examples for $h(\cdot)$ are $h(x) = \mathbf{1}_{x \in R^+}$ and $h(x) = e^x$.

Remark 4.1 *Many other generalizations of Hawkes processes have been proposed. They include mixed diffusion-Hawkes models [24], Hawkes models with shot noise exogenous events [18] and Hawkes processes with generation dependent kernels [37], to name a few.*

4.3 Compound Hawkes Processes

In this section we define non-linear compound Hawkes process with N-state dependent orders. The dependent orders means the dependency of both, the type of a book event and its corresponding inter-arrival times, on the type of the previous book event. We also consider special cases of this general compound Hawkes process.

Definition 4.7 (NLCHPnSDO) *(Non-Linear Compound Hawkes Process with n-state Dependent Orders (NLCHPnSDO) in Limit Order Books)*

Consider the mid-price process S_t

$$S_t = S_0 + \sum_{k=1}^{N(t)} a(X_k) \tag{4.12}$$

where X_k is a continuous time n-state Markov chain, $a(x)$ is a continuous and bounded function on the state space $X := \{1, 2, \ldots, n\}$, $N(t)$ is the non-linear Hawkes process (see, e.g., [48] defined by the intensity function in the following form (see Equation (4.11)*):*

$$\lambda(t) = h\Big(\lambda + \int_0^t \mu(t-s)dN(s)\Big)$$

where $h(\cdot)$ is a non-linear increasing function with support in R^+. We note that in [4] it was shown that if $h(\cdot)$ is α-Lipschitz (see [4]) such that $\alpha||h||_{L^1} < 1$ then there exists a unique stationary and ergodic Hawkes process satisfying the dynamics of Equation (4.11)*. We shall refer to the process in Equation* (4.12) *as a Non-linear Compound Hawkes Process with n-State Dependent Orders (NLCHPnSDO).*

This non-linear compound Hawkes process will be the foundation for our studies throughout this paper. In the following subsection we will introduce four specific examples, which will be used for our empirical investigations of the mid-price processes.

4.3.1 Special Cases of Compound Hawkes Processes in Limit Order Books

Definition 4.8 (GCHPnSDO) *(General Compound Hawkes Process With N-state Dependent Orders (GCHPnSDO))*

Suppose that X_k is an ergodic continuous-time Markov chain, independent of $N(t)$, with state space $X = \{1, 2, \ldots, n\}$, $N(t)$ is a one-dimensional Hawkes process defined in Definition 4.4 and $a(x)$ is any bounded and continuous function on X. We define the General Compound Hawkes process with N-state Dependent Orders (GCHPnSDO) by the following process

$$S_t = S_0 + \sum_{k=1}^{N(t)} a(X_k) \tag{4.13}$$

Note that this process can be recovered from Equation (4.12) *by letting $h(x) = x$.*

Definition 4.9 (GCHP2SDO) *(General Compound Hawkes Process with Two-State Dependent Orders (GCHP2SDO))*

Suppose that X_k is an ergodic continuous time Markov chain, independent of $N(t)$, with two states $\{1, 2\}$. Then Equation (4.13) *becomes*

$$S_t = S_0 + \sum_{k=1}^{N(t)} a(X_k) \tag{4.14}$$

where $a(X_k)$ takes only the values $a(1)$ and $a(2)$. Of course we can view this as a special case of the n-state case, where $n = 2$. This model was used in [45] for the mid-price process in limit order books with non-fixed tick δ and two-valued price changes.

Definition 4.10 (GCHPDO) *(General Compound Hawkes Process with Dependent Orders (GCHPDO))*

Suppose that $X_k \in \{-\delta, \delta\}$ and that $a(x) = x$, then S_t in Equation (4.13) *becomes*

$$S_t = S_0 + \sum_{k=1}^{N(t)} X_k \tag{4.15}$$

This type of process can be a model for the mid-price in limit order books, where δ is a fixed tick size and $N(t)$ is the number of order arrivals up to time t. We shall call this process a General Compound Hawkes Process with Dependent Orders (GCHPDO). This is a generalization of the previous process, obtained by letting $a(1) = -\delta$ and $a(2) = \delta$.

Having defined several mid-price processes, we now prove diffusion limit theorems and LLNs for each price process in the following section. These diffusion processes will be used for our exploration of the applicability of this model to real world limit order book data.

4.4 Limit Theorems for Hawkes Processes: LLN and FCLT

4.4.1 Law of Large Numbers (LLN) for Hawkes Processes

Let $0 < \hat{\mu} := \int_0^{+\infty} \mu(s)ds < 1$. Then

$$\frac{N(t)}{t} \to_{t\to+\infty} \frac{\lambda}{1-\hat{\mu}}.$$

4.4.2 Functional Central Limit Theorems (FCLT) for Hawkes Processes

Under LLN and $\int_0^{+\infty} s\mu(s)ds < +\infty$ conditions

$$P\Big(\frac{N(t) - \lambda t/(1-\hat{\mu})}{\sqrt{\lambda t/(1-\hat{\mu})^3}} < y\Big) \to_{t\to+\infty} \Phi(y),$$

where $\Phi(\cdot)$ is the c.d.f. of the standard normal distribution.

Remark 4.2 *For exponential decaying intensity* $\hat{\mu} = \alpha/\beta$.

4.5 Limit Theorems for Poisson Processes: LLN and FCLT

For complicity of the picture, we present here LLN and FCLT for Poisson process. We note that if $\mu(t) = 0$ in (2), then $\lambda(t) = \lambda$, and Hawkes process is simply Poisson process. Thus, we can easily get LLN and FCLT for Poisson process from LLN and FCLT for Hawkes process in Sections 3.1 and 3.2, respectively. In this case, $\hat{\mu} = 0$.

4.5.1 Law of Large Numbers (LLN) for Poisson Processes

$$\frac{N(t)}{t} \to_{t\to+\infty} \lambda.$$

4.5.2 Functional Central Limit Theorems (FCLT) for Hawkes Processes

$$P\Big(\frac{N(t) - \lambda t}{\sqrt{\lambda t}} < y\Big) \to_{t\to+\infty} \Phi(y),$$

where $\Phi(\cdot)$ is the c.d.f. of the standard normal distribution.

4.6 Stylized Properties of Hawkes Process

In this section we present justification of using Hawkes processes based on their stylized properties: non-exponential distribution of inter-arrival times, clustering affect of trades, and non-independency of the mid-price changes. In the present section we use the set of real data, namely EBAY, FB, MU, PCAR, SMH, CSCO, (provided by Cartea et al. (2015), [9], book), to check, justify and illustrate our approach in using Hawkes process. For more details, see Chapter 7.

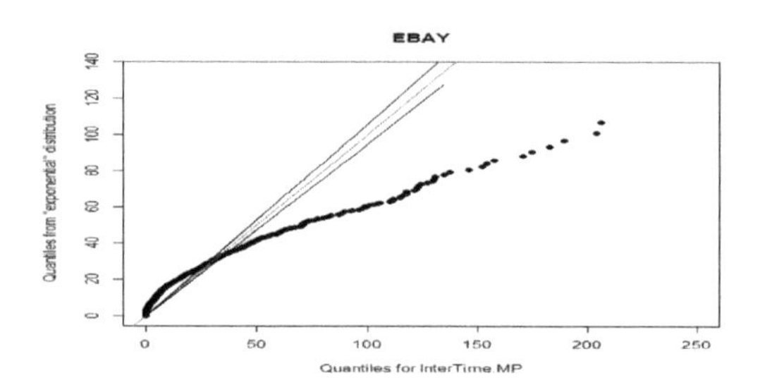

FIGURE 4.1
EBAY-QQplot.

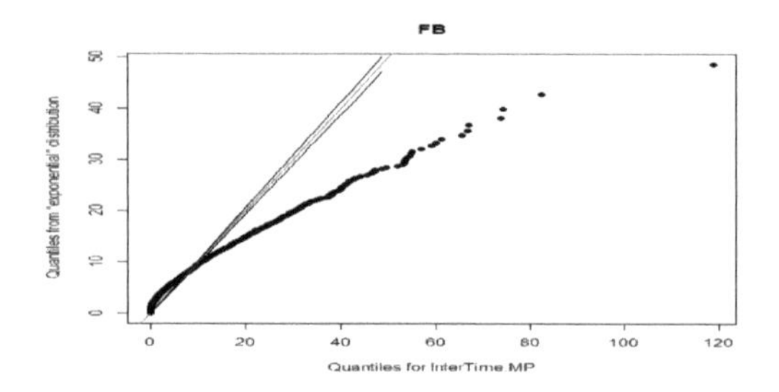

FIGURE 4.2
FB-QQplot.

4.6.1 Non-exponential Inter-arrival Times

In Figures 4.1–4.6, the QQ-plot rejects Poisson Process since the inter-arrival time does not follow the exponential distribution for all 6 stocks (EBAY, FB, MU, PCAR, SMH, CSCO). In Figures 4.1–4.6, the two black straight lines are the confidence boundary under 95% confidence level; the red straight line is $y = x$.

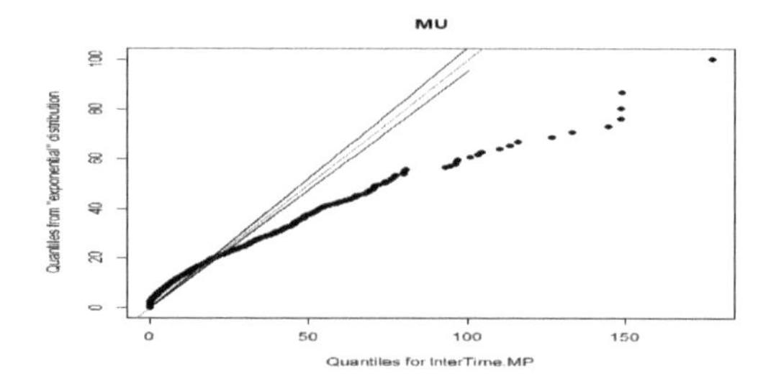

FIGURE 4.3
MU-QQplot.

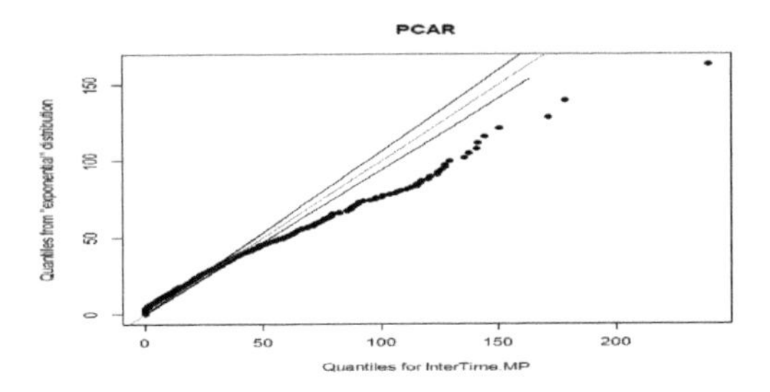

FIGURE 4.4
PCAR-QQplot.

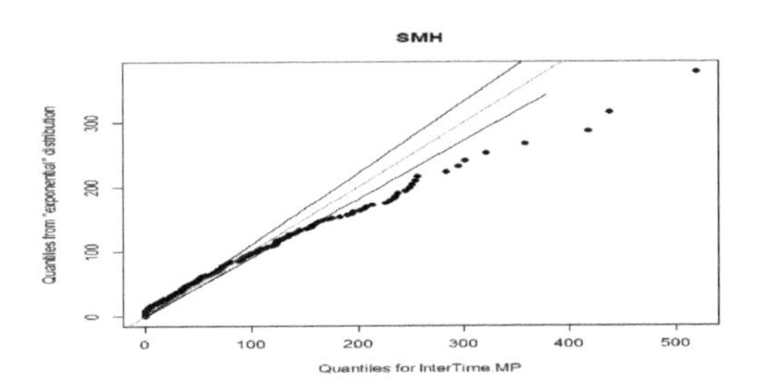

FIGURE 4.5
SMH-QQplot.

Furthermore, if the exponential distribution does not fit the inter-arrival times well, we want know which distribution will fit better. We tried normal distribution, Gamma distribution, Weibull distribution, and then compared the theoretical Cumulative Density Function (CDF) with the empirical CDF (see Figures 4.7–4.12). From Figures 4.7–4.12, it is obvious that in most cases, Gamma or Weibull distribution performs much better for all 6 stocks, EBAY, FB, MU, PCAR, SMH, CSCO, than the others since the curve of the theoretical CDF nearly coincides with that of empirical CDF.

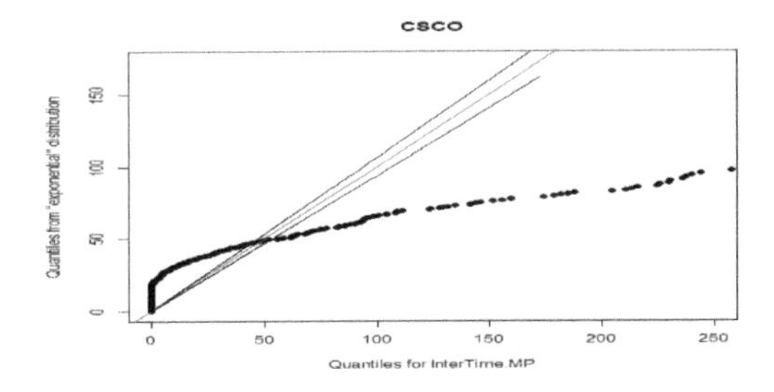

FIGURE 4.6
CSCO-QQplot.

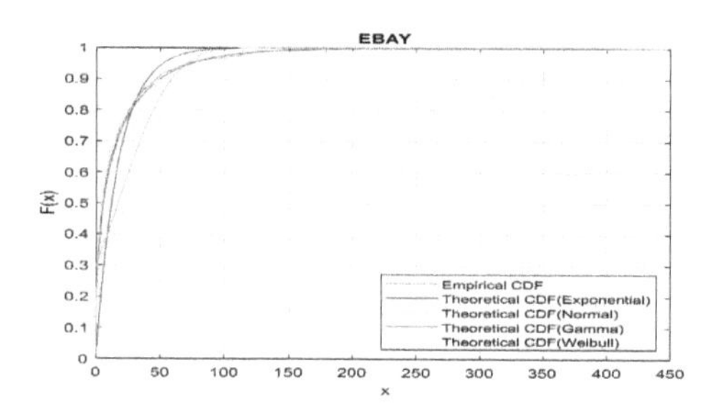

FIGURE 4.7
EBAY.

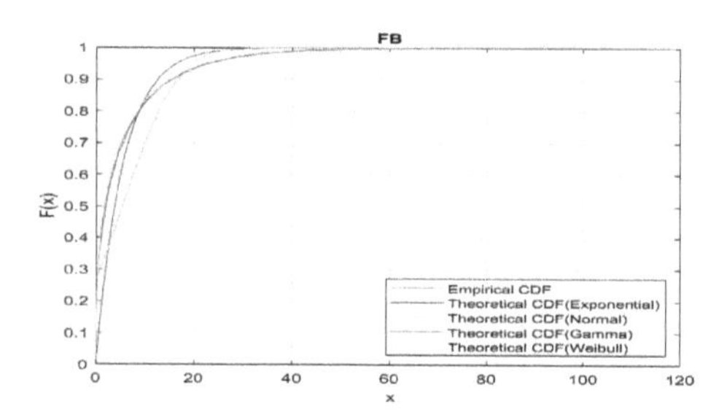

FIGURE 4.8
FB.

4.6.2 Clustering Effect of Trades

We also plot the number of mid price changes occurring in every minute during the trading day. The clustering feature is obvious to be discovered in Figures 4.13–4.18:

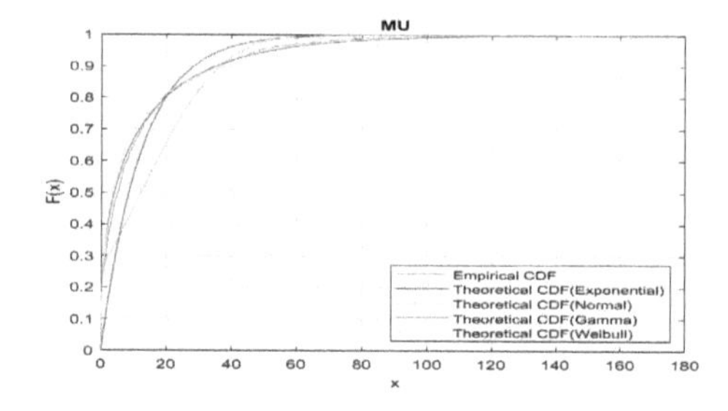

FIGURE 4.9
MU.

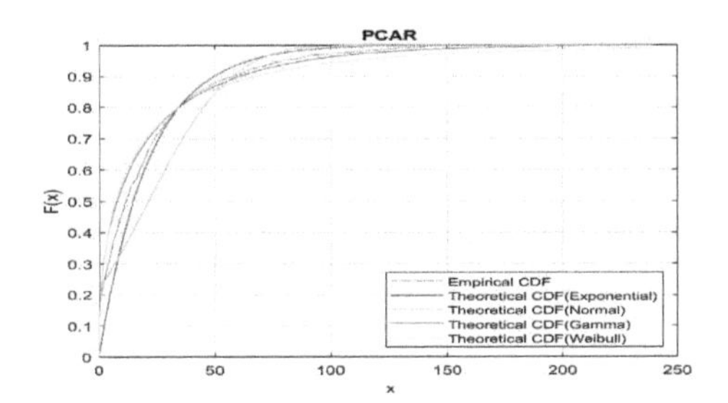

FIGURE 4.10
PCAR.

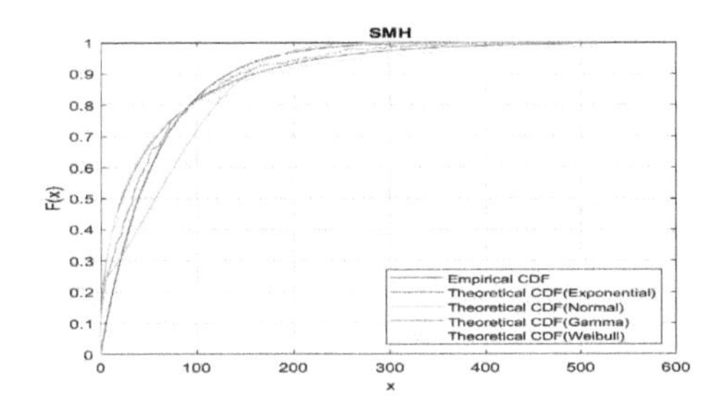

FIGURE 4.11
SMH.

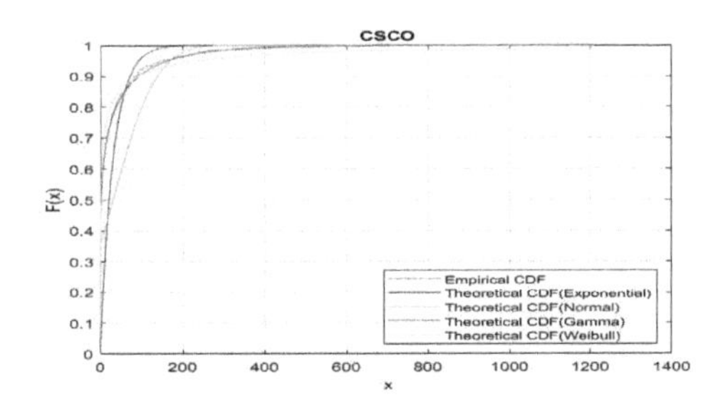

FIGURE 4.12
CSCO-20141107.

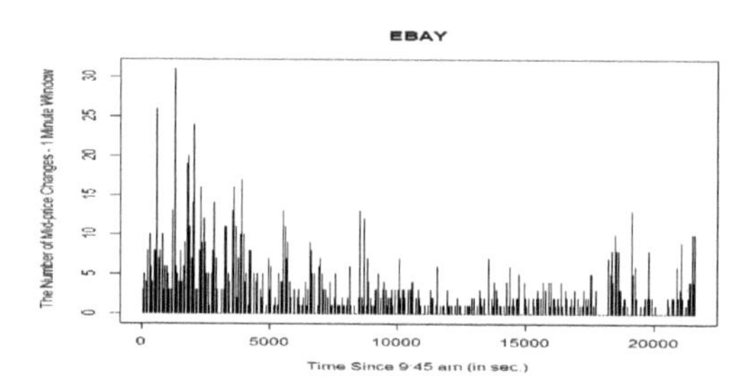

FIGURE 4.13
EBAY-clustering.

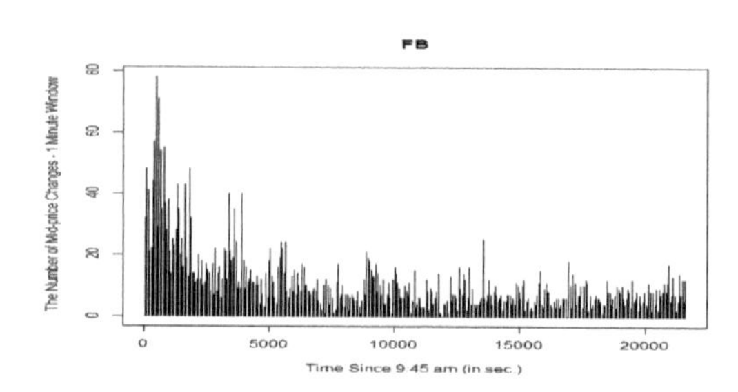

FIGURE 4.14
FB(clustering).

4.6.3 Non-independency of Mid-price Changes

To further confirm that the mid price changes are not independent, we calculated the autocorrelation function for one FB stock with different length of the time interval (see Figures 4.19–4.22).

Remark 4.3 *In the next chapters, we prove a Law of Larges Numbers (LLN) and a Functional Central Limit Theorem (FCLT) for specific cases of these*

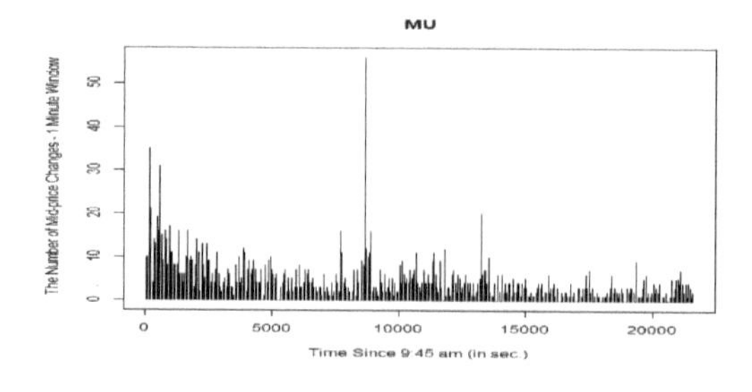

FIGURE 4.15
MU-clustering.

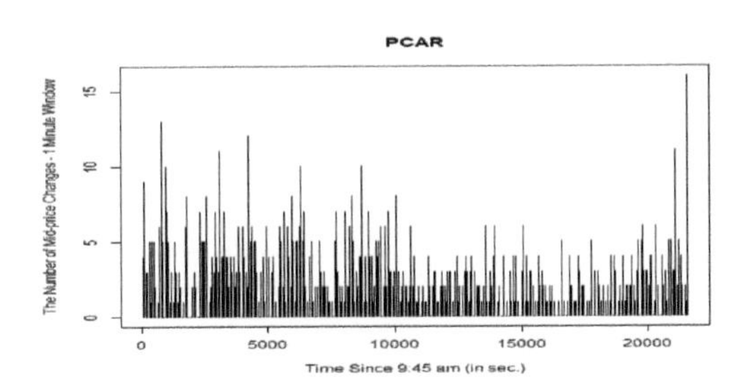

FIGURE 4.16
PCAR-clustering.

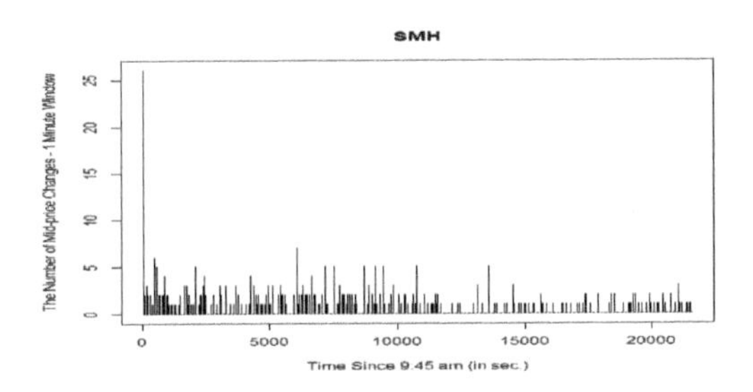

FIGURE 4.17
SMH-clustering.

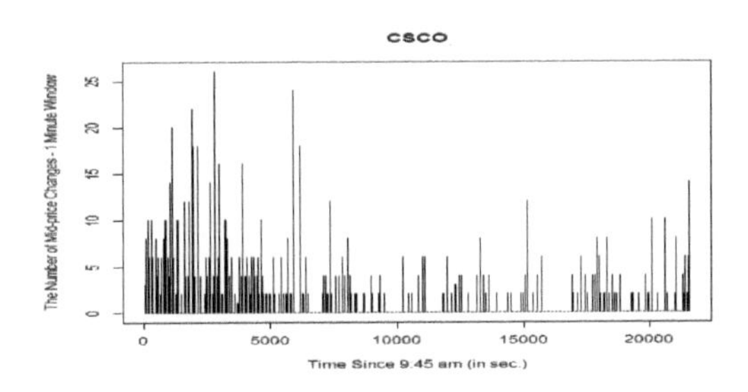

FIGURE 4.18
CSCO-20141107-clustering.

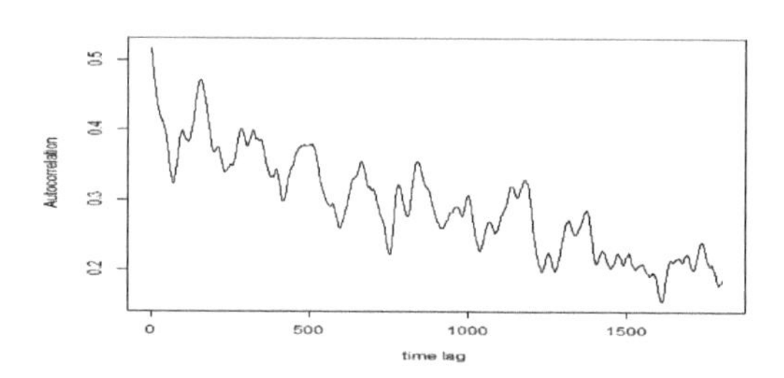

FIGURE 4.19
tao=20.

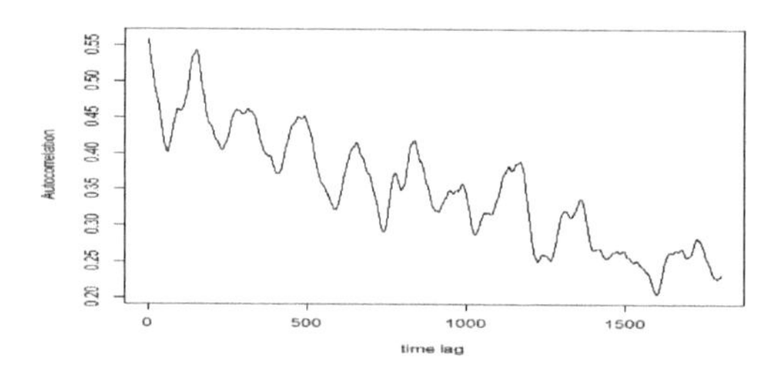

FIGURE 4.20
tao=30.

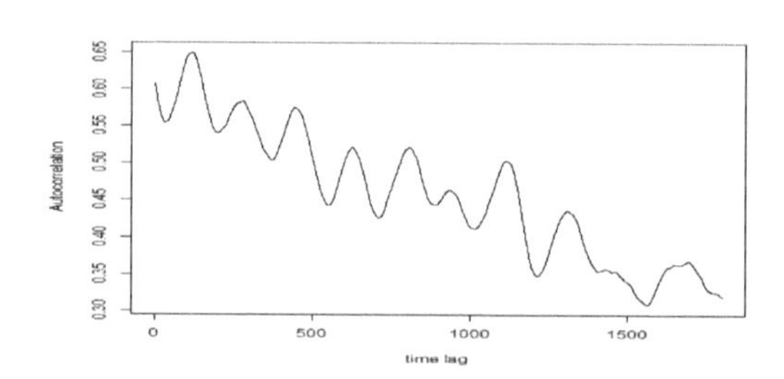

FIGURE 4.21
tao=60.

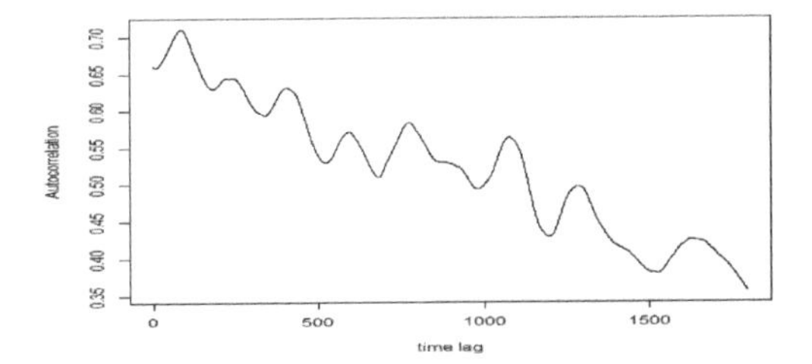

FIGURE 4.22
tao=90.

processes. Several of these FCLTs are applied to limit order books where we use asymptotic methods to study the link between price volatility and order flow in our models. The volatility of the price changes is expressed in terms of parameters describing the arrival rates and price changes. We also present some numerical examples.

4.7 Conclusion

In this introductory chapter, we presented some basics definitions of HPs, their stylized properties, and the descriptions of CHP and GCHP. The stylized properties include non-exponential distribution of inter-arrival times, clustering affect of trades, and non-independency of the mid-price changes. We used the set of real data, namely EBAY, FB, MU, PCAR, SMH, CSCO, (provided by Cartea et al. (2015), [9], book), to check, justify and illustrate our approach in using Hawkes process.

Bibliography

[1] Ait-Sahalia, Y., Cacho-Diaz, J. and Laeven, R. (2010): Modelling of financial contagion using mutually exciting jump processes. *Tech. Rep.*, 15850, Nat. Bureau of Ec. Res., USA.

[2] Buffington, J., Elliott, R. J. (2000): Regime Switching and European Options. Lawrence, K.S. (ed.) *Stochastic Theory and Control.* Proceedings of a Workshop, 73-81. Berlin Heidelberg New York: Springer. (2002).

[3] Buffington, J., Elliott, R.J. (2002): American Options with Regime Switching. *International Journal of Theoretical and Applied Finance* 5, 497–514.

[4] Brémaud, P. and Massoulié, L. (1996): Stability of nonlinear Hawkes processes. *The Annals of Probab.*, 24(3), 1563.

[5] Bacry, E., Mastromatteo, I. and Muzy, J.-F. (2015): Hawkes processes in finance. arXiv:1502.04592v2 [q-fin.TR] 17 May 2015.

[6] Bowsher, C. (2007): Modelling security market events in continuous time: intensity based, multivariate point process models. *J. Econometrica*, 141 (2), 876–912.

[7] Bauwens, L. and Hautsch, N. (2009): *Modelling Financial High Frequency Data Using Point Processes.* Springer.

[8] Bartlett, M. (1963): The spectral analysis of point processes. *J. R. Stat.* Soc., ser. B, 25 (2), 264–296.

[9] Cartea, A., Jaimungal, S. and Ricci, J. (2011): Buy low, sell high: a high-frequency trading prospective. *SIAM J. Fin. Math.*, 5(1), pp. 415–444.

[10] Cartea, Á., Jaimungal, S. and Penalva, J. (2015): *Algorithmic and High-Frequency Trading.* Cambridge University Press.

[11] Cartensen, L. (2010): *Hawkes processes and combinatorial transcriptional regulation.* PhD Thesis, University of Copenhagen.

[12] Chavez-Demoulin, V. and McGill, J. (2012): High-frequency financial data modelling using Hawkes processes. *J. Banking and Finance*, 36(12), 3415–3426.

[13] Chavez-Casillas, J., Elliott, R., Remillard, B. and Swishchuk, A. (2017): A level-1 limit order book with time dependent arrival rates. Proceed. IWAP, Toronto, June-20-25. Also available on arXiv: https://arxiv.org/submit/1869858

[14] Cohen, S. and Elliott, R. (2014): Filters and smoothness for self-exciting Markov modulated counting process. *IEEE Trans. Aut. Control.*, Papers 1311.6257, arXiv.org.

[15] Cont, R. and de Larrard, A. (2013): A Markovian modelling of limit order books. *SIAM J. Finan. Math.*, 4(1), pp. 1–25.

[16] Cox, D. (1955): Some statistical methods connected with series of events. *J. R. Stat. Soc.*, ser. B, 17 (2), 129–164.

[17] Daley, D.J. and Vere-Jones, D. (1988): *An Introduction to the Theory of Point Processes.* Springer.

[18] Dassios, A. and Zhao, H. (2011): A dynamic contagion process. *Advances in Applied Probab.*, 43(3), 814–846.

[19] Engel, R. and Russel, J. (1998): Autoregressive conditional duration: A new model for irregulary spaced transaction data. *Econometrica*, 66, 1127–1162.

[20] Engle, R. (2000): The econometrics of ultra-high-frequency data. *Econometrica*, 68: 1–20.

[21] Engle, R. and Large, J. (2001): Predicting vnet: a model of the dynamics of market depth. *J. Finan. Markets*, 4, 113–142.

[22] Engle, R. and Lunde, A. (2003): Trades and quotes: a bivariate point process. *J. Finan. Econom.*, 1 (2), 159–188.

[23] Embrechts, P., Liniger, T. and Lin, L. (2011): Multivariate Hawkes processes: an application to financial data. *J. Appl. Prob.*, 48, A: 367– 378.

[24] Errais, E., Giesecke, K. and Goldberg, L. (2010): Affine point processes and portfolio credit risk. *SIAM J. Fin. Math.* 1: 642–665.

[25] Fillimonov, V., Sornette, D., Bichetti, D. and Maystre, N. (2013): Quantifying of the high level of endogeneity and of structural regime shifts in comodity markets, 2013.

[26] Fillimonov, V. and Sornette, D. (2012): Quantifying reflexivity in financial markets: Toward a prediction of flash crashes. *Physical Review E*, 85(5):056108.

[27] Hawkes, A. (1971): Spectra of some self-exciting and mutually exciting point processes. *Biometrica*, 58, 83–90.

[28] Hawkes, A. and Oakes, D. (1974): A cluster process representation of a self-exciting process. *J. Applied Probab.*, 11: 493–503.

[29] Hewlett, P. (2006): Clustering of order arrivales, price impact and trade-path optimization.

[30] Hasbrouch, J. (1999):Trading fast and slow: security market events in real time. *Tech. Report*, Working paper series 99–112, New York University.

[31] Korolyuk, V. S. and Swishchuk, A. V. (1995): *Semi-Markov Random Evolutions*. Kluwer Academic Publishers, Dordrecht, The Netherlands.

[32] Large, J. (2007): Measuring the resiliency of an electronic limit order book. *J. Fin. Markets*, 10(1):1–25.

[33] Liniger, T. (2009): Multivariate Hawkes Processes. PhD thesis, Swiss Fed. Inst. Tech., Zurich.

[34] Lewis, P. (1964): J. R. *Stat. Soc.*, ser. B, 26 (3), 398.

[35] Laub, P., Taimre, T. and Pollett, P. (2015): Hawkes Processes.arXiv: 1507.02822v1[math.PR]10 Jul 2015.

[36] McNeil, A., Frey, R. and Embrechts, P. (2015): *Quantitative Risk Management: Concepts, Techniques and Tools.* Princeton University Press.

[37] Mehdad, B. and Zhu, L. (2014): On the Hawkes process with different exciting functions. *arXiv: 1403.0994.*

[38] Norris, J. R. (1997): Markov Chains. In Cambridge Series in Statistical and Probabilistic Mathematics. UK: Cambridge University Press.

[39] Skorokhod, A. (1965): Studies in the Theory of Random Processes, Addison-Wesley, Reading, Mass., (Reprinted by Dover Publications, NY).

[40] Swishchuk, A. (2017a): Risk model based on compound Hawkes process. Abstract, IME 2017, Vienna (https://fam.tuwien.ac.at/ ime2017/ program.php).

[41] Swishchuk, A. (2017b): Risk model based on compound Hawkes process. arXiv: http://arxiv.org/abs/1706.09038

[42] Swishchuk, A. (2017c): General Compound Hawkes Processes in Limit Order Books. Working Paper, U of Calgary, 32 pages, June 2017. Available on arXiv: http://arxiv.org/abs/1706.07459

[43] Swishchuk, A., Chavez-Casillas, J., Elliott, R. and Remillard, B. (2017): Compound Hawkes processes in limit order books. Available on SSRN: https://papers.ssrn.com/sol3/papers.cfm?abstract_id=2987943

[44] Swishchuk, A. and Vadori, N. (2017): A semi-Markovian modelling of limit order markets. SIAM J. Finan. Math., 8, 240–273.

[45] Swishchuk, A., Cera, K., Hofmeister, T. and Schmidt, J. (2017): General semi-Markov model for limit order books. Intern. *J. Theoret. Applied Finance*, 20, 1750019.

[46] Swishchuk, A. and Vadori, N. (2015): Strong law of large numbers and central limit theorems for functionals of inhomogeneous Semi-Markov processes. *Stochastic Analysis and Applications*, 13 (2), 213–243.

[47] Vinkovskaya, E. (2014): *A Point Process Model for the Dynamics of LOB.* PhD thesis, Columbia University.

[48] Zheng, B., Roueff, F. and Abergel, F. (2014): Ergodicity and scaling limit of a constrained multivariate Hawkes process. *SIAM J. Finan.* Math., 5, 2014.

[49] Zhu, L. (2013): Central limit theorem for nonlinear Hawkes processes. *J. Appl. Prob.*, 50(3), 760–771.

[50] G.O. Mohler, M.B. Short, P.J. Brantingham, F.P. Schoenberg, G.E. Tita. (2011): *Journal of the American Statistical Association* 106(493), 100.

[51] S. Azizpour, K. Giesecke, G. Schwenkler. Exploring the sources of default clustering. http://web.stanford.edu/dept/MSandE/cgi-bin/people/faculty/giesecke/pdfs/exploring.pdf (2010).

[52] Y. Ogata, (1988) *Journal of the American Statistical Association* 83(401), 9.

[53] LOBSTER, limit order book data. https://lobsterdata.com/

[54] E. Bacry, S. Delattre, M. Hoffmann, and J. F. Muzy. (2013) Modelling microstructure noise with mutually exciting point processes. *Quantitative Finance*, 13(1):65–77.

[55] Paul Embrechts, Thomas Liniger, and Lu Lin. (2011) Multivariate Hawkes processes: an application to financial data. *Journal of Applied Probability*, 48(A):367–378.

5

Stochastic Modelling of Big Data in Finance with CHP

This chapter introduces two new Hawkes processes, namely, compound and regime-switching compound Hawkes processes, to model the price processes in limit order books. We prove Law of Large Numbers and Functional Central Limit Theorems (FCLT) for both processes. The two FCLTs are applied to limit order books where we use these asymptotic methods to study the link between price volatility and order flow in our two models by using the diffusion limits of these price processes. The volatilities of price changes are expressed in terms of parameters describing the arrival rates and price changes. We also present some numerical examples.

5.1 Introduction

We recall that the Hawkes process (HP) is named after its creator Alan Hawkes [1971, 1974] [24]. The HP is a so-called "self-exciting point process" which means that it is a point process with a stochastic intensity which, through its dependence on the history of the process, captures the temporal and cross-sectional dependence of the event arrival process as well as the "self-exciting" property observed in empirical analysis. HPs have been used for many applications, such as modelling neural activity, genetics [11], occurrence of crime [Louie et al., 2010][27], bank defaults and earthquakes.

The most recent application of HPs is in financial analysis, in particular, to model limit order books, (e.g., high frequency data on price changes or arrival times of trades). In this paper we study two new Hawkes processes, namely, compound and regime-switching compound Hawkes processes to model the price processes in the limit order books. We prove a Law of Large Numbers and Functional Central Limit Theorems (FCLT) for both processes. The latter two FCLTs are applied to limit order books where we use these asymptotic methods to study the link between price volatility and order flow in our two models by using the diffusion limits of these price processes. The volatilities of price changes are expressed in terms of parameters describing the arrival rates

DOI: 10.1201/9781003265986-5

and price changes. The general compound Hawkes process was first introduced in [Swishchuk, 2017][33] to model a risk process in insurance.

Bowsher [4] was the first who applied a HP to financial data modelling. Cartea et al. [9] applied an HP to model market order arrivals. Filimonov and Sornette [23] and Filimonov et al. [22] apply a HP to estimate the percentage of price changes caused by endogenous self-generated activity, rather than the exogenous impact of news or novel information. Bauwens and Hautsch [5]use a 5-D HP to estimate multivariate volatility, between five stocks, based on price intensities. We note, that Brémaud et al. [1996] [6] generalized the HP to its nonlinear form. Also, a functional central limit theorem for the nonlinear Hawkes process was obtained in [Zhu, 2013][40]. The "Hawkes diffusion model" was introduced in Ait-Sahalia et al., [1] in an attempt to extend previous models of stock prices and include financial contagion. Chavez-Demoulin et al. [13] used Hawkes processes to model high-frequency financial data. Some applications of Hawkes processes to financial data are also given in Embrechts et al., [20].

Cohen et al. [15] derived an explicit filter for Markov modulated Hawkes process. Vinkovskaya [38] considered a regime-switching Hawkes process to model its dependency on the bid-ask spread in limit order books. Regime-switching models for the pricing of European and American options were considered in Buffington and Elliott, [7] and Buffington and Elliott, [8], respectively. A semi-Markov process was applied to limit order books in Swishchuk and Vadori, [34] to model the mid-price. We note, that a level-1 limit order books with time dependent arrival rates $\lambda(t)$ were studied in Chavez-Casillas et al., [14], including the asymptotic distribution of the price process. General semi-Markovian models for limit order books were considered in Swishchuk et al., 2017 [35].

The paper by Bacry et al. [2]proposes an overview of the recent academic literature devoted to the applications of Hawkes processes in finance. The book by Cartea et al. [10] develops models for algorithmic trading in contexts such as executing large orders, market making, trading pairs or collecting of assets, and executing in dark pool. That book also contains link to a website from which many datasets from several sources can be downloaded, and MATLAB code to assist in experimentation with the data. A detailed description of the mathematical theory of Hawkes processes is given in Liniger, citeL. The paper by Laub et al. [27] provides a background, introduces the field and historical developments, and touches upon all major aspects of Hawkes processes.

This chapter is organized as follows. Section 5.2 gives the definitions of a Hawkes process (HP), definitions of compound Hawkes process (CHP) and regime-switching compound Hawkes process (RSCHP). These definitions are new ones from the following point of view: summands associated in a Markov chain but not are i.i.d.r.v. Section 5.3 contains Law of Large Numbers and diffusion limits for CHP and RSCHP. Numerical examples are presented in Section 5.4.

5.2 Definitions of HP, CHP and RSCHP

In this section we give definitions of one-dimensional, compound and regime-switching compound Hawkes processes. Some properties of Hawkes process can be found in the existing literature. (See, e.g., [Hawkes, 1971] [24] and [Hawkes and Oakes, 1974], [25] [Embrechts et al., 2011] [20], [Zheng et al., 2014 [39]], to name a few.) However, the notions of compound and regime-switching compound Hawkes processes are new.

5.2.1 One-dimensional Hawkes Process

Definition 5.1 (Counting Process) *A counting process is a stochastic process $N(t), t \geq 0$, taking positive integer values and satisfying: $N(0) = 0$. It is almost surely finite, and is a right-continuous step function with increments of size $+1$.*

Denote by $\mathcal{F}^N(t), t \geq 0$, the history of the arrivals up to time t, that is, $\{\mathcal{F}^N(t), t \geq 0\}$, is a filtration, (an increasing sequence of σ-algebras).

A counting process $N(t)$ can be interpreted as a cumulative count of the number of arrivals into a system up to the current time t. The counting process can also be characterized by the sequence of random arrival times $(T_1, T_2, ...)$ at which the counting process $N(t)$ has jumped. The process defined by these arrival times is called a point process (see [Daley and Vere-Jones, 1988][18]).

Definition 5.2 (Point Process) *If a sequence of random variables $(T_1, T_2, ...)$, taking values in $[0, +\infty)$, has $P(0 \leq T_1 \leq T_2 \leq ...) = 1$, and the number of points in a bounded region is almost surely finite, then, $(T_1, T_2, ...)$ is called a point process.*

Definition 5.3 (Conditional Intensity Function) *Consider a counting process $N(t)$ with associated histories $\mathcal{F}^N(t), t \geq 0$. If a non-negative function $\lambda(t)$ exists such that*

$$\lambda(t) = \lim_{h \to 0} \frac{E[N(t+h) - N(t) | \mathcal{F}^N(t)]}{h}, \tag{5.1}$$

then it is called the conditional intensity function of $N(t)$ (see [Laub et al., 2015]). We note, that sometimes this function is called the hazard function (see [Cox, 1955][17]).

Definition 5.4 (One-dimensional Hawkes Process) *The one-dimensional Hawkes process (see [Hawkes, 1971][24] and [Hawkes and Oakes, 1974][25]) is a point process $N(t)$ which is characterized by its intensity $\lambda(t)$ with respect to its natural filtration:*

$$\lambda(t) = \lambda + \int_0^t \mu(t-s) dN(s), \tag{5.2}$$

where $\lambda > 0$, *and the response function* $\mu(t)$ *is a positive function and satisfies* $\int_0^{+\infty} \mu(s)ds < 1$.

The constant λ is called the background intensity and the function $\mu(t)$ is sometimes also called the excitation function. We suppose that $\mu(t) \neq 0$ to avoid the trivial case, which is, a homogeneous Poisson process. Thus, the Hawkes process is a non-Markovian extension of the Poisson process.

With respect to definitions of $\lambda(t)$ in (1) and $N(t)$ (2), it follows that

$$P(N(t+h) - N(t) = m|\mathcal{F}^N(t)) = \begin{cases} \lambda(t)h + o(h), & m = 1 \\ o(h), & m > 1 \\ 1 - \lambda(t)h + o(h), & m = 0. \end{cases}$$

The interpretation of equation (2) is that the events occur according to an intensity with a background intensity λ which increases by $\mu(0)$ at each new event then decays back to the background intensity value according to the function $\mu(t)$. Choosing $\mu(0) > 0$ leads to a jolt in the intensity at each new event, and this feature is often called a self-exciting feature, in other words, because an arrival causes the conditional intensity function $\lambda(t)$ in (1)–(2) to increase then the process is said to be self-exciting.

We should mention that the conditional intensity function $\lambda(t)$ in (1)–(2) can be associated with the compensator $\Lambda(t)$ of the counting process $N(t)$, that is:

$$\Lambda(t) = \int_0^t \lambda(s)ds. \tag{5.3}$$

Thus, $\Lambda(t)$ is the unique $\mathcal{F}^N(t), t \geq 0$, predictable function, with $\Lambda(0) = 0$, and is non-decreasing, such that

$$N(t) = M(t) + \Lambda(t) \quad a.s.,$$

where $M(t)$ is an $\mathcal{F}^N(t), t \geq 0$, local martingale (This is the Doob-Meyer decomposition of N.)

A common choice for the function $\mu(t)$ in (2) is one of exponential decay (see [24]):

$$\mu(t) = \alpha e^{-\beta t}, \tag{5.4}$$

with parameters $\alpha, \beta > 0$. In this case the Hawkes process is called the Hawkes process with exponentially decaying intensity.

Thus, the equation (2) becomes

$$\lambda(t) = \lambda + \int_0^t \alpha e^{-\beta(t-s)} dN(s), \tag{5.5}$$

We note, that in the case of (4), the process $(N(t), \lambda(t))$ is a continuous-time Markov process, which is not the case for the choice (2).

With some initial condition $\lambda(0) = \lambda_0$, the conditional density $\lambda(t)$ in (5) with the exponential decay in (4) satisfies the following stochastic differential equation (SDE):

$$d\lambda(t) = \beta(\lambda - \lambda(t))dt + \alpha dN(t), \quad t \geq 0,$$

which can be solved (using stochastic calculus) as

$$\lambda(t) = e^{-\beta t}(\lambda_0 - \lambda) + \lambda + \int_0^t \alpha e^{-\beta(t-s)} dN(s),$$

which is an extension of (5).

Another choice for $\mu(t)$ is a power law function:

$$\lambda(t) = \lambda + \int_0^t \frac{k}{(c + (t - s))^p} dN(s) \tag{5.6}$$

for some positive parameters c, k, p. This power law form for $\lambda(t)$ in (6) was applied in the geological model called Omori's law, and used to predict the rate of aftershocks caused by an earthquake.

Remark 5.1 *Many generalizations of Hawkes processes have been proposed. They include, in particular, multi-dimensional Hawkes processes [Embrechts et al., 2011] [20], non-linear Hawkes processes [Zheng et al., 2014] [39], mixed diffusion-Hawkes models [Errais et al., 2010] [21], Hawkes models with shot noise exogenous events [Dassios and Zhao, 2011][19], Hawkes processes with generation dependent kernels [Mehdad and Zhu, 2011] [30].*

5.2.2 Compound Hawkes Process (CHP)

In this section we give definitions of compound Hawkes process (CHP) and regime-switching compound Hawkes process (RSCHP). These definitions are new ones from the following point of view: summands are not i.i.d.r.v., as in classical compound Poisson process, but associated in a Markov chain.

Definition 5.5 (Compound Hawkes Process (CHP)) *Let $N(t)$ be a one-dimensional Hawkes process defined as above. Let also X_t be ergodic continuous-time finite state Markov chain, independent of $N(t)$, with space state X. We write τ_k for jump times of $N(t)$ and $X_k := X_{\tau_k}$. The compound Hawkes process is defined as*

$$S_t = S_0 + \sum_{k=1}^{N(t)} X_k. \tag{5.7}$$

Remark 5.2 *If we take X_k as i.i.d.r.v. and $N(t)$ as a standard Poisson process in (10) ($\mu(t) = 0$), then S_t is a compound Poisson process. Thus, the name of S_t in (10)-compound Hawkes process.*

Remark 5.3 (Limit Order Books: Fixed Tick, Two-values Price) (Change, Independent Orders *If Instead of Markov chain we take the sequence of i.i.d.r.v. X_k, then (10) becomes*

$$S_t = S_0 + \sum_{i=1}^{N(t)} X_k. \tag{5.8}$$

In the case of Poisson process $N(t)$ ($\mu(t) = 0$) this model was used in[Cont and Larrard, 2013][16] to model the limit order books with $X_k = \{-\delta, +\delta\}$, where δ is the fixed tick size.

5.2.3 Regime-switching Compound Hawkes Process

Let Y_t be an N-state Markov chain, with rate matrix A_t. We assume, without loss of generality, that Y_t takes values in the standard basis vectors in R^N. Then, Y_t has the representation

$$Y_t = Y_0 + \int_0^t A_s Y_s ds + M_t, \tag{5.9}$$

for M_t an R^N -valued P-martingale (see [Buffington and Elliott, 2000][7] for more details).

Definition 5.6 (One-dimensional Regime-switching Hawkes Process) *A one-dimensional regime-switching Hawkes Process N_t is a point process characterized by its intensity $\lambda(t)$ in the following way:*

$$\lambda_t =< \lambda, Y_t > + \int_0^t < \mu(t-s), Y_s > dN_s, \tag{5.10}$$

where $< \cdot, \cdot >$ is an inner product and Y_t is defined in (12).

Definition 5.7 (Regime-switching Compound Hawkes Process) (RSHP) *Let N_t be any one-dimensional regime-switching Hawkes process as defined in (13), Definition 6. Let also X_n be an ergodic continuous-time finite state Markov chain, independent of N_t, with space state X. The regime-switching compound Hawkes process is defined as*

$$S_t = S_0 + \sum_{i=1}^{N_t} X_k, \tag{5.11}$$

where N_t is defined in (13).

Remark 5.4 *In similar way, as in Definition 6, we can define regime-switching Hawkes processes with exponential kernel, (see (4)), or power law kernel (see (6)).*

Remark 5.5 *Regime-switching Hawkes processes were considered in [Cohen and Elliott, 2014] [15](with exponential kernel) and in [Vinkovskaya, 2014][38], (multi-dimensional Hawkes process). Paper [Cohen and Elliott, 2014][15] discussed a self-exciting counting process whose parameters depend on a hidden finite-state Markov chain, and the optimal filter and smoother based on observations of the jump process are obtained. Thesis [Vinkovskaya, 2014] [38]considers a regime-switching multi-dimensional Hawkes process with an exponential kernel which reflects changes in the bid-ask spread. The statistical properties, such as maximum likelihood estimations of its parameters, etc., of this model were studied.*

5.3 Diffusion Limits and LLNs for CHP and RSCHP in Limit Order Books

In this section, we consider LLNs and diffusion limits for the CHP and RSCHP, defined above, as used in the limit order books. In the limit order books, high-frequency and algorithmic trading, order arrivals and cancellations are very frequent and occur at the millisecond time scale (see, e.g., [Cont and Larrard, 2013] [16], [Cartea et al., 2015]) [10]. Meanwhile, in many applications, such as order execution, one is interested in the dynamics of order flow over a large time scale, typically tens of seconds or minutes. It means that we can use asymptotic methods to study the link between price volatility and order flow in our model by studying the diffusion limit of the price process. Here, we prove functional central limit theorems for the price processes and express the volatilities of price changes in terms of parameters describing the arrival rates and price changes. In this section, we consider diffusion limits and LLNs for both CHP, Section 3.1, and RSCHP, Section 3.2, in the limit order books. We note, that level-1 limit order books with time dependent arrival rates $\lambda(t)$ were studied in [Chavez-Casillas et al., 2016][14], including the asymptotic distribution of the price process.

5.3.1 Diffusion Limits for CHP in Limit Order Books

We consider here the mid-price process S_t (CHP) which was defined in (10), as,

$$S_t = S_0 + \sum_{k=1}^{N(t)} X_k. \tag{5.12}$$

Here, $X_k \in \{-\delta, +\delta\}$ is continuous-time two-state Markov chain, δ is the fixed tick size, and $N(t)$ is the number of price changes up to moment t, described by the one-dimensional Hawkes process defined in (2), Definition 4. It means that we have the case with a fixed tick, a two-valued price change and dependent orders.

Theorem 5.8 (Diffusion Limit for CHP) *Let X_k be an ergodic Markov chain with two states $\{-\delta, +\delta\}$ and with ergodic probabilities $(\pi^*, 1-\pi^*)$. Let also S_t be defined in (15). Then*

$$\frac{S_{nt} - N(nt)s^*}{\sqrt{n}} \rightarrow_{n\to+\infty} \sigma\sqrt{\lambda/(1-\hat{\mu})}W(t), \tag{5.13}$$

where $W(t)$ is a standard Wiener process, $\hat{\mu}$ is given by

$$0 < \hat{\mu} := \int_0^{+\infty} \mu(s)ds < 1 \quad \text{and} \quad \int_0^{+\infty} \mu(s)sds < +\infty, \tag{5.14}$$

$$s^* := \delta(2\pi^* - 1) \quad \text{and} \quad \sigma^2 := 4\delta^2\Big(\frac{1-p'+\pi^*(p'-p)}{(p+p'-2)^2} - \pi^*(1-\pi^*)\Big). \tag{5.15}$$

Here, (p, p') are the transition probabilities of the Markov chain X_k. We note that λ and $\mu(t)$ are defined in (2).

Proof *From (15) it follows that*

$$S_{nt} = S_0 + \sum_{k=1}^{N(nt)} X_k, \tag{5.16}$$

and

$$S_{nt} = S_0 + \sum_{k=1}^{N(nt)} (X_k - s^*) + N(nt)s^*.$$

Therefore,

$$\frac{S_{nt} - N(nt)s^*}{\sqrt{n}} = \frac{S_0 + \sum_{k=1}^{N(nt)}(X_k - s^*)}{\sqrt{n}}. \tag{5.17}$$

Since $\frac{S_0}{\sqrt{n}} \rightarrow_{n\to+\infty} 0$, we have to find the limit for

$$\frac{\sum_{k=1}^{N(nt)}(X_k - s^*)}{\sqrt{n}}$$

when $n \to +\infty$.

Consider the following sums

$$R_n := \sum_{k=1}^{n}(X_k - s^*) \tag{5.18}$$

and

$$U_n(t) := n^{-1/2}[(1-(nt-\lfloor nt \rfloor))R_{\lfloor nt \rfloor} + (nt - \lfloor nt \rfloor))R_{\lfloor nt \rfloor)+1}], \tag{5.19}$$

where $\lfloor \cdot \rfloor$ is the floor function.

Following the martingale method from [Swishchuk and Vadori, 2015], we have the following weak convergence in the Skorokhod topology (see [Skorokhod, 1965][32]):

$$U_n(t) \to_{n\to+\infty} \sigma \mathcal{W}_t, \tag{5.20}$$

where σ is defined in (18), and $\mathcal{W}_t$ is a standard Brownian motion.

We note that w.r.t LLN for Hawkes process $N(t)$ (see, e.g., [Daley and Vee-Jones, 2010][18]) we have:

$$\frac{N(t)}{t} \to_{t\to+\infty} \frac{\lambda}{1-\hat{\mu}} := \bar{\lambda},$$

or

$$\frac{N(nt)}{n} \to_{n\to+\infty} \frac{t\lambda}{1-\hat{\mu}} = \bar{\lambda}t, \tag{5.21}$$

where $\hat{\mu}$ is defined in (17).

Using a change of time in (23), $t \to N(nt)/n$, we can find from (23) and (24):

$$U_n(N(nt)/n) \to_{n\to+\infty} \sigma \mathcal{W}\Big(t\lambda/(1-\hat{\mu})\Big),$$

or

$$U_n(N(nt)/n) \to_{n\to+\infty} \sigma\sqrt{\lambda/(1-\hat{\mu})}W(t), \tag{5.22}$$

where $W_t = \mathcal{W}_{\bar{\lambda}t}/\sqrt{\bar{\lambda}}$. The Brownian motion $W(t)$ in (25) is equivalent by distribution to Brownian motion $\mathcal{W}$ in (23) by scaling property. The result (16) now follows from (20)-(25).

Remark 5.6 *In the case of exponential decay, $\mu(t) = \alpha e^{-\beta t}$ (see (4)), the limit in (16) is $[\sigma/\sqrt{\lambda/(1-\alpha/\beta)}]W(t)$, because $\hat{\mu} = \int_0^{+\infty} \alpha e^{-\beta s} ds = \alpha/\beta$.*

5.3.2 LLN for CHP

Lemma 5.1 (LLN for CHP) *The process S_{nt} in (19) satisfies the following weak convergence in the Skorokhod topology (see [Skorokhod, 1965][32]):*

$$\frac{S_{nt}}{n} \to_{n\to+\infty} s^* \frac{\lambda}{1-\hat{\mu}} t, \tag{5.23}$$

where s^* and $\hat{\mu}$ are defined in (18) and (17), respectively.

Proof *From (19) we have*

$$S_{nt}/n = S_0/n + \sum_{k=1}^{N(nt)} X_k/n. \tag{5.24}$$

The first term goes to zero when $n \to +\infty$. From the other side, using the strong LLN for Markov chains (see, e.g., [Norris, 1997])

$$\frac{1}{n}\sum_{k=1}^{n} X_k \to_{n\to+\infty} s^*, \tag{5.25}$$

where s^* is defined in (18).

Finally, taking into account (24) and (28), we obtain:

$$\sum_{k=1}^{N(nt)} X_k/n = \frac{N(nt)}{n}\frac{1}{N(nt)}\sum_{k=1}^{N(nt)} X_k \to_{n\to+\infty} s^*\frac{\lambda}{1-\hat{\mu}}t,$$

and the result in (26) follows.

Remark 5.7 *In the case of exponential decay, $\mu(t) = \alpha e^{-\beta t}$ (see (4)), the limit in (26) is $s^* t(\lambda/(1-\alpha/\beta))$, because $\hat{\mu} = \int_0^{+\infty} \alpha e^{-\beta s} ds = \alpha/\beta$.*

5.3.3 Corollary: Extension to a Point Process

The price process S is expressed as

$$S_t = S_0 + \sum_{i=1}^{N(t)} X_i, \qquad t \geq 0,$$

where N is a point process, and Markov chain X_i is defined in (10).

Assumption C1: As $n \to \infty$, $N(nt)/n \xrightarrow{Pr} \bar{\lambda}t$, where $\bar{\lambda} := \lambda/(1-\hat{\mu})$.

Note that if $N(t) = \max\{n : V_n \leq t\}$, then $N(nt)/n \xrightarrow{Pr} \bar{\lambda}t = \frac{1}{\bar{v}}$ iff $V_n/n \xrightarrow{Pr} \bar{v}$. This representation is useful in particular for renewal processes where $V_n = \sum_{k=1}^{n} \tau_k$, with the τ_k i.i.d. with mean $\bar{v}$.

Assumption C2: $U_n(t) \rightsquigarrow W$, where W is a Brownian motion, and $U_n(t)$ is defined in (22).

It then follows from Assumptions C1 and C2 that

$$n^{-1/2}\{S_{nt} - S_0 - s^* N(nt)\}\} = \sigma U_n(N(nt)/n) = n^{-1/2}\sum_{i=1}^{N(nt)}\{X_i - s^*\} \rightsquigarrow \sigma\sqrt{\bar{\lambda}}\, W_t,$$

where W is a Brownian motion, and s^* is denied in (18). In fact, for any $t \geq 0$, $W_t = \mathcal{W}_{\bar{\lambda}t}/\sqrt{\bar{\lambda}}$.

The limiting variance $\sigma^2\bar{\lambda}$ can probably be approximated by summing the square of the increments $S_{nt_i} - S_{nt_{i-1}} - s^*(N(nt_i) - N(nt_{i-1}))$. In any cases, $\bar{\lambda}$ cab be easily estimated by $N(T)/T$, and σ can be estimated from the distribution of the price increments.

Suppose now that there is also a CLT for the point process N. More precisely,

Assumption C3: $n^{1/2}\left(\frac{N(nt)}{n} - t\lambda\right) \rightsquigarrow \bar{\sigma}\bar{W}_t$, where $\bar{W}$ is a Brownian motion independent of W.

Then under Assumptions C1–C3,

$$n^{-/2}\left\{S_{nt} - nt\bar{\lambda}s^*\right\} \rightsquigarrow \tilde{\sigma}\mathbb{W}_t,$$

where $\mathbb{W} = \left\{\sigma\sqrt{\bar{\lambda}}W + s^*\bar{\sigma}\bar{W}\right\}/\tilde{\sigma}$ is a Brownian motion, and

$$\tilde{\sigma} = \left[\sigma^2\bar{\lambda} + \{s^*\}^2\bar{\sigma}^2\right]^{1/2}.$$

This follows from Assumptions and the fact that

$$n^{-1/2}\left\{S_{nt} - S_0 - nt\bar{\lambda}s^*\right\} = n^{-1/2}\sum_{i=1}^{N(nt)}\{X_i - s^*\} + s^*n^{1/2}\left(\frac{N(nt)}{n} - t\lambda\right).$$

Remark 5.8 *Assumption C3 is true in many interesting cases. For renewal processes, if σ_τ is the standard deviation of τ_k, then $\bar{\sigma} = \sigma_\tau\bar{\lambda}^{3/2}$. This is also true for Hawkes processes [Bacry et al., 2013] with $\lambda(t) = \lambda_0 + \int_0^t \mu(t-s)dN_s$, provided $\hat{\mu} = \int_0^\infty \mu(s)ds < 1$. Then $\bar{\lambda} = \frac{\lambda}{1-\hat{\mu}}$ and $\bar{\sigma} = \sqrt{\bar{\lambda}}/(1-\hat{\mu})$.*

5.3.4 Diffusion Limits for RSCHP in Limit Order Books

Consider now the mid-price process S_t (RSCHP) in the form

$$S_t = S_0 + \sum_{k=1}^{N_t} X_k, \tag{5.26}$$

where $X_k \in \{-\delta, +\delta\}$ is continuous-time two-state Markov chain, δ is the fixed tick size, and N_t is the number of price changes up to the moment t, described by a one-dimensional regime-switching Hawkes process with intensity given by:

$$\lambda_t = <\lambda, Y_t> + \int_0^t \mu(t-s)dN_s, \tag{5.27}$$

(compare with (11), Definition 6).

Here we would like to relax the model for one-dimensional regime-switching Hawkes process, considering only the case of a switching the parameter λ, background intensity, in (20), which is reasonable from a limit order book's point of view. For example, we can consider a three-state Markov chain $Y_t \in \{e_1, e_2, e_3\}$ and interpret $<\lambda, Y_t>$ as the imbalance, where $\lambda_1, \lambda_2, \lambda_3$, represent high, normal and low imbalance, respectively (see [Cartea et al., 2015] [10] for imbalance notion and discussion). Of course, a more general case (13) can be considered as well, where the excitation function $<\mu(t), Y_t>$,, can take three values, corresponding to high imbalance, normal imbalance, and low imbalance, respectively.

Theorem 5.9 (Diffusion Limit for RSCHP) *Let X_k be an ergodic Markov chain with two states $\{-\delta, +\delta\}$ and with ergodic probabilities $(\pi^*, 1-\pi^*)$. Let also S_t be defined in (29) with λ_t as in (30). We also consider Y_t to be an ergodic Markov chain with ergodic probabilities $(p_1^*, p_2^*, ..., p_N^*)$. Then*

$$\frac{S_{nt} - N_{nt}s^*}{\sqrt{n}} \rightarrow_{n\to+\infty} \sigma\sqrt{\hat{\lambda}/(1-\hat{\mu})}W(t), \tag{5.28}$$

where $W(t)$ is a standard Wiener process with s^* and σ defined in (18),

$$\hat{\lambda} := \sum_{i=1}^{N} p_i^* \lambda_i \neq 0, \quad \lambda_i :=< \lambda, i >, \tag{5.29}$$

and $\hat{\mu}$ is defined in (17).

Proof *From (29) it follows that*

$$S_{nt} = S_0 + \sum_{i=1}^{N_{nt}} X_k, \tag{5.30}$$

and

$$S_{nt} = S_0 + \sum_{i=1}^{N_{nt}} (X_k - s^*) + N_{nt} s^*,$$

where N_{nt} is an RGCHP with regime-switching intensity λ_t as in (30). Then,

$$\frac{S_{nt} - N_{nt} s^*}{\sqrt{n}} = \frac{S_0 + \sum_{i=1}^{N_{nt}} (X_k - s^*)}{\sqrt{n}}. \tag{5.31}$$

As long as $\frac{S_0}{\sqrt{n}} \to_{n \to +\infty} 0$, we wish to find the limit of

$$\frac{\sum_{i=1}^{N_{nt}} (X_k - s^*)}{\sqrt{n}}$$

when $n \to +\infty$.

Consider the following sums, similar to (21) and (22):

$$R_n := \sum_{k=1}^{n} (X_k - s^*) \tag{5.32}$$

and

$$U_n(t) := n^{-1/2} [(1 - (nt - \lfloor nt \rfloor)) R_{\lfloor nt \rfloor)} + (nt - \lfloor nt \rfloor)) R_{\lfloor nt \rfloor)+1}], \tag{5.33}$$

where $\lfloor \cdot \rfloor$ is the floor function.

Following the martingale method from [Swishchuk and Vadori, 2015], we have the following weak convergence in the Skorokhod topology (see [Skorokhod, 1965][32]):

$$U_n(t) \to_{n \to +\infty} \sigma W(t), \tag{5.34}$$

where σ is defined in (18).

We note that with respect to the LLN for the Hawkes process N_t in (34) with regime-switching intensity λ_t as in (30) we have (see [Korolyuk and Swishchuk, 1995][26] for more details):

$$\frac{N_t}{t} \to_{t\to+\infty} \frac{\hat{\lambda}}{1-\hat{\mu}},$$

or

$$\frac{N_{nt}}{n} \to_{n\to+\infty} \frac{t\hat{\lambda}}{1-\hat{\mu}}, \tag{5.35}$$

where $\hat{\mu}$ is defined in (17) and $\hat{\lambda}$ in (32).

Using a change of time in (37), $t \to N_{nt}/n$, we can find from (37) and (38):

$$U_n(N_{nt}/n) \to_{n\to+\infty} \sigma W\Big(t\hat{\lambda}/(1-\hat{\mu})\Big),$$

or

$$U_n(N_{nt}/n) \to_{n\to+\infty} \sigma\sqrt{\hat{\lambda}/(1-\hat{\mu})}W(t), \tag{5.36}$$

The result (31) now follows from (33)-(39).

Remark 5.9 *In the case of exponential decay,* $\mu(t) = \alpha e^{-\beta t}$ *(see (4)), the limit in (31) is* $[\sigma\sqrt{\hat{\lambda}/(1-\alpha/\beta)}]W(t)$, *because* $\hat{\mu} = \int_0^{+\infty} \alpha e^{-\beta s} ds = \alpha/\beta$.

5.3.5 LLN for RSCHP

Lemma 5.2 (LLN for RSCHP) *The process* S_{nt} *in (33) satisfies the following weak convergence in the Skorokhod topology (see [Skorokhod, 1965][32]):*

$$\frac{S_{nt}}{n} \to_{n\to+\infty} s^* \frac{\hat{\lambda}}{1-\hat{\mu}} t, \tag{5.37}$$

where s^*, $\hat{\lambda}$ *and* $\hat{\mu}$ *are defined in (13), (27) and (12), respectively.*

Proof *From (33) we have*

$$S_{nt}/n = S_0/n + \sum_{i=1}^{N_{nt}} X_k/n, \tag{5.38}$$

where N_{nt} *is a Hawkes process with regime-switching intensity* λ_t *in (30).*

The first term goes to zero when $n \to +\infty$.

From the other side, with respect to the strong LLN for Markov chains (see, e.g., [Norris, 1997])

$$\frac{1}{n}\sum_{k=1}^{n} X_k \to_{n\to+\infty} s^*, \tag{5.39}$$

where s^* is defined in (18).

Finally, taking into account (38) and (42), we obtain:

$$\sum_{i=1}^{N_{nt}} X_k/n = \frac{N_{nt}}{n}\frac{1}{N_{nt}}\sum_{i=1}^{N_{nt}} X_k \to_{n\to+\infty} s^*\frac{\hat{\lambda}}{1-\hat{\mu}}t.$$

The result in (40) follows.

Remark 5.10 *In the case of exponential decay, $\mu(t) = \alpha e^{-\beta t}$ (see (4)), the limit in (40) is $s^* t(\hat{\lambda}/(1-\alpha/\beta))$, because $\hat{\mu} = \int_0^{+\infty} \alpha e^{-\beta s} ds = \alpha/\beta$.*

5.4 Numerical Examples and Parameters Estimations

Formula (16) in Theorem 1 (Diffusion Limit for CHP) relates the volatility of intraday returns at lower frequencies to the high-frequency arrival rates of orders. The typical time scale for order book events are milliseconds. Formula (16) states that, observed over a larger time scale, e.g., 5, 10 or 20 minutes, the price has a diffusive behaviour with a diffusion coefficient given by the coefficient at $W(t)$ in (16):

$$\sigma\sqrt{\lambda/(1-\hat{\mu})}, \tag{5.40}$$

where all the parameters here are defined in (17)-(18). We mention, that this formula (43) for volatility contains all the initial parameters of the Hawkes process, Markov chain transition and stationary probabilities and the tick size. In this way, formula (43) links properties of the price to the properties of the order flow.

Also, the left hand side of (16) represents the variance of price changes, whereas the right hand side in (16) only involves the tick size and Hawkes process and Markov chain quantities. From here it follows that an estimator for price volatility may be computed without observing the price at all. As we shall see below, the error of estimation of comparison of the standard deviation of the LNS of (16) and the RHS of (16) multiplied by $\sqrt{n}$ is approximately 0.08, indicating that approximation in (16) for diffusion limit for CHP in Theorem 1, is pretty good.

Section 5.4.1 below presents parameters estimation for our model using CISCO Data (5 Days, 3-7 November 2014 (see [Cartea et al., 2015][10])). Section 5.4.2 contains the errors of estimation of comparison of of the standard deviation of the LNS of (16) and the RHS of (16) multiplied by $\sqrt{n}$. Section 5.4.3 depicts some graphs based on parameters estimation from Section 5.4.1. And Section 5.4.4 presents some ideas of how to implement the regime switching case from Section 4.4.

5.4.1 Parameters Estimation for CISCO Data

(5 Days, 3-7 November 2014 (see [Cartea et al., 2015]))

We have the following estimated parameters for 5 days, 3-7 November 2014, from Formula (16):

$$
\begin{aligned}
s^* &= 0.0001040723; 0.0002371220; 0.0002965143;\\
&\quad 0.0001263690; 0.0001554404;\\
\sigma &= 1.066708e-04; 1.005524e-04; 1.165201e-04; 1.134621e-04;\\
&\quad 9.954487e-05;\\
\lambda &= 0.03238898; 0.02643083; 0.02590728; 0.02530517; 0.02417804;\\
\alpha &= 438.2557; 401.0505; 559.1927; 418.7816; 449.8632;\\
\beta &= 865.9344; 718.0325; 1132.0741; 834.2553; 878.9675;\\
\hat{\lambda} := \lambda/(1-\alpha/\beta) &= 0.06560129; 0.059801686; 0.051181133; 0.050801432; 0.04957073.
\end{aligned}
$$

Volatility Coefficient $\sigma\sqrt{\lambda/(1-\alpha/\beta)}$ (volatility coefficient for the Brownian Motion in the right hand-side (RHS) of (16)):

$$0.04033114; 0.04098132; 0.04770726; 0.04725449; 0.04483260.$$

Transition Probabilities p :

Day1:

uu	ud
0.5187097	0.4812903
du	dd
0.4914135	0.5085865

Day2:

0.4790503	0.5209497
0.5462555	0.4537445

Day3:

0.6175041	0.3824959
0.4058722	0.5941278

Day4:

0.5806988	0.4193012
0.4300341	0.5699659

Day5:

0.4608844	0.5391156
0.5561404	0.4438596

We note, that stationary probabilities $\pi_i^*, i = 1, ..., 5$, are, respectively: $0.5525; 0.6195; 0.6494; 0.5637; 0.5783$. Here, we assume that the tick δ size is $\delta = 0.01$.

The following set of parameters are related to the following expression

$$S_{nt} - N(nt)s^* = S_0 + \sum_{k=1}^{N(nt)} (X_k - s^*),$$

-LHS of the expression in (16) multiplied by $\sqrt{n}$.

The first set of numbers are for the 10 minutes time horizon ($nt = 10$ minutes, for 5 days, the 7 sampled hours, total 35 numbers):

Table 1

[1]24.50981; [2]24.54490; [3]24.52375; [4]24.59209; [5]24.47209; [6]24.57042; [7]24.61063; [8]24.76987; [9]24.68749; [10]24.81599; [11]24.77026; [12]24.79883; [13]24.80073; [14]24.90121; [15]24.87772; [16]24.98492; [17]25.09788; [18]25.09441; [19]24.99085; [20]25.18195; [21]25.15721; [22]25.04236; [23]25.18323; [24]25.15222; [25]25.20424; [26]25.14171; [27]25.18323; [28]25.25348; [29]25.10225; [30]25.29003; [31]25.28282; [32]25.33267; [33]25.30313; [34]25.27407; [35]25.30438;

The standard deviation (SD) is: 0.2763377. The Standard Error (SE) for SD for the 10 min is: 0.01133634 (for standard error calculations see [Casella and Berger, 2002, page 257][12].

The second set of numbers are for the 5 minutes time horizon ($nt = 5$ minutes, for 5 days, the 7 sampled hours):

Table 2

[1]24.49896; [2]24.52906; [3]24.50417; [4]24.53417; [5]24.53500; [6]24.51458; [7]24.55479; [8]24.93026; [9]24.66931; [10]24.74263; [11]24.79358; [12]24.80310; [13]24.84500; [14]24.88405; [15]24.85729; [16]24.98907; [17]25.08085; [18]25.07500; [19]24.99322; [20]25.13381; [21]25.15144; [22]25.15197; [23]25.12475; [24]25.15449; [25]25.18475; [26]25.20348; [27]25.20500; [28]25.25348; [29]25.21251; [30]25.35376; [31]25.30407; [32]25.30469; [33]25.30469; [34]25.27500; [35]25.30469;

The standard deviation for those numbers is: 0.2863928. The SE for SD for the 5 min is: 0.01233352.

The third and last set of numbers are for the 20 minutes time horizon ($nt = 20$ minutes, for 5 days, the 7 sampled hours):

Table 3

[1]24.48419; [2]24.53970; [3]24.56292; [4]24.57105; [5]24.48938; [6]24.52751; [7]24.50751; [8]24.76465; [9]24.59753; [10]24.82935; [11]24.76552; [12]24.81741; [13]24.75409; [14]24.84077; [15]24.92942; [16]24.99721; [17]25.05551; [18]25.04848; [19]25.08492; [20]25.09780; [21]25.09551; [22]24.95124; [23]25.24222; [24]25.19096; [25]25.18273; [26]25.14070; [27]25.20171; [28]25.26785; [29]25.23013; [30]25.38661; [31]25.32127; [32]25.34065; [33]25.30313; [34]25.25251; [35]25.24972;

The standard deviation is: 0.2912967. The SE for SD for the 20 min is: 0.01234808.

As we can see, the SE is approximately 0.01 for all three cases.

5.4.2 Error of Estimation

Here, we would like to calculate the error of estimation comparing the standard deviation for

$$S_{nt} - N(nt)s^* = S_0 + \sum_{k=1}^{N(nt)} (X_k - s^*)$$

and standard deviation in the right-hand side of (16) multiplied by $\sqrt{n}$, namely,

$$\sqrt{n}\sigma\sqrt{\lambda/(1-\alpha/\beta)}.$$

We calculate the error of estimation with respect to the following formula:

$$ERROR = (1/m)\sum_{k=1}^{m}(sd - \hat{sd})^2,$$

where $\hat{sd} = \sqrt{n}Coef$, where $Coef$ is the volatility coefficient in the right-hand side of equation (16). In this case $n = 1000$, and $Coef = 0.3276$.
We take observations of $S_{nt} - N(tn)s^*$ every 10 min and we have 36 samples per day for 5 days.

Using the first approach with formula above we take $m = 5$ and for computing the standard deviation "sd" we take 36 samples of the first day. In that case, we have
$ERROR = 0.07617229$.

Using the second approach with formula above, we take $m = 36$ and for computing "sd" we take samples of 5 elements (the same time across 5 days). In that case we have
$ERROR = 0.07980041$.

As we can see, the error of estimation in both cases is approximately 0.08, indicating that approximation in (16) for diffusion limit for CHP, Theorem 1, is pretty good.

5.4.3 Graphs based on Parameters Estimation for CISCO Data (5 Days, 3-7 November 2014 ([Cartea et al., 2015])) from Section 4.1

The following graphs, Figure 5.1, contain the empirical intensity for the point process for those 5 days vs a simulated path using the above-estimated parameters.

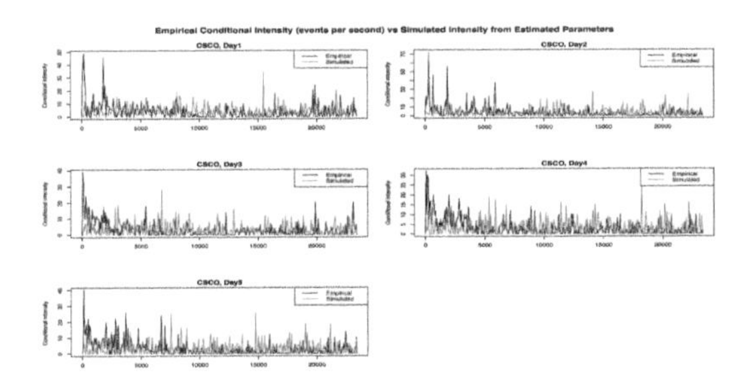

FIGURE 5.1
CondSimInt.

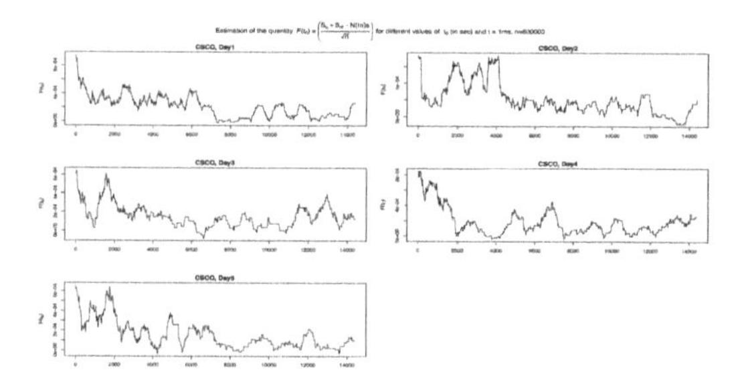

FIGURE 5.2
LimEst.

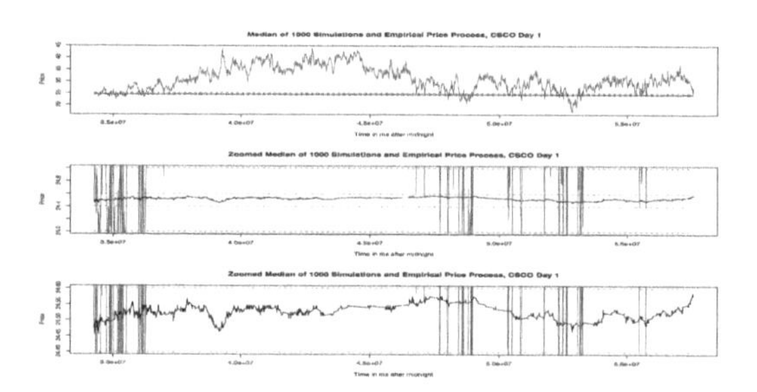

FIGURE 5.3
SimMedianEmp.

In the next graphs, Figure 5.2, we estimate the left hand-side (LHS) of (16). The time horizon is $nt = 10$ min. We took the time from which the start time measuring the 10 min. as the independent variable or x-axis. The dependent variable or y-axis is

$$F(t_0) = (S_{t_0} + S_{tn} - N(tn)s^*)/\sqrt{n}.$$

The following graphs, Figure 5.3, are the same as above but just considering the median of the 1000 simulations and zoomed in the range so that it is easy to compare.

The next graphs, Figure 5.4, contain information on the quantiles of simulations of the price process according to equation (16). That is, for a fixed big n and fixed t_0 and t. We use 1000 simulations of the process (with the parameters estimated for $N(t)$). The time horizon is a trading day.

The following graph, Figure 5.5, is the same as above but the time horizon is 5 minutes (e.g., $nt = 5$ minutes now, n is the same).

The last graph, Figure 5.6 is the same as above but the time horizon is 60 minutes (e.g., $nt = 60$ minutes now, n is the same).

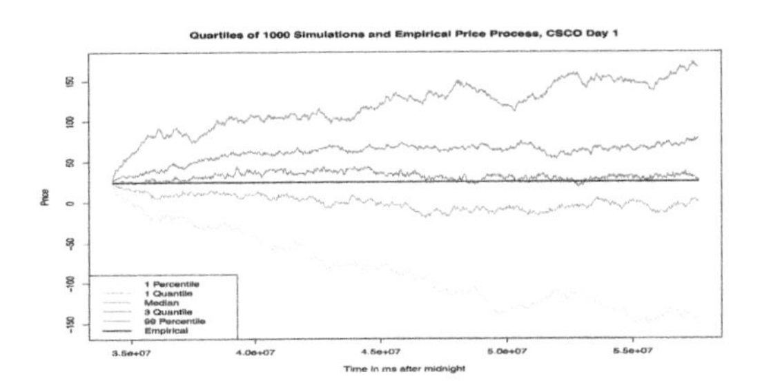

FIGURE 5.4
SimQuanEmp.

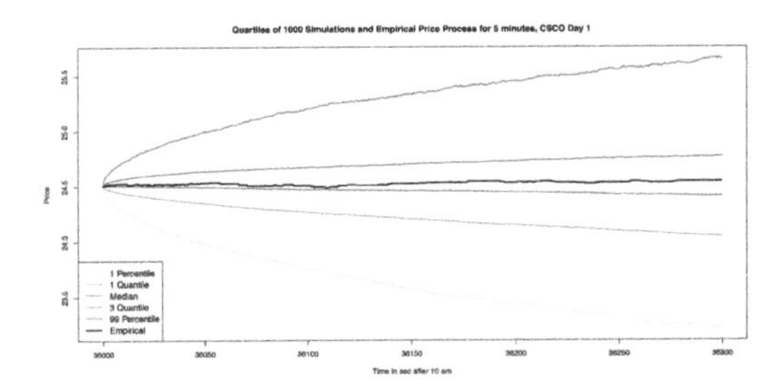

FIGURE 5.5
SimQuanEmp5min.

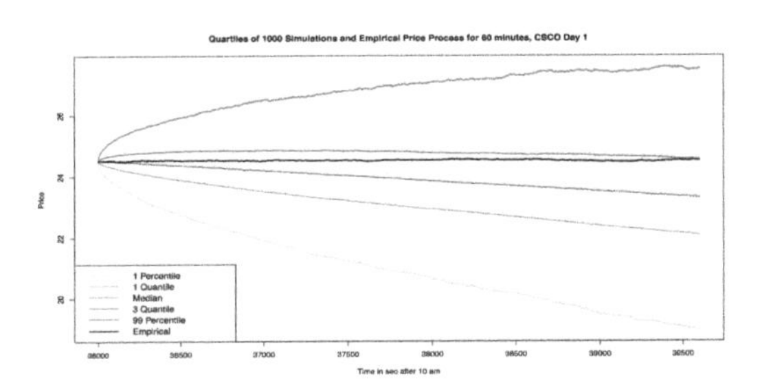

FIGURE 5.6
SimQuanEmp60min.

5.4.4 Remark on Regime-switching Case (Section 3.4)

We present here some ideas of how to implement the regime-switching case from Section 5.3.2. We take a look at the case of two states for intensity λ. The first state is constructed as the intensity that is above the intensities average, and the second state is constructed as the intensity that is below the intensities average. The transition probabilities matrix P are calculated using the relative frequencies of the intensities, and the stationary probabilities $\vec{p} = (p_1, p_2)$ are calculated from the equation $\vec{p}P = \vec{p}$. Then $\hat{\lambda}$ can be calculated from formula (32). For example, for the case of 5 days CISCO data [10], we have $\lambda_1 = 0.03238898$, $\lambda_2 = 0.02545533$ and $(p_1, p_2) = (0.2, 0.8)$. In this way, the value for $\hat{\lambda}$ in (32) is $\hat{\lambda} = 0.02688$. As we could see from the data for λ in Section 4.1 and the latter number, the error does not exceed 0.0055. It means that the errors of estimation for our standard deviations in Section 4.2 is almost the same. This is the evidence that in the case of regime-switching CHP the diffusion limit gives a very good approximation as well.

5.5 Conclusion

This chapter introduced two new Hawkes processes, namely, compound and regime-switching compound Hawkes processes, to model the price processes in limit order books. We proved Law of Large Numbers and Functional Central Limit Theorems (FCLT) for both processes. The two FCLTs are applied to limit order books, where we used these asymptotic methods to study the link between price volatility and order flow in our two models by using the diffusion limits of these price processes. The volatilities of price changes were expressed in terms of parameters describing the arrival rates and price changes. We also presented some numerical examples.

Bibliography

[1] Ait-Sahalia, Y., Cacho-Diaz, J. and Laeven, R. (2010). Modelling of financial contagion using mutually exciting jump processes. *Tech. Rep.*, 15850, Nat. Bureau of Ec. Res., USA.

[2] Bacry, E., Delattre, S., Hoffman, M. and Muzy, J.-F. (2013). Some limit theorems for Hawkes processes and application to financial statistics. *Stochastic Processes and their Applications*, 123(7), pp. 2475–2499.

[3] Bacry, E., Mastromatteo, I. and Muzy, J.-F. (2015). Hawkes processes in finance. arXiv:1502.04592v2 [q-fin.TR] 17, May.

[4] Bowsher, C. (2007). Modelling security market events in continuous time: intensity based, multivariate point process models. *J. Econometrica*, 141 (2), pp. 876–912.

[5] Bauwens, L. and Hautsch, N. (2009). *Modelling Financial High Frequency Data Using Point Processes.* Springer.

[6] Brémaud, P. and Massoulié, L. (1996). Stability of nonlinear Hawkes processes. *The Annals of Probab.*, 24(3), 1563.

[7] Buffington, J., Elliott, R.J. (2002). American Options with Regime Switching. *International Journal of Theoretical and Applied Finance*, 5, pp. 497–514.

[8] Buffington, J. and Elliott, R. J. (2000). Regime Switching and European Options. Lawrence, K.S. (ed.) *Stochastic Theory and Control.* Proceedings of a Workshop, 73–81. Berlin Heidelberg New York: Springer.

[9] Cartea, A., Jaimungal, S. and Ricci, J. (2011). Buy low, sell high: a high-frequency trading prospective. *Tech. Report.*

[10] Cartea, Á., Jaimungal, S. and Penalva, J. (2015). *Algorithmic and High-Frequency Trading.* Cambridge University Press.

[11] Cartensen, L. (2010). *Hawkes Processes and Combinatorial Transcriptional Regulation.* PhD Thesis, University of Copenhagen.

[12] Casella, G. and Berger, R. (2002). *Statistical Inference.* Duxbury-Thompson Learning Inc.

[13] Chavez-Demoulin, V. and McGill, J. (2012). High-frequency financial data modelling using Hawkes processes. *J. Banking and Finance*, 36(12), pp. 3415–3426.

[14] Chavez-Casillas, J., Elliott, R., Remillard, B. and Swishchuk, A. (2017). A level-1 limit order book with time dependent arrival rates. *Proceed. IWAP*, Toronto, June-20-25. Also available on arXiv: https://arxiv.org/submit/1869858

[15] Cohen, S. and Elliott, R. (2014). Filters and smoothness for self-exciting Markov modulated counting process. *IEEE Trans. Aut. Control.* Papers 1311.6257, arXiv.org.

[16] Cont, R. and de Larrard, A. (2013). A Markovian modelling of limit order books. *SIAM J. Finan. Math.*, 4 (1), pp. 1–25.

[17] Cox, D. (1955). Some statistical methods connected with series of events. *J. R. Stat.Soc.*, ser. B, 17 (2), pp. 129–164.

[18] Daley, D.J. and Vere-Jones, D. (1988). *An Introduction to the theory of Point Processes.* Springer.

[19] Dassios, A. and Zhao, H. (2011). A dynamic contagion process. *Advances in Applied Probab.*, 43 (3), pp. 814–846.

[20] Embrechts, P., Liniger, T. and Lin, L. (2011). Multivariate Hawkes processes: an application to financial data. *J. Appl. Prob.*, 48, A, pp. 367–378.

[21] Errais, E., Giesecke, K. and Goldberg, L. (2010). Affine point processes and portfolio credit risk. *SIAM J. Fin. Math.* 1, pp. 642–665.

[22] Fillimonov, V., Sornette, D., Bichetti, D. and Maystre, N. (2013). Quantifying of the high level of endogeneity and of structural regime shifts in comodity markets.

[23] Fillimonov, V. and Sornette, D. (2012). Quantifying reflexivity in financial markets: Toward a prediction of flash crashes. *Physical Review E*, 85 (5):056108.

[24] Hawkes, A. (1971). Spectra of some self-exciting and mutually exciting point processes. *Biometrica*, 58, pp. 83–90.

[25] Hawkes, A. and Oakes, D. (1974). A cluster process representation of a self-exciting process. *J. Applied Probab.*, 11, pp. 493–503.

[26] Korolyuk, V. S. and Swishchuk, A. V. (1995). Semi-Markov Random Evolutions. *Kluwer Academic Publishers*, Dordrecht, The Netherlands.

[27] Laub, P., Taimre, T. and Pollett, P. (2015). Hawkes Processes.arXiv: 1507.02822v1[math.PR]10 Jul 2015.

[28] Liniger, T. (2009). *Multivariate Hawkes Processes.* PhD thesis, Swiss Fed. Inst. Tech., Zurich.

[29] McNeil, A., Frey, R. and Embrechts, P. (2015). *Quantitative Risk Management: Concepts, Techniques and Tools.* Princeton University Press.

[30] Mehdad, B. and Zhu, L. (2014). On the Hawkes process with different exciting functions. *arXiv: 1403.0994.*

[31] Norris, J. R. (1997). Markov Chains. In Cambridge Series in Statistical and Probabilistic Mathematics. UK: Cambridge University Press.

[32] Skorokhod, A. (1965). Studies in the Theory of Random Processes, Addison-Wesley, Reading, Mass., (Reprinted by Dover Publications, NY).

[33] Swishchuk, A. (2017). Risk model based on compound Hawkes process. Abstract, IME 2017, Vienna.

[34] Swishchuk, A. and Vadori, N. (2017). A semi-Markovian modelling of limit order markets. *SIAM J. Finan. Math.*, 8, pp. 240–273.

[35] Swishchuk, A., Cera, K., Hofmeister, T. and Schmidt, J. (2017). General semi-Markov model for limit order books. *Intern. J. Theoret. Applied Finance*, 20, 1750019.

[36] Swishchuk, A., Chavez-Casillas, J., Elliott, R. and Remillard, B. (2018): Compound Hawkes processes in limit order books. *Financial Mathematics, Volatility and Covariance Modelling.* Routledge: Taylor & Francis Group. Available also on SSRN: https://papers.ssrn.com/sol3/papers.cfm?abstract_id=2987943

[37] Swishchuk, A. and Vadori, N. (2015). Strong law of large numbers and central limit theorems for functionals of inhomogeneous Semi-Markov processes. *Stochastic Analysis and Applications*, 13 (2), pp. 213–243.

[38] Vinkovskaya, E. (2014). *A Point Process Model for the Dynamics of LOB.* PhD thesis, Columbia University.

[39] Zheng, B., Roueff, F. and Abergel, F. (2014). Ergodicity and scaling limit of a constrained multivariate Hawkes process. *SIAM J. Finan. Math.*, 5.

[40] Zhu, L. (2013). Central limit theorem for nonlinear Hawkes processes, *J. Appl. Prob.*, 50(3), pp. 760–771.

6

Stochastic Modelling of Big Data in Finance with GCHP

This chapter provides the Law of Large Numbers (LLN) and a Functional Central Limit Theorems (FCLT) for several specific variations of GCHP introduced in Chapter 4. We apply these FCLTs to limit order books to study the link between price volatility and order flow, where the volatility in mid-price changes is expressed in terms of parameters describing the arrival rates and mid-price process.

6.1 A Brief Introduction and Literature Review

The Hawkes process (HP), named after its creator Alan Hawkes (1971, 1974), [27], [28], is a simple point process equipped with a self-exciting property, clustering effect and long run memory, as we have seen these in Chapter 4. Through its dependence on the history of the process, the HP captures the temporal and cross sectional dependence of the event arrival process as well as the "self-exciting" property observed in our empirical data on limit order books. Self-exciting point processes have recently been applied to high frequency data for price changes [55] or order arrival times [56]. HPs have seen their application in many areas, like genetics (2010) [11], occurrence of crime (2010) [51], bank defaults [52] and earthquakes [53].

Point processes gained a significant amount of attention in statistics during the 1950s and 1960s. Cox (1955) [16] introduced the notion of a doubly stochastic Poisson process (called the Cox process now) and Bartlett (1963) [8] investigated statistical methods for point processes based on their power spectral densities. Lewis (1964) [34] formulated a point process model (for computer power failure patterns) which was a step in the direction of the HP. A nice introduction to the theory of point processes can be found in Daley et al. (1988) [17]. The first type of point process in the context of market microstructure is the autoregressive conditional duration (ACD) model introduced by Engel et al. (1998) [19].

A recent application of HP is in financial analysis, in particular limit order books. In this paper, we study various new Hawkes processes, namely general

DOI: 10.1201/9781003265986-6

compound Hawkes processes to model the price process in limit order books. We prove a Law of Larges Numbers (LLN) and a Functional Central Limit Theorem (FCLT) for specific cases of these processes. Several of these FCLTs are applied to limit order books where we use asymptotic methods to study the link between price volatility and order flow in our models. The volatility of the price changes is expressed in terms of parameters describing the arrival rates and price changes. We also present some numerical examples. The general compound Hawkes process was first introduced in [40] to model the risk process in insurance and studied in detail in [41]. In the paper [43] we obtained functional CLTs and LLNs for general compound Hawkes processes with dependent orders and regime-switching compound Hawkes processes.

Bowsher (2007) [6] was the first one who applied the HP to financial data modelling. Cartea et al. (2011) [9] applied HP to model market order arrivals. Filimonov and Sornette (2012) [25] and Filimonov et al. (2013) [26] applied the HPs to estimate the percentage of price changes caused by endogenous self-generated activity rather than by the exogenous impact of news or novel information. Bauwens and Hautsch (2009) [7] used a five dimensional HP to estimate multivariate volatility between five stocks, based on price intensities. Hewlett (2006) [29] used the instantaneous jump in the intensity caused by the occurrence of an event to qualify the market impact of that event, taking into account the cascading effect of secondary events causing further events. Hewlett (2006) [29] also used the Hawkes model to derive optimal pricing strategies for market makers and optimal trading strategies for investors given that the rational market makers have the historic trading data. Large (2007) [32] applied a Hawkes model for the purpose of investigating market impact, with a specific interest in order book resiliency. Specifically, he considered limit orders, market orders and cancellations on both the buy and sell side, and further categorizes these events based on their level of aggression, resulting in a ten dimensional Hawkes process. Other econometric models based on marked point processes with stochastic intensity include autoregressive conditional intensity (ACI) models with the intensity depending on its history. Hasbrouck (1999) [30] introduced a multivariate point process to model the different events of an order book but did not parametrize the intensity. We note that Brémaud et al. (1996) [4] generalized the HP to its nonlinear form. Also, a functional central limit theorem for nonlinear Hawkes processes was obtained in Zhu (2013) [50]. The "Hawkes diffusion model" introduced in Ait-Sahalia et al. (2013) [1] attempted to extend previous models of stock prices to include financial contagion. Chavez-Demoulin et al. (2012) [12] used Hawkes processes to model high-frequency financial data. An application of affine point processes to portfolio credit risk may be found in Errais et al. (2010) [24]. Some applications of Hawkes processes to financial data are also given in Embrechts et al. (2011) [23].

Cohen et al. (2014) [14] derived an explicit filter for Markov modulated Hawkes processes. Vinkovskaya (2014) [48] considered a regime-switching Hawkes process to model its dependency on the bid-ask spread in limit order book. Regime-switching models for pricing of European and American options were considered in Buffington et al. (2000) [2] and Buffington et al. (2002) [3], respectively. Semi-Markov processes were applied to limit order books in [44] to model the mid-price. We also note that level-1 limit order books with time dependent arrival rates $\lambda(t)$ were studied in [13], including the asymptotic distribution of the price process.

The paper by Bacry et al. (2015) [5] proposes an overview of the recent academic literature devoted to the applications of Hawkes processes in finance. It is a nice survey of applications of Hawkes processes in finance. In general, the main models in high-frequency finance can be divided into univariate models, price models, impact models, order-book models and some systemic risk models, models accounting for news, high-dimensional models and clustering with graph models. The book by Cartea et al. (2015) [10] developed models for algorithmic trading such as methods for executing large orders, market making, trading pairs of collections of assets, and executing in the dark pool. This book also contains a link from which several datasets can be downloaded, along with MATLAB code to assist in experimentation with the data.

The results of this chapter were first announced in [42].

6.2 Diffusion Limits and LLNs

We prove in this chapter Law of Large Numbers (LLN) and a Functional Central Limit Theorems (FCLT) for several specific variations of GCHP introduced in Chapter 4.

6.2.1 Diffusion Limit and LLN for NLCHPnSDO

We consider the mid-price process S_t defined in Definition 4.7, namely

$$S_t = S_0 + \sum_{k=1}^{N(t)} a(X_k)$$

where X_k is a continuous time n-state Markov chain and a(x) is a continuous bounded function on the state space $X = \{1, 2, ..., n\}$. N_t is the number of price changes up to time t, described by the non-linear Hawkes process given in Equation (4.11).

Theorem 6.1 (Diffusion Limit for NLCHPnSDO) *Let X_k be an ergodic Markov chain with n states $\{1, 2, ..., n\}$ and with ergodic probabilities $(\pi_1^*, \pi_2^*, ..., \pi_n^*)$. Let also S_t be as defined in Definition 4.7, then*

$$\frac{S_{nt} - N(nt) \cdot \hat{a}^*}{\sqrt{n}} \xrightarrow{n \to \infty} \hat{\sigma}^* \sqrt{E[N[0,1]]} W_t \tag{6.1}$$

where W_t is a standard Wiener process and $E[N[0,1]]$ is the mean of the number of arrivals on a unit interval under the stationary and ergodic measure. Furthermore

$$0 < \hat{\mu} := \int_0^\infty \mu(s)ds < 1 \text{ and } \int_0^\infty s\mu(s)ds < \infty \tag{6.2}$$

$$\begin{aligned}
(\hat{\sigma}^*)^2 &:= \sum_{i\in X} \pi_i^* v(i) \\
v(i) &= b(i)^2 + \sum_{j\in X} (g(j) - g(i))^2 P(i,j) - 2b(i) \sum_{j\in X} (g(j) - g(i))P(i,j) \\
b &= (b(1),\ b(2), ...,\ b(n))' \\
b(i) &:= a(X_i) - a^* := a(i) - a^* \\
g &:= (P + \Pi^* - I)^{-1} b \\
\hat{a}^* &:= \sum_{i\in X} \pi_i^* a(X_i)
\end{aligned} \tag{6.3}$$

P is the transition probability matrix for X_k, i.e. $P(i,j) = P(X_{k+1} = j \mid X_k = i)$. Π^* denotes the matrix of stationary distributions of P and $g(j)$ is the jth entry of g.

Proof

From Equation (4.12) we have

$$S_{nt} = S_0 + \sum_{k=1}^{N(nt)} a(X_k) \tag{6.4}$$

and

$$S_{nt} = S_0 + \sum_{i=1}^{N(nt)} (a(X_k) - \hat{a}^*) + N(nt)\hat{a}^* \tag{6.5}$$

therefore

$$\frac{S_{nt} - N(nt)\hat{a}^*}{\sqrt{n}} = \frac{S_0 + \sum_{k=1}^{N(nt)} (a(X_k) - \hat{a}^*)}{\sqrt{n}}. \tag{6.6}$$

As long as $\frac{S_0}{\sqrt{n}} \xrightarrow{n\to\infty} 0$, we need only find the limit for

$$\frac{a(X_k) - \hat{a}^*}{\sqrt{n}}$$

when $n \to +\infty$. Consider the following sums

$$\hat{R}_n^* := \sum_{k=1}^{n} (a(X_k) - \hat{a}^*) \tag{6.7}$$

and

$$\hat{U}_n^*(t) := n^{-1/2}[(1-(nt-\lfloor nt \rfloor))]\hat{R}^*_{\lfloor nt \rfloor} + (nt - \lfloor nt \rfloor)\hat{R}^*_{\lfloor nt \rfloor + 1}] \tag{6.8}$$

where $\lfloor \cdot \rfloor$ *is the floor function. Following the martingale method from [46], we have the following weak convergence in the Skorokhod topology (see [39]):*

$$\hat{U}_n^*(t) \xrightarrow{n\to\infty} \hat{\sigma}^* W(t) \tag{6.9}$$

We note that the results from [4] imply by the ergodic theorem that

$$\frac{N_t}{t} \xrightarrow{t\to\infty} E[N[0,1]] \tag{6.10}$$

or

$$\frac{N_{nt}}{nt} \xrightarrow{n\to\infty} tE[N[0,1]]. \tag{6.11}$$

Using the change of time $t \to N_{nt}/n$*, we find that*

$$\hat{U}_n^*(N_{nt}/n) \xrightarrow{n\to\infty} \hat{\sigma}^* W(tE[N[0,1]]) \tag{6.12}$$

or

$$\hat{U}_n^*(N_{nt}/n) \xrightarrow{n\to\infty} \hat{\sigma}^* \sqrt{E[N[0,1]]}W(t) \tag{6.13}$$

The result now follows from Equations (6.4)-(6.6)

Lemma 6.1 (LLN for NLCHPnSDO) *The process* S_{nt} *in Equation* (6.4) *satisfies the following weak convergence in the Skorokhod topology (see [40]):*

$$\frac{S_{nt}}{n} \xrightarrow{n\to\infty} \hat{a}^* E[N[0,1]]t \tag{6.14}$$

where $\hat{a}^*$ *is defined in Equation* (6.3) *respectively.*

Proof *From Equation* (4.12) *we have*

$$\frac{S_{nt}}{n} = \frac{S_0}{n} + \sum_{k=1}^{N(nt)} \frac{a(X_k)}{n} \tag{6.15}$$

The first term goes to zero when $n \to \infty$*. On the right hand side, with respect to the strong LLN for Markov chains (see, e.g. [38])*

$$\frac{1}{n}\sum_{k=1}^{n} a(X_k) \xrightarrow{n\to\infty} \hat{a}^* \tag{6.16}$$

Then taking into account Equation (6.11)*, we obtain*

$$\sum_{i=1}^{N_{nt}} \frac{a(X_k)}{n} = \frac{N_{nt}}{n}\frac{1}{N_{nt}}\sum_{k=1}^{N(nt)} a(X_k) \xrightarrow{n\to\infty} \hat{a}^* E[N[0,1]]t \tag{6.17}$$

from which the desired result follows.

6.2.2 Diffusion Limit and LLN for GCHPnSDO

We consider here the mid-price process S_t which was defined in Definition 4.8, namely

$$S_t = S_0 + \sum_{k=1}^{N(t)} a(X_k) \tag{6.18}$$

where X_k is a continuous time n-state Markov chain, a(x) is a continuous and bounded function on the state space $X = \{1,\ 2,\ ...,\ n\}$, and $N(t)$ is the number of price changes up to time t, described by a one-dimensional Hawkes process defined in Definition 4.4. This can be interpreted as the case of non-fixed tick sizes, n-valued price changes and dependent orders.

Theorem 6.2 (Diffusion limit for GCHPnSDO) *Let X_k be an ergodic Markov chain with n states $\{1,\ 2,\ ...,\ n\}$ and with ergodic probabilities $(\pi_1^*,\ \pi_2^*,\ ...,\ \pi_n^*)$. Let also S_t be as defined in Definition 4.8, then*

$$\frac{S_{nt} - N(nt)\hat{a}^*}{\sqrt{n}} \xrightarrow{n\to\infty} \hat{\sigma}^* \sqrt{\lambda/(1-\hat{\mu})} W(t) \tag{6.19}$$

where W(t) is a standard Wiener process,

$$0 < \hat{\mu} := \int_0^\infty \mu(s)ds < 1 \ and \int_0^\infty s\mu(s)ds < \infty \tag{6.20}$$

$$\begin{aligned}
(\hat{\sigma}^*)^2 &: = \sum_{i\in X} \pi_i^* v(i) \\
v(i) &= b(i)^2 + \sum_{j\in X}(g(j) - g(i))^2 P(i,j) - 2b(i)\sum_{j\in X}(g(j) - g(i))P(i,j) \\
b &= (b(1),\ b(2), ...,\ b(n))' \\
b(i) &: = a(X_i) - a^* := a(i) - a^* \\
g &: = (P + \Pi^* - I)^{-1} b \\
\hat{a}^* &: = \sum_{i\in X} \pi_i^* a(X_i)
\end{aligned} \tag{6.21}$$

P is the transition probability matrix for X_k, i.e. $P(i,j) = P(X_{k+1} = j \mid X_k = i)$. Π^* denotes the matrix of stationary distributions of P and $g(j)$ is the jth entry of g.

Proof *As in the previous theorem we have that*

$$S_{nt} = S_0 + \sum_{k=1}^{N(nt)} a(X_k) \tag{6.22}$$

and

$$S_{nt} = S_0 + \sum_{k=1}^{N(nt)} (a(X_k) - \hat{a}^*) + N(nt)\hat{a}^* \tag{6.23}$$

where $\hat{a}^*$ *is as defined above.*

Therefore, we can conclude that

$$\frac{S_{nt} - N(nt)\hat{a}^*}{\sqrt{n}} = \frac{S_0 + \sum_{k=1}^{N(nt)} (a(X_k) - \hat{a}^*)}{\sqrt{n}} \tag{6.24}$$

as long as $\frac{S_0}{\sqrt{n}} \xrightarrow{n \to \infty} 0$, *we need only find the limit for*

$$\frac{\sum_{k=1}^{N(nt)} (a(X_k) - \hat{a}^*)}{\sqrt{n}}$$

as $n \to \infty$. *We consider the following sums*

$$\hat{R}_n^* = \sum_{k=1}^{n} (a(X_k) - \hat{a}^*) \tag{6.25}$$

and

$$\hat{U}_n^*(t) := n^{-1/2}[(1 - (nt - \rfloor nt \lfloor))\hat{R}^*_{\rfloor nt \lfloor} + (nt - \rfloor nt \lfloor)\hat{R}^*_{\rfloor nt \lfloor +1}] \tag{6.26}$$

where $\lfloor \cdot \rfloor$ *is the floor function.*

Then following the martingale method from [45], we have the following weak convergence in the Skorokhod topology (see [39])

$$\hat{U}_n^* \xrightarrow{n \to \infty} \hat{\sigma}^* W(t) \tag{6.27}$$

where $\hat{\sigma}^*$ *is as defined above.*

We note again that with respect to the LLN for Hawkes processes (see, e.g., [17]) we have

$$\frac{N(t)}{t} \xrightarrow{t \to \infty} \frac{\lambda}{1 - \hat{\mu}} \tag{6.28}$$

or

$$\frac{N(nt)}{n} \xrightarrow{n \to \infty} \frac{t\lambda}{1 - \hat{\mu}} \tag{6.29}$$

where $\hat{\mu}$ *is as defined above. Then using the change of time defined before where* $t \to N(nt)/n$ *we can find from Equations* (6.27)-(6.29)*:*

$$\hat{U}_n^* \left(\frac{N(nt)}{n} \right) \xrightarrow{n \to \infty} \hat{\sigma}^* W \left(\frac{t\lambda}{1 - \hat{\mu}} \right)$$

or

$$\hat{U}_n^* \left(\frac{N(nt)}{n} \right) \xrightarrow{n \to \infty} \hat{\sigma}^* \sqrt{\frac{\lambda}{1 - \hat{\mu}}} W(t)$$

The result in Equation (6.19) *now follows from Equations* (6.22)-(6.29)

Lemma 6.2 (LLN for GCHPnSDO) *The process S_{nt} defined in Definition 4.8, satisfies the following weak convergence in the Skorokhod topology (see [40])*

$$\frac{S_{nt}}{n} \xrightarrow{n\to\infty} \hat{a}^* \frac{\lambda}{1-\hat{\mu}} t \tag{6.30}$$

where $\hat{\mu}$ and $\hat{a}^*$ are defined in Equations (6.20) *and* (6.21) *respectively.* **Proof.** *From Equation* (4.13) *we have*

$$\frac{S_{nt}}{n} = \frac{S_0}{n} + \sum_{k=1}^{N(nt)} \frac{a(X_k)}{n}. \tag{6.31}$$

The first term goes to zero when $n \to \infty$. On the right hand side, with respect to the strong LLN for Markov chains (see, e.g., [38])

$$\frac{1}{n}\sum_{k=1}^{n} a(X_K) \xrightarrow{n\to\infty} \hat{a}^* \tag{6.32}$$

Then taking into account Equation (6.29) *we obtain*

$$\sum_{k=1}^{N(nt)} \frac{a(X_k)}{n} = \frac{N(nt)}{n}\frac{1}{N(nt)}\sum_{k=1}^{N(nt)} a(X_k) \xrightarrow{n\to\infty} \hat{a}^* \frac{\lambda}{1-\hat{\mu}} t \tag{6.33}$$

from which the desired result follows.

6.2.3 Diffusion Limits and LLNs for Special Cases of GCHPnSDO

Theorem 6.2 can be reduced to some of the special cases we outlined previously in Subsection 4.3.1. Specifically in Definitions 4.9 and 4.10, we consider the case of a 2-state Markov chain for which we provide the diffusion limit and LLN result as Corollaries below.

We begin by considering the mid-price process S_t (GCHP2SD) which was defined in 4.9, namely

$$S_t = S_0 + \sum_{k=1}^{N(t)} a(X_k) \tag{6.34}$$

where X_k is a continuous-time 2-state Markov chain, a(x) is a continuous and bounded function on $X = \{1,\ 2\}$ and $N(t)$ is the number of price changes up to moment t, described by a one-dimensional Hawkes process defined in Definition 4.4. This can be interpreted as the case of non-fixed tick sizes, two-valued price changes and dependent orders.

Corollary 6.1 (Diffusion Limit for GCHP2SDO) *Let X_k be an ergodic Markov chain with two states $\{1,\ 2\}$ and with ergodic probabilities $(\pi_1^*,\ \pi_2^*)$. Further, let S_t be defined as in Definition 4.9. Then*

$$\frac{S_{nt} - N(nt)a^*}{\sqrt{n}} \xrightarrow{n\to\infty} \sigma^* \sqrt{\lambda/(1-\hat{\mu})} W(t) \tag{6.35}$$

where $W(t)$ is standard Wiener process,

$$0 < \hat{\mu} := \int_0^\infty \mu(s)ds < 1 \text{ and } \int_0^\infty \mu(s)ds < \infty \tag{6.36}$$

$$\begin{aligned}
(\sigma^*)^2 : &= \pi_1^* + \pi_2^* a_2^2 + (\pi_1^* a_1 + \pi_2^* a_2) \\
&\quad \times \left[-2a_1\pi_1^* - 2a_2\pi_2^* + (\pi_1^* a_1 + \pi_2^* a_2)(\pi_1^* + \pi_2^*)\right] \\
&\quad + \frac{\pi_1^*(1-p) + \pi_2^*(1-p')(a_1-a_2)^2}{(p+p'-2)^2} \\
&\quad + 2(a_2 - a_1)\left[\frac{\pi_2^* a_2(1-p') - \pi_1^* a_1(1-p)}{p+p'-2}\right. \\
&\quad \left. + \frac{(\pi_1^* a_1 + \pi_2^* a_2)(\pi_1^* - p\pi_1^* - \pi_2^* + p'\pi_2^*)}{p+p'-2}\right] \\
a^* : &= a_1\pi_1^* + a_2\pi_2^*
\end{aligned} \tag{6.37}$$

where (p, p') are the transition probabilities of the Markov chain.

Corollary 6.2 (LLN for GCHP2SDO) *The process S_{nt} defined in Definition 4.9 satisfies the following weak convergence in the Skorokhod topology (see [40])*

$$\frac{S_{nt}}{n} \xrightarrow{n\to\infty} \hat{a}^* \frac{\lambda}{1-\hat{\mu}} t \tag{6.38}$$

where $\hat{\mu}$ and a^* are defined in Equations (6.36) *and* (6.37) *respectively.*

Now let us consider the process S_t defined in Definition 4.10, specifically

$$S_t = S_0 + \sum_{k=1}^{N(t)} X_k \tag{6.39}$$

where $X_k \in \{-\delta, \delta\}$ is a continuous-time 2-state Markov chain, δ is the fixed tick size, and $N(t)$ is the number of price changes up to moment t, described by a one-dimensional Hawkes process defined in Definition 4.4. This case we have a fixed tick size, two-valued price changes and dependent orders.

Corollary 6.3 (Diffusion Limit for CHPDO) *Let X_k be an ergodic Markov chain with two states $\{\delta, -\delta\}$ and with ergodic probabilities $(\pi^*, 1-\pi^*)$. Taking S_t, as defined in Definition 4.10, we have*

$$\frac{S_{nt} - N(nt)a^*}{\sqrt{n}} \xrightarrow{n\to\infty} \sigma \sqrt{\frac{\lambda}{1-\hat{\mu}}} W(t), \tag{6.40}$$

where $W(t)$ is a standard Wiener process,

$$0 < \hat{\mu} := \int_0^\infty \mu(s)ds < 1 \text{ and } \int_0^\infty \mu(s)ds < \infty \tag{6.41}$$

$$\begin{aligned} a^* &:= \delta(2\pi^* - 1) \\ \sigma^2 &:= 4\delta^2 \left(\frac{1 - p' + \pi^*(p' - p)}{(p + p' - 2)^2} - \pi^*(1 - \pi^*) \right), \end{aligned} \tag{6.42}$$

and (p, p') are the transition probabilities of the Markov chain X_k. We note that λ and $\mu(t)$ are defined in Equation (4.2).

We also note that the LLN for both GCHP2SDO and CHPDO is identical to the result given in Lemma 6.2, after some simplification.

6.3 Empirical Results

In order to test the validity of our models and determine which best fits empirical data, we have considered level 1 LOB data (meaning the limit orders sitting at the best bid and ask) for Apple, Amazon, Google, Microsoft and Intel on June 21, 2012 [54]. We first verify that the data is reasonable for our model by checking its liquidity. That is illustrated in Table 6.1.

The high number of daily price changes motivates the idea that we can use asymptotic analysis in order to approximate long-run volatility using order flow by finding the diffusion limit of the price process. Because we do not want to include opening and closing auctions we omit the first and last fifteen minutes of our data. We motivate the arrival process by analyzing the inter-arrival times and clustering to a ensure the arrival process is not Poisson and exhibits the characteristics of a Hawkes Process. That is illustrated in Figures 6.1 and 6.2.

TABLE 6.1
Stock liquidity of AAPL, AMZN, GOOG, MSFT, and INTC for June 21, 2012.

Ticker	Avg # of Orders per Second	Price Changes in 1 Day
AAPL	51	64,350
AMZN	25	27,557
GOOG	21	24,084
MSFT	173	3,217
INTC	176	4,060

Furthermore, the relationship between our diffusion coefficient and arrival process is limited to the expected number of arrivals on a unit interval. This implies that results for a simple exponential model can be easily generalized to a non-linear one. This makes it possible to work with a simplified model for our Hawkes process which is still rich enough to capture our observations. Keeping this in mind, we restrict ourselves to an exponential kernel and estimate parameters using a MLE [35]. We provide these estimates in Table 6.2 and compare the empirical expected number of arrivals and compare with the MLE estimate in Table 6.3.

TABLE 6.2
Each parameter was estimated using a particle swarm optimization method in an attempt to globally optimize the negative log-likelihood function. The values for λ, α and β for each data set are as provided.

	λ	α	β
AAPL	1.4683	1045.2676	2556.1844
AMZN	0.6443	653.7524	1556.1702
GOOG	0.4985	865.8553	1980.4409
MSFT	0.0659	479.3482	908.0032
INTC	0.0471	399.6389	760.4991

Obviously the MLE method accurately estimates the expected number of arrivals on the unit interval. This means we can confidently say for our data that our parameters will work reasonably with our models. We provide the estimated parameters in Table 6.2.

TABLE 6.3
Expected number of arrivals on a unit interval using estimated parameters from an MLE method is compared against the empirical arrivals.

	Emp. $E[N([0,1])]$	MLE
AAPL	2.4840	2.4841
AMZN	1.1110	1.1110
GOOG	0.8857	0.8857
MSFT	0.1395	0.1396
INTC	0.0991	0.0992

6.3.1 CHPDO

We first consider a compound Hawkes process with dependent orders defined in Definition 4.10, namely

$$S_t = S_0 + \sum_{k=1}^{N(t)} X_k \tag{6.43}$$

where $X_k \in \{+\delta, -\delta\}$ is a continuous-time two state Markov chain, δ is of fixed size and $N(t)$ is the number of mid-price movements up to time t described by a Hawkes process.

We have opted to study the mid-price changes of our model. Thus S_t can be computed by averaging the best bid and ask price. Noting that the price is recorded in cents the smallest possible jump in the mid-price is a half a cent which we will use as δ. Furthermore, in order to estimate the transition matrix for the Markov chain X_k we count the absolute frequencies of upward and downward price movements and from this calculate the relative frequencies giving an estimate for p_{uu} and p_{dd} which represent the conditional probabilities of an up/down movement given an up/down movement. Later we will consider several different sizes of mid-price movements and will work with the convention that each movement will be assigned a state based off of its ordering in the reals. In this case, $-\delta$ will be state one, and δ with be state two. This results in the transition matrix P given below.

$$P = \begin{bmatrix} p_{dd} & 1 - p_{dd} \\ 1 - p_{uu} & p_{uu} \end{bmatrix}$$

After determining our parameters and transition probabilities we calculate a^* and σ in Table 6.4 together with p_{uu} and p_{dd}.

TABLE 6.4
Provided above are the values for s^*, σ as well as the probabilities of an upward/downward movement given an upward/downward movement for each of the 5 stocks in question.

	p_{dd}	p_{uu}	σ	a^*
AAPL	0.4956	0.4933	0.0049	-1.1463e-5
AMZN	0.4635	0.4576	0.0046	-2.7373e-5
GOOG	0.4769	0.4461	0.0046	-1.4301e-4
MSFT	0.6269	0.5827	0.0062	-2.7956e-4
INTC	0.6106	0.5588	0.0059	-3.1185e-4

Provided these values, we can test our claim that our model accurately describes the mid-price process. We will use the diffusion limit proved considered earlier, namely

$$\frac{S_{nt} - N(nt)a^*}{\sqrt{n}} \xrightarrow{n\to\infty} \sigma\sqrt{\frac{\lambda}{1-\alpha/\beta}}W(t). \tag{6.44}$$

If the data satisfies our proposed model, then after considering large windows of time (5min, 10min, 20min), we would expect to see the empirical and theoretical standard deviations to follow each other closely. To test this we compare the equivalent process, constructed by multiplying the LHS and RHS by $\sqrt{n}$. Then cutting our data into disjoint windows of size n, specifically

$[in, (i+1)n]$ with $t = 1$ and by setting the left bound as our starting time we can calculate $S_{nt} - N(nt)a^*$ for each individual window and give a generalized formula for this below.

$$S_i^* = S_{(i+1)nt} - S_{int} - (N((i+1)nt) - N(int))a^* \tag{6.45}$$

This gives a collection of values $\{S_i^*\}$ over which we compute the standard deviation. If our model is accurate, we would would expect that

$$\text{std}\{S_i^*\} \approx \sqrt{nt}\sigma\sqrt{\frac{\lambda}{1-\alpha/\beta}}, \text{ where } t = 1. \tag{6.46}$$

We plot the empirical standard deviation against the theoretical one for various window sizes starting at 10 seconds and increasing in steps of 10 seconds until we reach 20 minutes, this is illustrated in Figure 6.3.

Several important remarks should be made at this point. It is clear that while the model accurately predicts the overall trend for MSFT and INTC, we severely underestimate the variability in the mid-price process for APPL, AMZN and GOOG. Furthermore, as the window size increases the overall spread in the data increases. We attribute this to the decreasing sample size imposed on us as we increase the window size. For example when we consider a 20 minute window, we can only construct 27 disjoint windows in the 9 hour trading day forcing us to deal with the problem of predicting a "population" standard deviation from a increasingly small sample. We remedy this by using a variance stabilizing transformation in later sections. Specifically, a popular method for a Poisson process is to take the square root of our empirical and theoretical standard deviations. This makes it possible to qualitatively view the overall trend in the data, gaining a clearer idea of goodness of fit from there.

6.3.2 GCHP2SDO

As of now we have considered a fixed δ related to the trading tick size. However, if we consider the mid-price changes for APPL, AMZN and GOOG the assumption of a fixed tick size is violated. In fact, we observe that approximately 61%, 53% and 71% of all mid-price changes are larger than half a tick size, which is opposed to what we observe for MSFT and INTC where all mid-price changes occur at the half tick size, we illustrate this for AAPL, AMZN and GOOG in Figure 6.4.

It is clear in Figure 6.4 additional considerations need to be made. A simple way to include the variability in mid-price movements in our model is to introduce $a(X_i)$ as described in Definition 4.9. It is of course necessary to determine the values of $a(\cdot)$ for each state of our Markov chain. A naive method is to take the mean of the downward and upward mid-price movements and assign them to $a(1)$ and $a(2)$ respectively. We provide these values in Table 6.5.

TABLE 6.5
$a(i)$ is the average of the upward or downward mid-price movements. Following our previous convention, the first state will be associated with the mean of all downward mid-price movements and the second state will be associated with the mean of all upward mid-price movements.

	$a(1)$	$a(2)$
AAPL	-0.0172	0.0170
AMZN	-0.0134	0.0133
GOOG	-0.0302	0.0308

In this step we have only endeavoured to better realize the actual price movements in our data. Therefore, when we observe a downward mid-price movement we continue to assign it to state one and similarly for an upward price movement we continue to assign it to state two. It follows that our transition matrix will remain the same. Then using these new state values we recalculate a^* and σ, providing them in Table 6.6. The effect of these changes is investigated in Figure 6.5. Note that in Figure 6.5 we have used the variance stabilizing transformation discussed earlier in order to better visualize the overall trend in our data.

TABLE 6.6
We have the values for a^*, σ, as well as the probabilities of an upwards/downwards movement, given an upwards/downwards movement for the 3 stocks of interest.

	p_{dd}	p_{uu}	σ	a^*
AAPL	0.4956	0.4933	0.0169	-1.5624e-4
AMZN	0.4635	0.4576	0.0123	-1.0475e-4
GOOG	0.4769	0.4461	0.0282	-5.5095e-4

Notice that there is a significant qualitative improvement in the fits for AAPL and GOOG in Figure 6.5 but the variability in mid-price movements for AMZN is still clearly underestimated by our model. The unexplained variance may be captured by investigating an n-state Markov chain since the additional transition probabilities could explain the variability missing in the 2-state case.

6.3.3 GCHPNSDO

We recall the N-state model described in Definition 4.8. The immediate question becomes how best to choose the state values. We modify the quantile based approach from [45]. After calculating the mid-price changes we separate the data into upward and downward price movements. Then we calculate evenly distributed quantiles for both data sets. Depending on the data, several quantiles may be identical, we reject any duplicates. We thus obtain a list of bounds which we complete by adding the minimum observed value if necessary.

To determine the state values $a(X_i)$, we take the average of all mid-price changes located between two neighbouring boundary values. Furthermore, we assign a mid-price change to state i if it is greater than or equal to the $(i-1)$th boundary and strictly less than the ith boundary. An exception is made for the largest upper bound where equality is permitted at both ends.

As we could not capture the full variability of mid-price changes for AMZN in the previous method we investigate for this case. Furthermore, for tractability we only consider 14 boundary values from which we obtain a 12-state Markov chain. Instead of providing the transition matrix, we provide the ergodic probabilities for the transition matrix and the associated states in Table 6.7.

TABLE 6.7
We have provided the state, associated ergodic probabilities and state values $a(i)$ for AMZN, given a 12 state Markov chain which was obtained from choosing a 16 quantile method.

AMZN

i	π_i^*	$a(X_i)$
1	0.0275	-0.0524
2	0.0281	-0.0318
3	0.0264	-0.0250
4	0.0382	-0.0200
5	0.0576	-0.0150
6	0.3249	-0.0064
7	0.2321	0.0050
8	0.0923	0.0100
9	0.0578	0.0150
10	0.0353	0.0200
11	0.0412	0.0271
12	0.0387	0.0476

In order to compare the two state and N-state approaches we first take a qualitative approach and plot the two theoretical and empirical standard deviations against each other in Figure 6.6. When we compare the mean squared residuals, the 2-state model discussed before has mean squared error 0.0208 while our 12-state Markov chain brings that down to 0.0125. Considering an even larger Markov chain with 24-states we are only able to obtain a meager improvement to 0.0123, suggesting that there is some underlying variance in the mid-price process not captured by our model. We investigate these more quantitative measures in the following subsection. First, comparing the models against a numerical best fit, and then investigating the mean squared error, we gain a better quantitative understanding of the overall improvement obtained from each model increasing the number of quantiles.

6.3.4 Quantitative Analysis

While our model does visually appear to fit the expected variability in four of the five cases, it still fails to capture the complete dynamics of mid-price changes seen in our AMZN data. We investigate the mean square error of our models with a varying number of quantiles in Table 6.8. This gives a good indication as to whether the N-state model is a better fit for our data. If we look closely at AAPL and GOOG we see the N-state case can still improve our results from the two state case. For AAPL we constructed a 17 state Markov chain by taking 16 quantiles on the downward movements and 16 quantiles upward movements. This resulted in a mean squared error of 0.0036 which is approximately a 28% improvement to the two state case where the mean squared error was 0.0050. Even more extreme, using a 25 state Markov chain for GOOG which was constructed similarly, we observed a mean squared error of 0.0046 which is a 60% improvement from the two state case with a mean squared error of 0.0115. We conclude with Table 6.8 which provides the mean residuals for AAPL, AMZN and GOOG with several of Markov chains constructed from various numbers of quantiles. We also include mean residuals for INTC and MSFT for comparison.

TABLE 6.8
We list the mean residuals for several Markov chains with varying numbers of states. These were generated using our modified quantile approach choosing to start with 2, 8, 16 or 32 quantiles. We see that in general the mean residual decreases to some lower limit where we can no longer perform any better. Recall that the only observed mid-price changes for INTC and MSFT were of a half tick size, any increase in the number of quantiles will result in the same performance.

	CHPDO	2	8	16	32
AAPL	0.2679	0.0050	0.0036	0.0036	0.0036
AMZN	0.1122	0.0208	0.0131	0.0124	0.0123
GOOG	0.4036	0.0115	0.0048	0.0045	0.0047
INTC	1.7917e-5	1.7917e-5	1.7917e-5	1.7917e-5	1.7917e-5
MSFT	1.0586e-4	1.0586e-4	1.0586e-4	1.0586e-4	1.0586e-4

Another quantitative measure of our fit would be to compare them to the best theoretical one available given the data. Notice that each of our models assumes that the standard deviation evolves according to the square root of the time step times some determinable coefficient. Therefore we can estimate the best possible coefficient by minimizing the mean squared residual. We provide plots of these hypothetical best fits against the empirical data and theoretical fits in Figure 6.7. We also provide the coefficients for the theoretical fits and hypothetical best fit in Table 6.9 in order to have a more quantitative comparison.

TABLE 6.9
The coefficients calculated for AAPL, AMZN and GOOG are generated using a Markov chain created by 16 quantiles on the upward and downward movements, while the coefficients for INTC and MSFT are obtained from the CHPDO case.

	Theoretical Coefficient	Regression Coefficient	Percent Error
AAPL	0.02868	0.02828	1.42%
AMZN	0.01450	0.01831	20.8%
GOOG	0.02883	0.03023	4.63%
INTC	0.00186	0.00193	3.4%
MSFT	0.00231	0.00246	6.4%

We notice that in each case the errors are close to, or under five percent with AMZN being the biggest offender. This is consistent with the discussion provided throughout our analysis and highlights the general applicability of our model.

6.3.5 Remarks

Overall the N state model outperforms the others, generating fits for four out of the five datasets that are reasonable and providing reductions in the mean squared error by upwards of 25%. While not able to capture the full dynamics observed in AMZN it appears to be a strong candidate for a simple model of price dynamics observed in our data. Further investigation would be necessary to determine what causes the additional volatility observed in AMZN, and potentially implement a more robust model which captures this.

6.3.6 Figures to Chapter 6

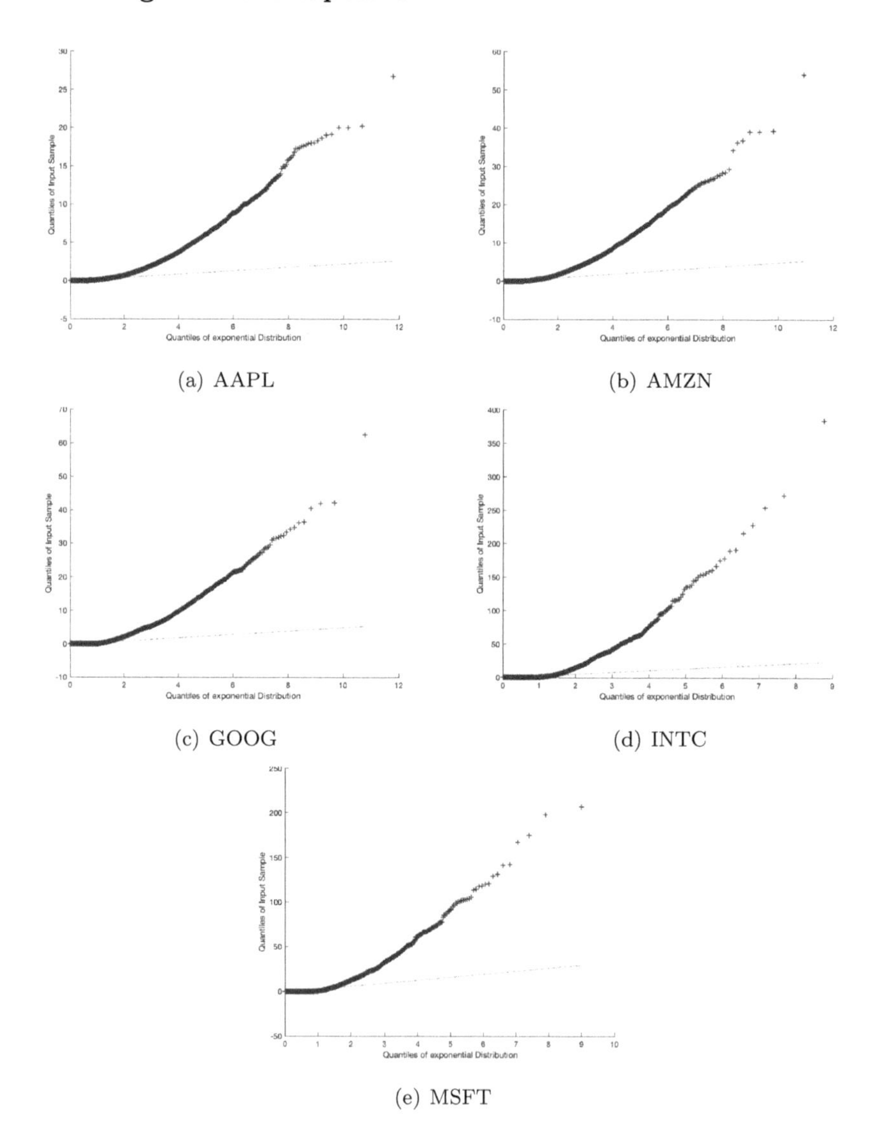

(a) AAPL (b) AMZN

(c) GOOG (d) INTC

(e) MSFT

FIGURE 6.1
Above we provide a quantile-quantile plot of our empirical inter-arrival times against a Poisson process for each of the five stocks. We see that the inter-arrival data does not fit the expected curve, providing evidence that the underlying arrival process is not Poisson.

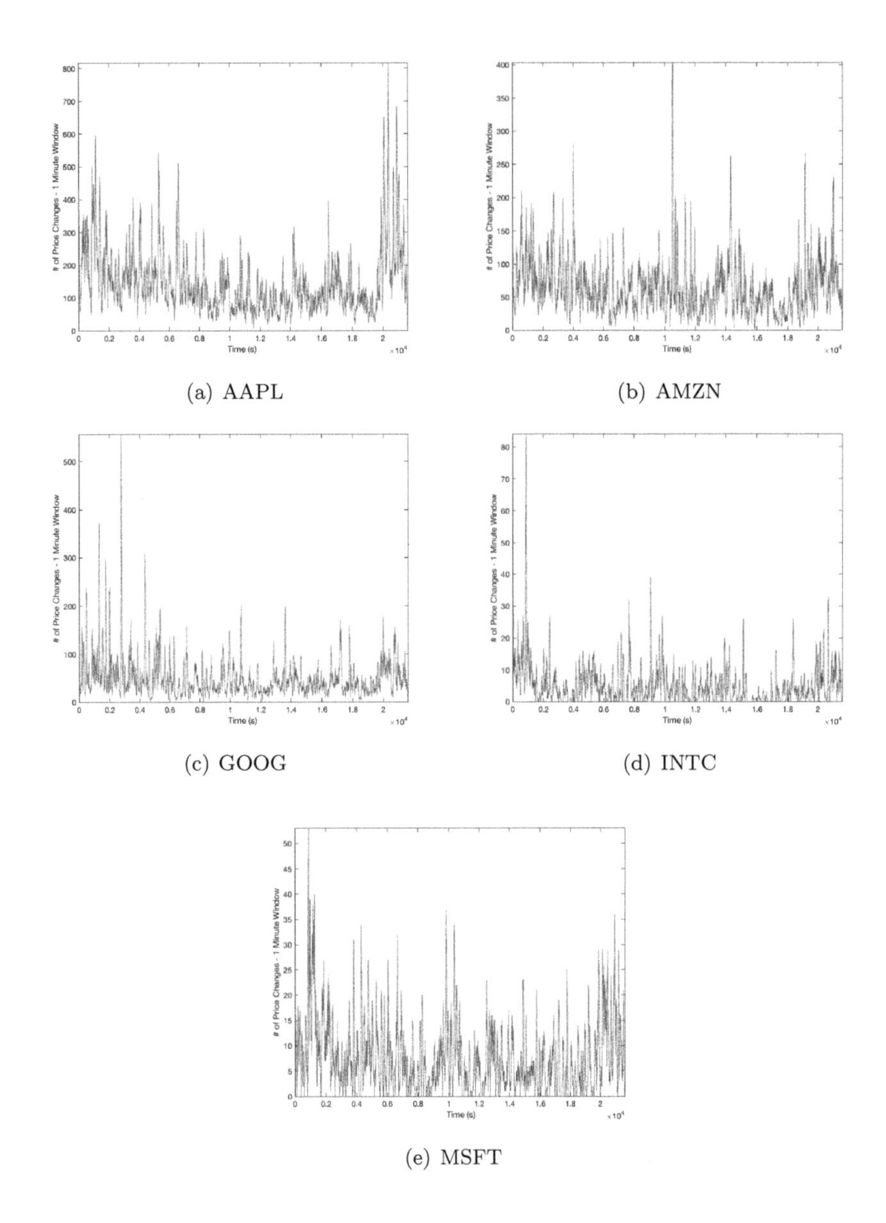

FIGURE 6.2
Each plot shows the number of arrivals for a moving one minute window. From this we can conclude that there is a significant amount of clustering in the arrival of mid-price changes, motivating the Hawkes model of our arrival process.

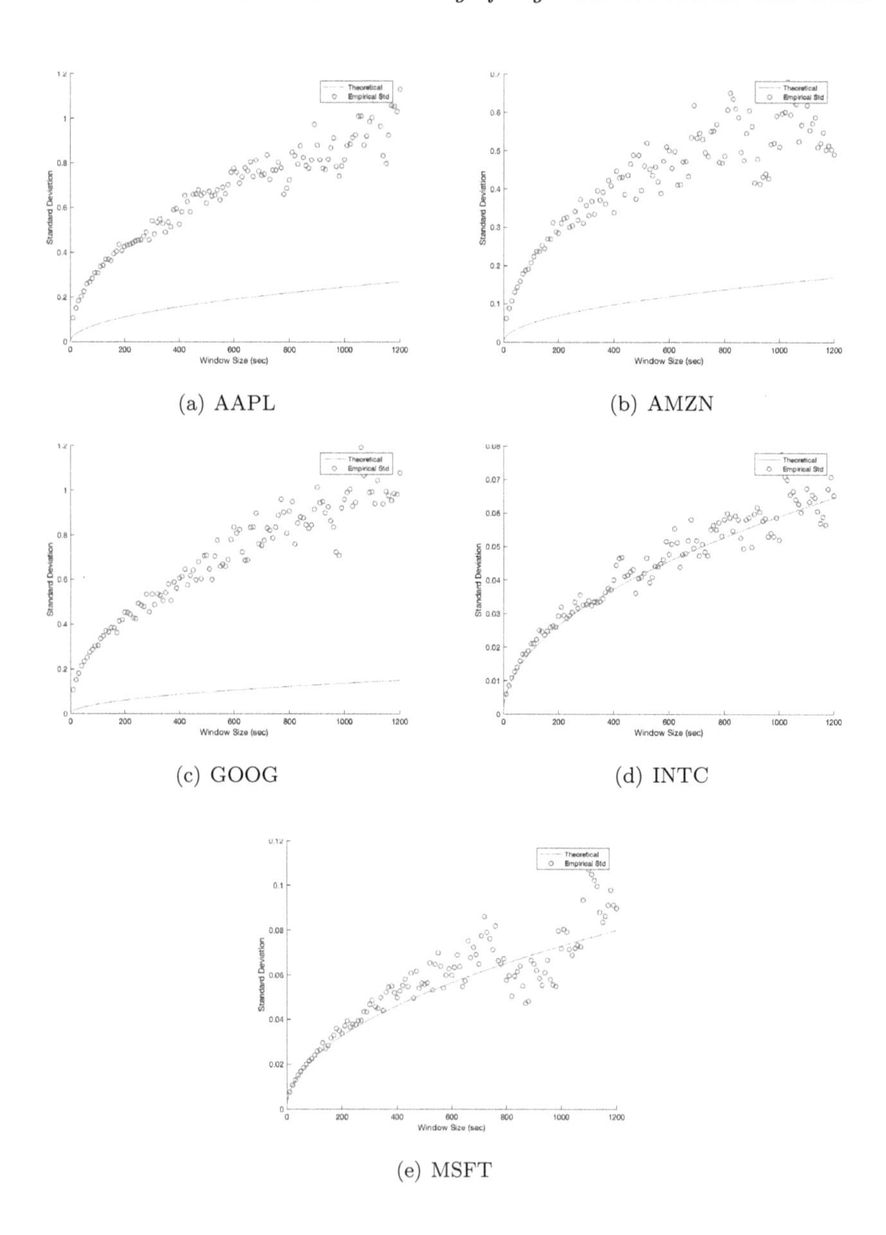

(a) AAPL

(b) AMZN

(c) GOOG

(d) INTC

(e) MSFT

FIGURE 6.3
Each figure compares the empirical standard deviation for a fixed window size to the theoretical standard deviation. We have plotted an empirical standard deviation for all n from 10 seconds to 20 minutes in step sizes of 10 seconds. Each empirical standard deviation corresponds to a single point in the scatter plot and the plotted curve corresponds to the predicted theoretical value.

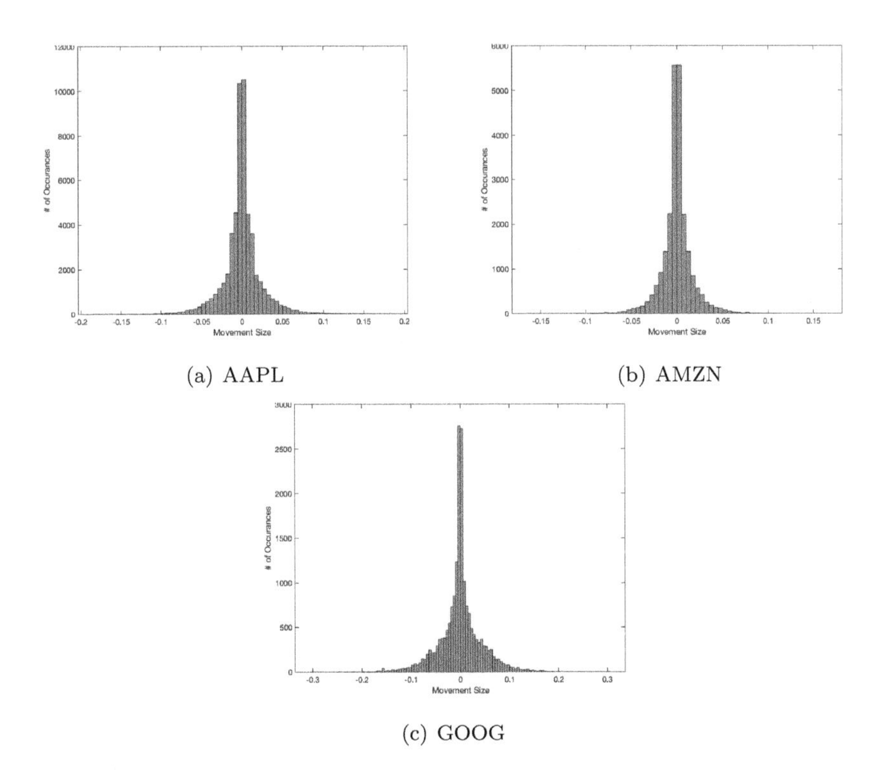

(a) AAPL

(b) AMZN

(c) GOOG

FIGURE 6.4
We can see clearly that the change in the mid-price is often larger than a half tick. These mid-price changes make up a significant portion of the actual data, contradicting the assumption needed for the CHPDO model that the mid-price changes occur on average at a half tick size.

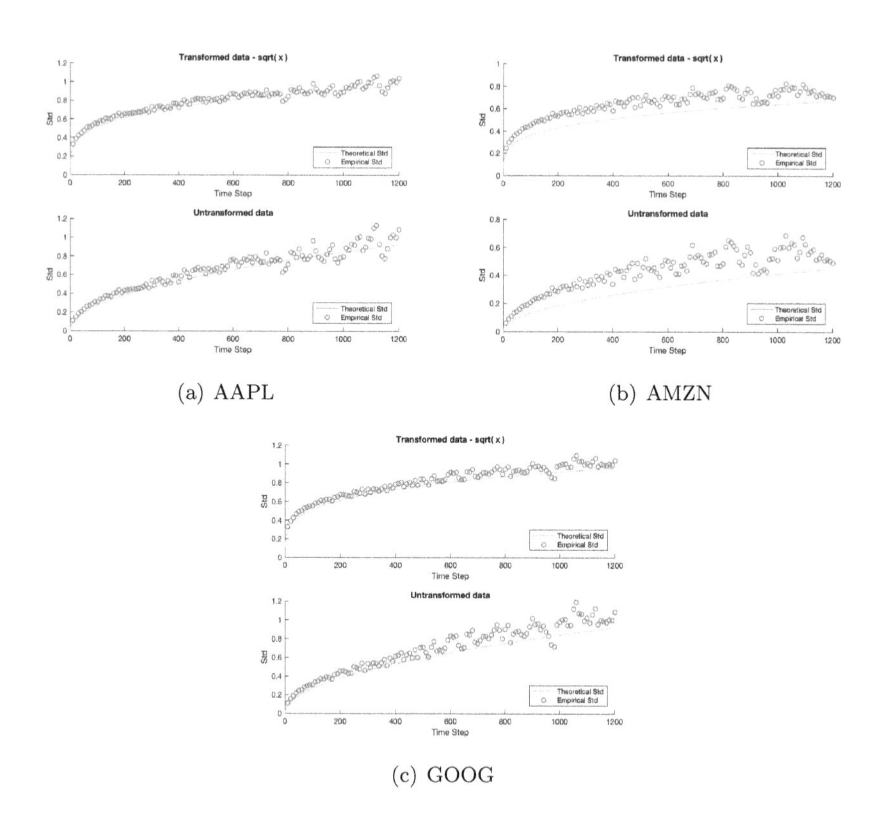

(a) AAPL

(b) AMZN

(c) GOOG

FIGURE 6.5
A comparison of the empirical standard deviation for a fixed window size n to the theoretical standard deviation for AAPL, AMZN and GOOG using the 2 state dependent order model. We have plotted the empirical standard deviation for all n from 10 seconds to 20 minutes in step sizes of 10 seconds. Each empirical standard deviation corresponds to a single point in the scatter plot and the plotted curve corresponds to the predicted theoretical value. Visually there is a significant improvement for all stocks, although the theoretical standard deviation for AMZN is still underestimating the empirical variability.

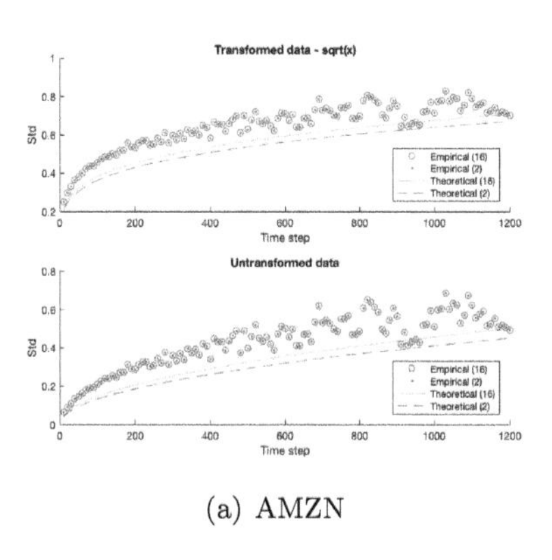

(a) AMZN

FIGURE 6.6

We consider the N-state model for AMZN discussed previously in the paper. While there is a slight improvement against the original fit, the model still struggles to perfectly predict the variability in the mid-price changes of our data.

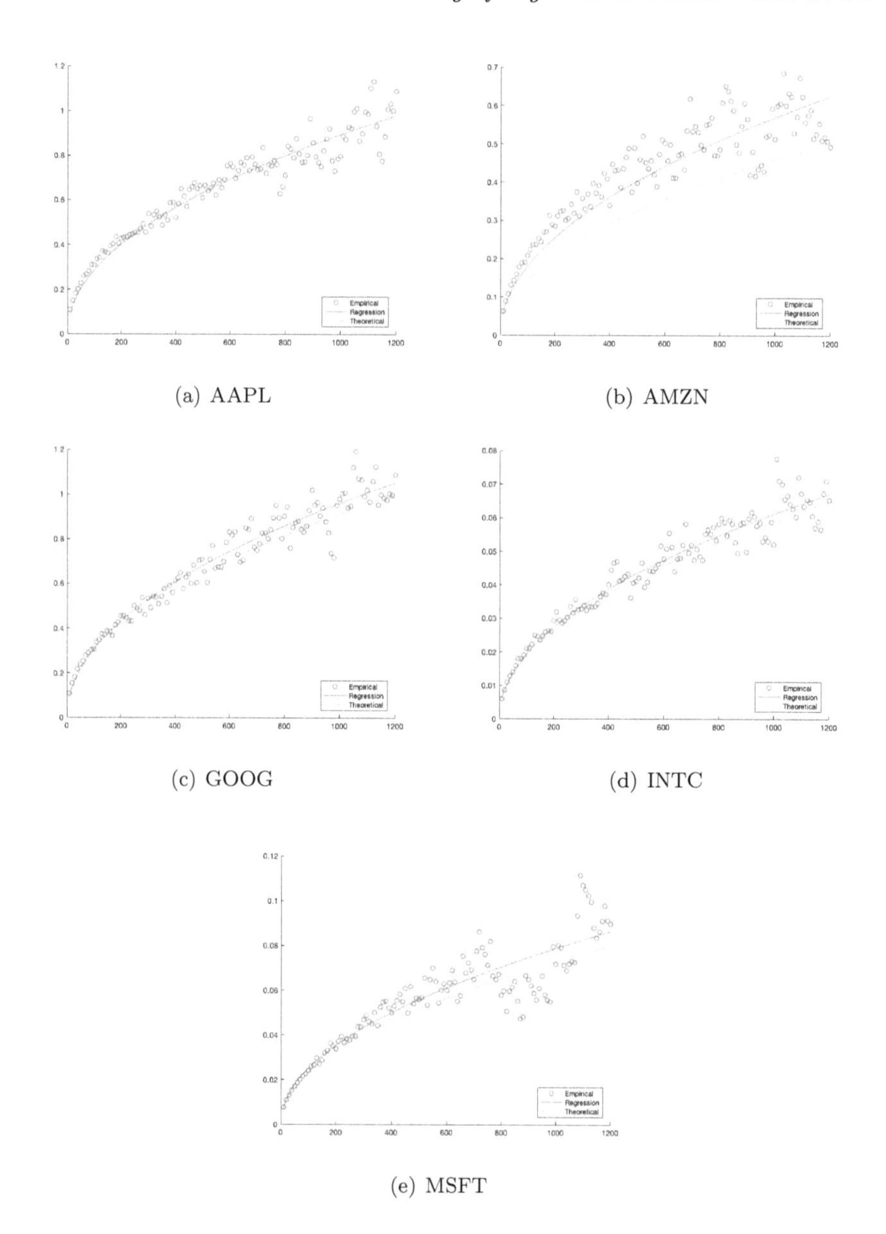

(a) AAPL

(b) AMZN

(c) GOOG

(d) INTC

(e) MSFT

FIGURE 6.7
A qualitative comparison of the regression to the theoretical model. For APPL, AMZN and GOOG we have used a Markov chain generated from 16 quantiles taken on the upward movements and downward movements. For INTC and MSFT we have taken the CHPDO coefficient since a different coefficient is not possible with the other models.

6.4 Conclusion

This chapter provided the Law of Large Numbers (LLN) and a Functional Central Limit Theorems (FCLT) for several specific variations of GCHP introduced in Chapter 4. We applied these FCLTs to limit order books to study the link between price volatility and order flow, where the volatility in mid-price changes is expressed in terms of parameters describing the arrival rates and mid-price process. We also presented empirical results based on real data.

Bibliography

[1] Ait-Sahalia, Y., Cacho-Diaz, J. and Laeven, R. (2010): Modelling of financial contagion using mutually exciting jump processes. *Tech. Rep.*, 15850, Nat. Bureau of Ec. Res., USA.

[2] Buffington, J., Elliott, R. J. (2000): Regime Switching and European Options. Lawrence, K.S. (ed.) *Stochastic Theory and Control.* Proceedings of a Workshop, 73–81. Berlin Heidelberg New York: Springer. (2002).

[3] Buffington, J., Elliott, R.J. (2002): American Options with Regime Switching. *International Journal of Theoretical and Applied Finance* 5, 497–514.

[4] Brémaud, P. and Massoulié, L. (1996): Stability of nonlinear Hawkes processes. The Annals of Probab., 24(3), 1563.

[5] Bacry, E., Mastromatteo, I. and Muzy, J.-F. (2015): Hawkes processes in finance. arXiv:1502.04592v2 [q-fin.TR] 17 May 2015.

[6] Bowsher, C. (2007): Modelling security market events in continuous time: intensity based, multivariate point process models. *J. Econometrica*, 141 (2): 876–912.

[7] Bauwens, L. and Hautsch, N. (2009): *Modelling Financial High Frequency Data Using Point Processes.* Springer.

[8] Bartlett, M. (1963): The spectral analysis of point processes. *J. R. Stat.* Soc., ser. B, 25 (2), 264–296.

[9] Cartea, A., Jaimungal, S. and Ricci, J. (2011): Buy low, sell high: a high-frequency trading prospective. *Tech. Report.*

[10] Cartea, Á., Jaimungal, S. and Penalva, J. (2015): *Algorithmic and High-Frequency Trading.* Cambridge University Press.

[11] Cartensen, L. (2010): *Hawkes processes and combinatorial transcriptional regulation.* PhD Thesis, University of Copenhagen.

[12] Chavez-Demoulin, V. and McGill, J. (2012): High-frequency financial data modelling using Hawkes processes. *J. Banking and Finance*, 36(12), 3415–3426.

[13] Chavez-Casillas, J., Elliott, R., Remillard, B. and Swishchuk, A. (2017): A level-1 limit order book with time dependent arrival rates. Proceed. IWAP, Toronto, June-20-25. Also available on arXiv: https://arxiv.org/submit/1869858

[14] Cohen, S. and Elliott, R. (2014): Filters and smoothness for self-exciting Markov modulated counting process. *IEEE Trans. Aut. Control.*

[15] Cont, R. and de Larrard, A. (2013): A Markovian modelling of limit order books. *SIAM J. Finan. Math.*

[16] Cox, D. (1955): Some statistical methods connected with series of events. *J. R. Stat.Soc.*, ser. B, 17 (2), 129–164.

[17] Daley, D.J. and Vere-Jones, D. (1988): *An Introduction to the theory of Point Processes.* Springer.

[18] Dassios, A. and Zhao, H. (2011): A dynamic contagion process. *Advances in Applied Probab.*, 43(3), 814–846.

[19] Engel, R. and Russel, J. (1998): Autoregressive conditional duration: A new model for irregulary spaced transaction data. *Econometrica*, 66, 1127–1162.

[20] Engle, R. (2000): The econometrics of ultra-high-frequency data. *Econometrica*, 68: 1–20.

[21] Engle, R. and Large, J. (2001): Predicting vnet: a model of the dynamics of market depth. *J. Finan. Markets*, 4, 113–142.

[22] Engle, R. and Lunde, A. (2003): Trades and quotes: a bivariate point process. *J. Finan. Econom.*, 1 (2), 159–188.

[23] Embrechts, P., Liniger, T. and Lin, L. (2011): Multivariate Hawkes processes: an application to financial data. *J. Appl. Prob.*, 48, A: 367– 378.

[24] Errais, E., Giesecke, K. and Goldberg, L. (2010): Affine point processes and portfolio credit risk. *SIAM J. Fin. Math.* 1: 642–665.

[25] Fillimonov, V., Sornette, D., Bichetti, D. and Maystre, N. (2013): Quantifying of the high level of endogeneity and of structural regime shifts in comodity markets, 2013.

[26] Fillimonov, V. and Sornette, D. (2012): Quantifying reflexivity in financial markets: Toward a prediction of flash crashes. *Physical Review E*, 85(5):056108.

[27] Hawkes, A. (1971): Spectra of some self-exciting and mutually exciting point processes. *Biometrica*, 58, 83–90.

[28] Hawkes, A. and Oakes, D. (1974): A cluster process representation of a self-exciting process. *J. Applied Probab.*, 11: 493–503.

[29] Hewlett, P. (2006): Clustering of order arrivales, price impact and trade-path optimization.

[30] Hasbrouch, J. (1999):Trading fast and slow: security market events in real time. *Tech. Report.*

[31] Korolyuk, V. S. and Swishchuk, A. V. (1995): Semi-Markov Random Evolutions. *Kluwer Academic Publishers,* Dordrecht, The Netherlands.

[32] Large, J. (2007): Measuring the resiliency of an electronic limit order book. *J. Fin. Markets*, 10(1):1–25.

[33] Liniger, T. (2009): Multivariate Hawkes Processes. PhD thesis, Swiss Fed. Inst. Tech., Zurich.

[34] Lewis, P. (1964): J. R. *Stat. Soc.*, ser. B, 26 (3), 398.

[35] Laub, P., Taimre, T. and Pollett, P. (2015): Hawkes Processes.arXiv: 1507.02822v1[math.PR]10 Jul 2015.

[36] McNeil, A., Frey, R. and Embrechts, P. (2015): *Quantitative Risk Management: Concepts, Techniques and Tools.* Princeton University Press.

[37] Mehdad, B. and Zhu, L. (2014): On the Hawkes process with different exciting functions. *arXiv: 1403.0994.*

[38] Norris, J. R. (1997): Markov Chains. In Cambridge Series in Statistical and Probabilistic Mathematics. UK: Cambridge University Press.

[39] Skorokhod, A. (1965): Studies in the Theory of Random Processes, Addison-Wesley, Reading, Mass., (Reprinted by Dover Publications, NY).

[40] Swishchuk, A. (2017a): Risk model based on compound Hawkes process. Abstract, IME 2017, Vienna (https://fam.tuwien.ac.at/ime2017/program.php).

[41] Swishchuk, A. (2017b): Risk model based on compound Hawkes process. arXiv: http://arxiv.org/abs/1706.09038

[42] Swishchuk, A. (2017c): General Compound Hawkes Processes in Limit Order Books. Working Paper, U of Calgary, 32 pages, June 2017. Available on arXiv: http://arxiv.org/abs/1706.07459

[43] Swishchuk, A., Chavez-Casillas, J., Elliott, R. and Remillard, B. (2017): Compound Hawkes processes in limit order books. Available on SSRN: https://papers.ssrn.com/sol3/papers.cfm?abstract_id=2987943

[44] Swishchuk, A. and Vadori, N. (2017): A semi-Markovian modelling of limit order markets. SIAM J. Finan. Math., 8, 240–273.

[45] Swishchuk, A., Cera, K., Hofmeister, T. and Schmidt, J. (2017): General semi-Markov model for limit order books. Intern. *J. Theoret. Applied Finance*, 20, 1750019.

[46] Swishchuk, A. and Vadori, N. (2015): Strong law of large numbers and central limit theorems for functionals of inhomogeneous Semi-Markov processes. *Stochastic Analysis and Applications*, 13 (2), 213–243.

[47] Swishchuk, A. and Huffman, A. (2018): General Compound Hawkes Process in Limit Order Books. *Risks* 2020, 8(1), 28; https://doi.org/10.3390/risks8010028

[48] Vinkovskaya, E. (2014): *A point process model for the dynamics of LOB.* PhD thesis, Columbia University.

[49] Zheng, B., Roueff, F. and Abergel, F. (2014): Ergodicity and scaling limit of a constrained multivariate Hawkes process. *SIAM J. Finan.* Math., 5.

[50] Zhu, L. (2013): Central limit theorem for nonlinear Hawkes processes. *J. Appl. Prob.*, 50(3), 760–771.

[51] G.O. Mohler, M.B. Short, P.J. Brantingham, F.P. Schoenberg, G.E. Tita, (2011). *Journal of the American Statistical Association* 106(493), 100.

[52] S. Azizpour, K. Giesecke, G. Schwenkler. (2010) Exploring the sources of default clustering. http://web.stanford.edu/ dept/MSandE/cgi-bin/people/faculty/giesecke/pdfs/exploring.pdf.

[53] Ogata, Y. (1988): *Journal of the American Statistical Association* 83(401), 9.

[54] LOBSTER, limit order book data. https://lobsterdata.com/

[55] E. Bacry, S. Delattre, M. Hoffmann, and J. F. Muzy. (2013) Modelling microstructure noise with mutually exciting point processes. *Quantitative Finance*, 13(1):65–77.

[56] Paul Embrechts, Thomas Liniger, and Lu Lin. (2011) Multivariate hawkes processes: an application to financial data. *Journal of Applied Probability*, 48(A):367–378.

7

Quantitative and Comparative Analyses of Big Data with GCHP

This chapter solves the problem of quantitative and comparative analysis of modelling big data arising in high-frequency and algorithmic trading using real data. Namely, we introduce different new types of General Compound Hawkes Processes (GCHPDO, GCHP2SDO, GCHPnSDO) and find their diffusive limits to model the mid price movements of 6 stocks, EBAY, FB, MU, PCAR,SMH, CSCO, (provided by Cartea et al. (2015) book [14]. We also define error rates to estimate the models fitting accuracy. Maximum Likelihood Estimation (MLE) and Particle Swarm Optimization (PSO) are used for Hawkes processes and models parameters' calibration. See [16].

7.1 Introduction

This chapter introduces new different types of general compound Hawkes processes (GCHP), namely GCHPDO (two fixed ticks, $+\delta$, $-\delta$, dependent orders), GCHP2SDO (two non-fixed ticks, $(a(1), a(2))$, dependent orders) and GCHP-nSDO (n non-fixed ticks, $(a(1), ..., a(n))$, dependent order) to model the mid prices S_t dynamics of the assets in HFT, namely, EBAY, FB, MU, PCAR, SMH, CSCO, (provided by Cartea et al. (2015) book [14].

As we mentioned, high-frequency trading happens in milliseconds, as order arrivals and cancellations are very frequent. How we can study and model the dynamics of the mid-prices? One of the ways is to look over a larger time scale, e.g., 5, 10 or 20 minutes, i.e., consider time scale nt instead of t, then n could be $n = 100, 1000, .., $ etc. It means that we consider the dynamics of order flow over large time scale. Thousands of order book events may occur over such large time scales (e.g., for CISCO data on one day in 2014 it is around $500,000$, see Cartea et al. [14]. Thus, we can use asymptotic methods to study the link between order flow and price volatility by considering the diffusive limit of the mid-price processes. More precisely, we use the functional central limit theorems for above-mentioned GCHP, and present the volatility of price changes in terms of parameters of initial models.

DOI: 10.1201/9781003265986-7

To estimate the models fitting accuracy we define an error rate for each model, and set the threshold value as 15%. The model with error rate less than the threshold is considered as well fitted. Thus, we define which of our model is the best fit for our real data.

We also use Maximum Likelihood Estimation (MLE) and Particle Swarm Optimization (PSO) for Hawkes processes and our parameters' calibration.

There are many papers devoted to the modelling of HFT and applications of Hawkes processes in finance. See Bacry et al. [11] for more details. Below we give an overview of the most relevant and close to the present paper literature.

Cont and de Larrard [10] proposed a simple Markovian stochastic model for the dynamics of a limit order book, in which arrivals of market orders, limit orders, and order cancellations are independent, and inter-arrival times have exponential distribution. They also studied the diffusion limit of the price process and expressed the volatility of price changes in terms of parameters describing the arrival rates of buy and sell orders and cancellations.

As suggested by empirical observations, Swishchuk and Vadori [6] extended their framework to: (1) arbitrary distributions for book events inter-arrival times (possibly non-exponential) and (2) both the nature of a new book event and its corresponding inter-arrival time depend on the nature of the previous book event. They did so by resorting to Markov renewal processes to model the dynamics of the bid and ask queues. They kept analytical tractability via explicit expressions for the Laplace transforms of various quantities of interest. They also justified and illustrated their approach by calibrating their model to the five stocks Amazon, Apple, Google, Intel, Microsoft on June 21, 2012, to the 15 stocks from Deutsche Boerse Group (September 23, 2013) and to CISCO asset (November 3, 2014). As in Cont and A. de Larrard [10], the bid-ask spread remains constant equal to one tick, only the bid and ask queues are modelled (they are independent from each other and get reinitialized after a price change), and all orders have the same size. We discussed possible extensions of our model for the case when the spread is not fixed, including the diffusion limit of the price dynamics in this case, and we also discussed stochastic optimal control, and market making problems.

The paper by Swishchuk, Hoffmeister, Cera and Schmidt [7] and Chapter 3 considered a general semi-Markov model for limit order books with two states that incorporates price changes that are not fixed to one tick. Furthermore, even more general case of the semi-Markov model for limit order books was introduced that incorporates an arbitrary number of states for the price changes. For both cases the justifications, diffusion limits, implementations and numerical results were presented for different limit order book data: Apple, Amazon, Google, Microsoft, Intel on June 21, 2012 and Cisco, Facebook, Intel, Liberty Global, Liberty Interactive, Microsoft, Vodafone from November 3, 2014 to November 7, 2014. Chavez-Casillas, Elliott, Remillard and Swishchuk [15] proposed a simple stochastic model for the dynamics of a limit order book, extending the recent work of Cont and de Larrard [10], where the price dynamics are endogenous, resulting from market transactions.

They also shown that the conditional diffusion limit of the price process is the so-called Brownian meander.

Trading activity is not a completely random and memoryless process. That is why the Poisson process is not suitable for modelling trade arrival times. Trading activity also show a clustering behaviour. These properties suggest the use of the Hawkes process, a point process mathematically defined by Hawkes (1971) [13], which is an extension of the classical Poisson process that possesses this clustering property. It explains the large number of works on trading activity and more generally high-frequency econometrics based on this process as a modelling framework. (See Bacry, Mastromatteo and Muzy [11] for applications of Hawkes processes in finance).

Da Fonseca and Zaatour [12] provided explicit formulas for the moments and the autocorrelation function of the number of jumps over a given interval for a self-excited Hawkes process. The estimation strategy was applied to trade arrival times for major stocks that show a clustering behaviour, a feature the Hawkes process can effectively handle. As the calibration is fast, the estimation was rolled to determine the stability of the estimated parameters. Also, the analytical results enable the computation of the diffusive limit in a simple model for the price evolution based on the Hawkes process. It determines the connection between the parameters driving the high-frequency activity to the daily volatility.

Swishchuk, Elliott, Remillard and Chavez-Casillas [8] introduced two new Hawkes processes, namely, compound and regime-switching compound Hawkes processes, to model the price processes in limit order books. They proved Law of Large Numbers and Functional Central Limit Theorems (FCLT) for both processes. The two FCLTs were applied to limit order books, where they used these asymptotic methods to study the link between price volatility and order flow in these two models by using the diffusion limits of these price processes. The volatilities of price changes were expressed in terms of parameters describing the arrival rates and price changes. They also presented some numerical examples based on CISCO data (November 3-7, 2014).

Swishchuk and Huffman [3] studied various new Hawkes processes. Specifically, they constructed general compound Hawkes processes and investigate their properties in limit order books. With regards to these general compound Hawkes processes, they proved a Law of Large Numbers (LLN) and a Functional Central Limit Theorems (FCLT) for several specific variations. Then they applied several of these FCLTs to limit order books in Lobster data (June 21, 2012) to study the link between price volatility and order flow, where the volatility in mid-price changes is expressed in terms of parameters describing the arrival rates and mid-price process. Quantitative and comparative analyses were performed for different models.

The rest of the chapter is organized as follows. Different type of Hawkes processes and diffusive limits for them are introduced in Section 7.2. Applications to limit order books are considered in Section 7.3, including data and their clustering features descriptions, QQ-plots and autocorrelations. Section 7.4 presents Hawkes process and different models' parameters calibration's results. Error measurements and comparative analysis are considered in Section 7.5. Section 7.6 concludes the chapter.

7.2 Theoretical Analysis

7.2.1 One-dimensional Hawkes Process

7.2.1.1 Definition

The one-dimensional Hawkes Process is a point process $N(t)$ which is characterized by its intensity function $\lambda(t)$

$$\lambda(t) = h(\lambda + \int_0^t \mu(t-s)dN(s))$$

where the constant λ is called the background intensity and the function $\mu(\cdot)$ is called the excitation function that satisfies $\int_0^\infty \mu(s)ds < 1$. In this equation, if $h(\cdot)$ is a non-linear function, then we call this Hawkes Process as One-dimensional Non-linear Hawkes Process.

Typically, in this paper, the point process $N(t)$ is represented as the cumulative number of mid-price changes up to current time t in the trading system. $h(\cdot)$ is set to be a linear function, i.e., $h(x) = x$, and $\mu(\cdot)$ is set to be an exponential function, i.e., $\mu(t) = \alpha e^{-\beta t}$. This typical setting is widely used in most of the academic research (see [1]). Therefore, in this paper, the intensity function of the One-dimensional Hawkes Process is

$$\lambda(t) = \lambda + \int_0^t \alpha e^{-\beta(t-s)}dN(s) \tag{7.1}$$

7.2.1.2 Calibration

In this paper, Maximum Likelihood Estimation (MLE) is used for callibation of One-dimensional Hawkes Process. Suppose that the Hawkes process is observed over some time period $[0,T] \supset [0,t_k]$. Then, the log-likelihood function of Hawkes process is (See [1],[2]),

$$l = \sum_{i=1}^{k} \log[\lambda + \alpha \sum_{j=1}^{i-1} e^{-\beta(t_i-t_j)}] - \lambda t_k + \frac{\alpha}{\beta}\sum_{i=1}^{k}[e^{-\beta(t_k-t_i)} - 1]$$

If we let $A(i) = \sum_{j=1}^{i-1} e^{-\beta(t_i-t_j)}$, for $i = \{2,3,...,k\}$, and $A(1) = 0$, then last equation can be transformed into

$$l = \sum_{i=1}^{k} \log[\lambda + \alpha A(i)] - \lambda t_k + \frac{\alpha}{\beta}\sum_{i=1}^{k}[e^{-\beta(t_k-t_i)} - 1]$$

However, in this case, the log-likelihood function is difficult to be solved manually and thus we need to use some computational methods (e.g., Partical Swarm Optimization (PSO)) to find the global optimization of the parameters λ, α, β.

7.2.2 General Compound Hawkes Process

7.2.2.1 Definition

Compound Poisson Process is a widely used model in risk management and finance, with some independently arriving jumps that follow the Poisson Process,

$$S_t = S_0 + \sum_{k=1}^{N(t)} a(X_k).$$

In this model, $N(t)$ is a Poisson Process, X_k is a continuous time n-state Markov Chain; $a(X_k)$ is a continuous and bounded mapping function on the state space $X := \{1, 2, ..., n\}$. If we change the model a little bit, we can get our new General Compound Hawkes Process

$$S_t = S_0 + \sum_{k=1}^{N(t)} a(X_k)$$

where $N(t)$ is a one-dimensional Hawkes Process with the intensity function mentioned in equation (7.1). In our case, this model is used for mid-price modelling and thus $N(t)$ represents the number of mir-price changes up to time t; $a(X_k)$ represents the price movements interval.

As we can see, the previous two models look the same at first sight, but they are different with respect to $N(t)$: it's Poisson process in the former, and it's Hawkes process in the latter.

From general to typical case, if X_k is an n-state Markov Chain, this model is called as General Compound Hawkes Process with n-state Dependent Orders (GCHPnSDO); if X_k is a 2-state Markov Chain, this model is called as General Compound Hawkes Process with 2-state Markov Chain (GCHP2SDO); if in the 2-state Markov Chain, the value of $a(X_k)$ is set to be the most typical case, i.e., the value of $a(X_k)$ is fixed, $a(X_k) = \delta$ or $a(X_k) = -\delta$, then this model is called as General Compound Hawkes Process with Dependent Orders (GCHPDO).

7.2.2.2 Diffusive Limit

Let X_k be an ergodic n-state Markov Chain with state space $X := \{1, ..., n\}$, i.e., $X_k = \{1, 2, ..., n\}$, and with ergodic probabilistic $(\pi_1^*, \pi_2^*, ..., \pi_n^*)$; S_t is defined in GCHPnSDO model. The diffusive limit of GCHPnSDO is derived in [3],

$$\frac{S_{nt} - N(nt)\hat{a}^*}{\sqrt{n}} \overset{n\to\infty}{\longrightarrow} \hat{\sigma}^* \sqrt{\lambda/(1-\hat{\mu})} W(t)$$

where $W(t)$ is a standard Wiener Process,

$$0 < \hat{\mu} := \int_0^{\infty} \mu(s)ds = \frac{\alpha}{\beta} < 1$$

$$\int_0^{\infty} s\mu(s)ds < \infty$$

$$\hat{a}^* := \sum_{i \in X} \pi_i^* a(i)$$

$$\hat{\sigma}^* := \sum_{i \in X} \pi_i^* v(i)$$

$$v(i) = b(i)^2 + \sum_{j \in X} (g(j) - g(i))^2 P(i,j) - 2b(i) \sum_{j \in X} (g(j) - g(i))P(i,j)$$

$$b(i) := a(i) - a^*$$

$$b = (b(1), b(2), ..., b(n))'$$

$$g := (P + \Pi - I)^{-1} b$$

In the last equation, P denotes the transition probability matrix of the Markov Chain, i.e., $P(i,j) = P(X_{k+1} = j \mid X_k = i)$; Π^* denotes the matrix of stationary distribution of the Markov Chain. When the power n of the transition probability matrix P is large enough, $P^n \stackrel{n \to \infty}{\longrightarrow} \Pi^*$, and all rows of Π^* will be the same.

The diffusive limit of GCHP2SDO and GCHPDO model can still use previous convergence for GCHPnSDO since they are the typical cases of GCHP-nSDO model. In Section 7.3, we will further introduce how to use the diffusive limit to estimate the correctness of the mid-price modelling by General Compound Hawkes Process.

7.3 Application

7.3.1 Limit Order Book

Limit order book (LOB) is just like a book that records financial big data in the market. Here, we consider two types of tradings, one is called market order (buy/sell) and the other one is called limit order (buy/sell). In the trading system, market order will be executed immediately when it arrives; limit order will wait for later execution and while waiting it can be cancelled. Therefore, limit order book is a collection of queued active limit orders awaiting execution or cancellation (see [4]).

Then we will give some definitions in LOB. Lot size means the smallest amount of the asset that can be traded within LOB. Tick size means the smallest permissible price interval between different orders within LOB (e.g.,

Suppose that the tick size in the trading system is 1 cent. Then, all the orders should be submitted in the exactly two decimal places). Bid price means the highest stated price among active buy orders at time t. Ask price means the lowest stated price among active sell orders at time t. Mid price is the price between the bid and ask price, i.e.,

$$Mid\ Price = \frac{Bid\ Price + Ask\ Price}{2}$$

7.3.2 Data

One remark should be made with respect to the choice of real data. In different papers we used different types of real data: LOBster (five stocks Amazon, Apple, Google, Intel, Microsoft on June 21, 2012), 15 stocks from Deutsche Boerse Group (September 23, 2013) and CISCO asset (November 3, 2014) in Swishchuk and Vadori [6], Swishchuk, Hofmeister, Cera and Schmidt [7], Chavez-Casillas, Elliott, Remillard and Swishchuk [15], respectively, to justify and illustrate our approach.

In the present chapter we decide to use another set of real data, namely EBAY, FB, MU, PCAR,SMH, CSCO, (provided by Cartea et al. (2015) book [14]), to check our methods and to justify and illustrate our approach. And as in our previous chapters, the outcomes of the present chapter show that the method is right and gives very good results.

Thus, in this chapter, the real LOB data is downloaded from http://sebastian.statistics.utoronto.ca/books/algo-and-hf-trading/data/.

For the mid price modelling in Section 3.5, the data we use in this chapter are EBAY20141110, FB20141110, MU20141110, PCAR20141110, SMH20141110, CSCO20141107. Consider for each trading day, it starts from 9:30 and ends at 16:00. In order to avoid the open and close auctions, we ignore the first and last 15 minutes, and thus we end with 6 trading hours per day, between 9:45 and 15:45. The number of mid price changes in a day for EBAY, FB, MU, PCAR, SMH, CSCO are 1255, 3988, 1756, 1008, 379, 943.

7.3.3 Descriptive Data Analysis

Before we work on our new General Compound Hawkes Process Model, we need to check the following questions first. Compared with Compound Poisson Process, we want to know whether we could reject Poisson Process; furthermore, we want to know whether there is a correlation between the mid-price changes; and then we plot the number of arrivals under a fixed time window to explore the features in the data.

7.3.3.1 QQ-plot

In Figures 7.1–7.6, the QQ-plot rejects Poisson Process since the inter-arrival time does not follow the exponential distribution. In Figures 7.1–7.6, the two

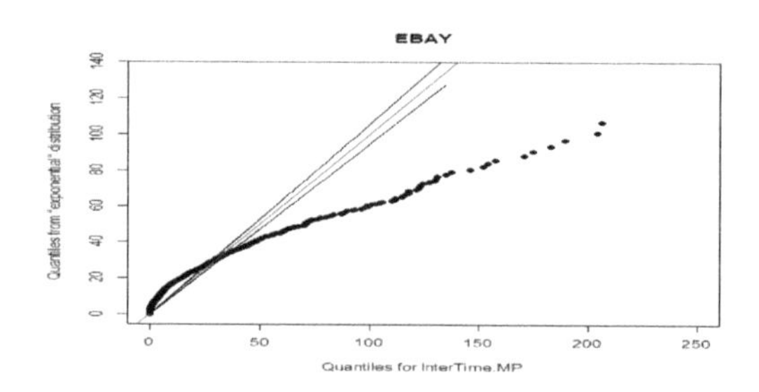

FIGURE 7.1
EBAY-QQplot.

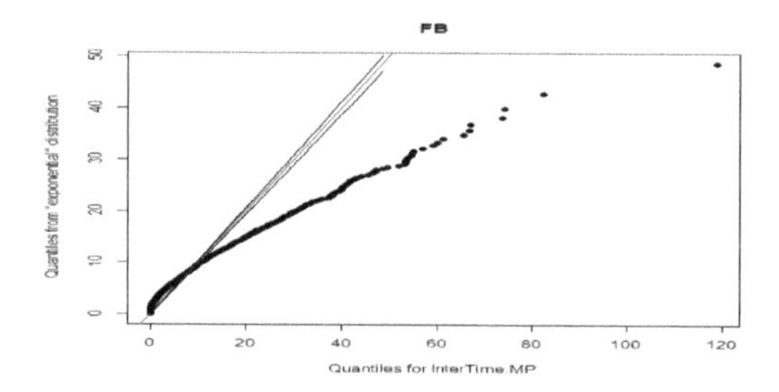

FIGURE 7.2
FB-QQplot.

black straight lines are the confidence boundary under 95% confidence level; the red straight line is $y = x$. If the inter-arrival time of the mid price changes follow the exponential distribution, all the black points should approximately be on the red line, within the confidence boundary. However, from the Figures 7.1–7.6, it is obvious to reject this assumption.

Furthermore, if the exponential distribution does not fit the inter-arrival times well, we want know which distribution will fit better. We tried normal

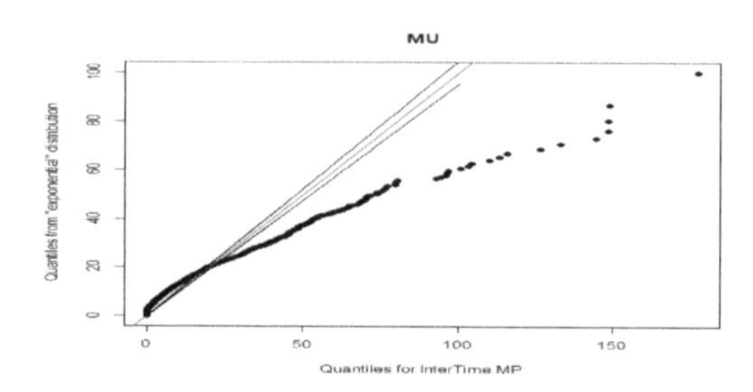

FIGURE 7.3
MU-QQplot.

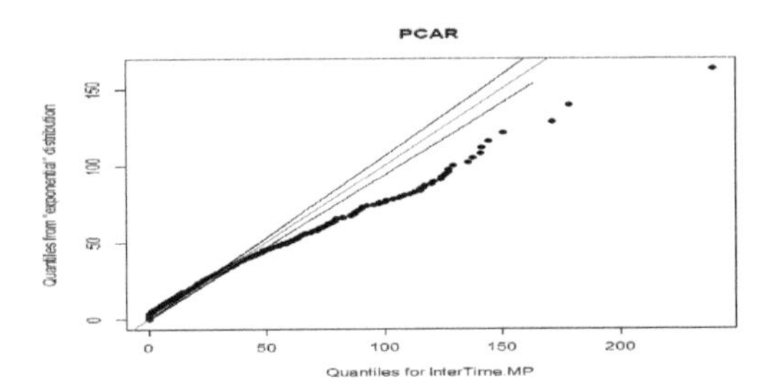

FIGURE 7.4
PCAR-QQplot.

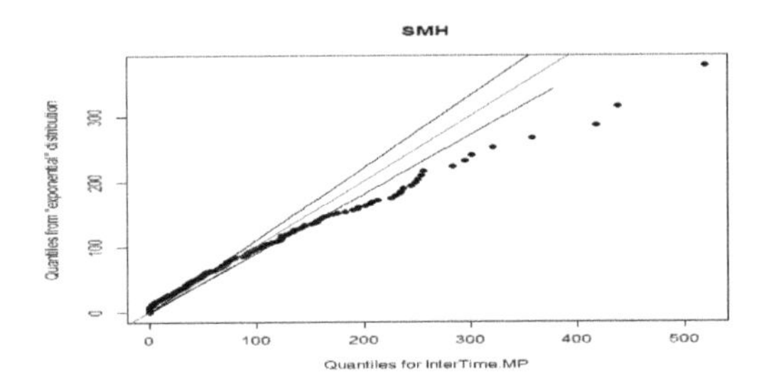

FIGURE 7.5
SMH-QQplot.

distribution, Gamma distribution, Weibull distribution, and then compared the theoretical Cumulative Density Function (CDF) with the empirical CDF (see Figures 7.7–7.12). From Figures 7.7–7.12, it is obvious that in most cases, Gamma or Weibull distribution performs much more better than the others since the curve of the theoretical CDF nearly coincides with that of empirical CDF.

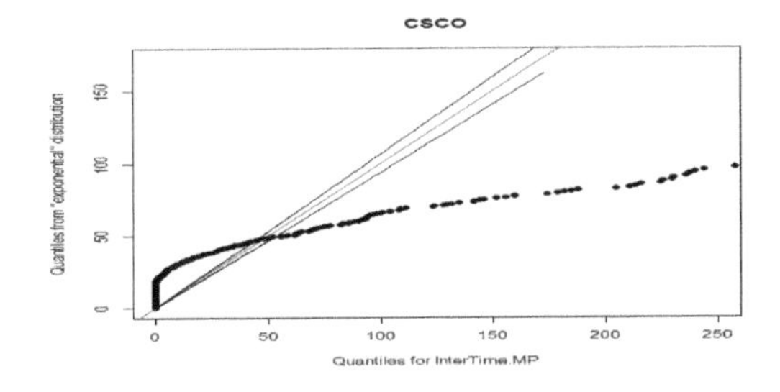

FIGURE 7.6
CSCO-QQplot.

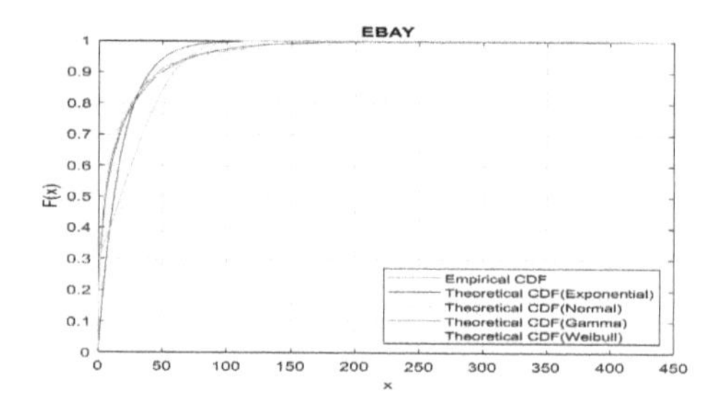

FIGURE 7.7
EBAY.

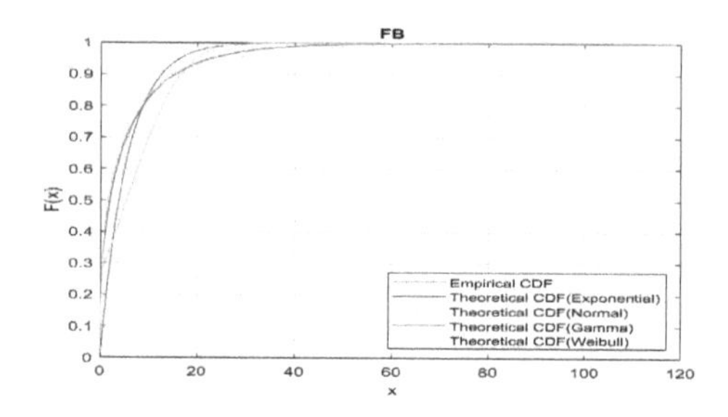

FIGURE 7.8
FB.

Of course, quantitative measures can be used as well to compare the different models, such as MLE of the Weibull and Gamma distributions' parameters as we did in Swishchuk and Vadori (2017), [5], for LOBster data or in Chavez-Cassilas et al (2019), [14])[15], for Facebook data. However, we gave here two different visualized comparisons, namely QQ-plots and CDFs, for 6 different stocks to illustrate our approach which proved to be right.

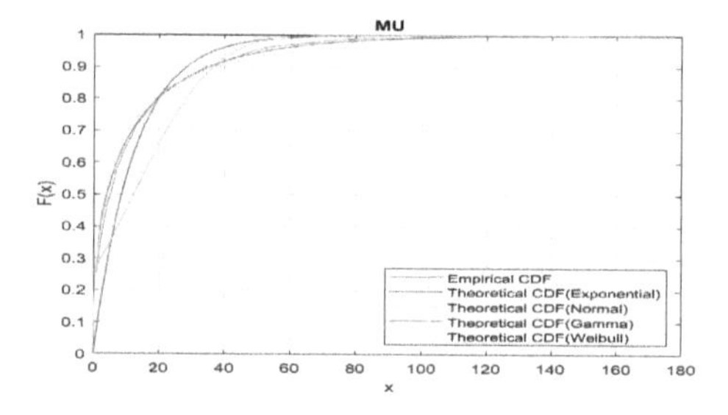

FIGURE 7.9
MU.

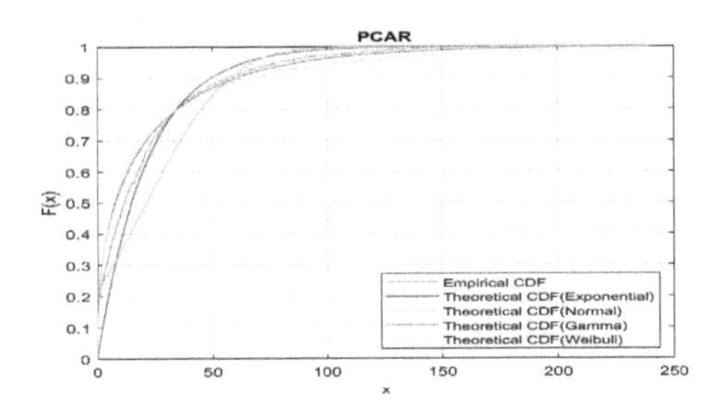

FIGURE 7.10
PCAR.

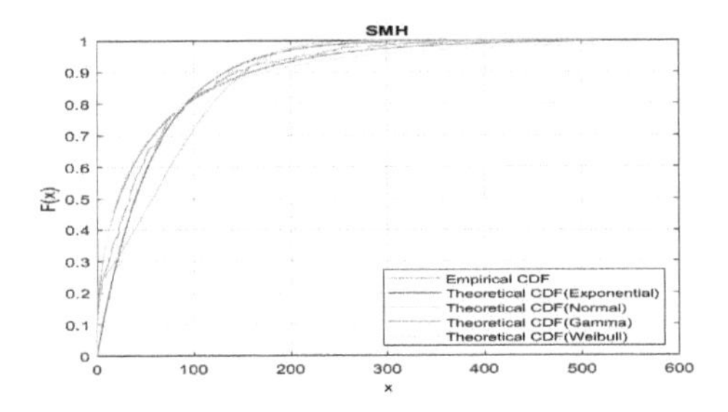

FIGURE 7.11
SMH.

7.3.3.2 Autocorrelation

To further confirm that the mid price changes are not independent, we calculated the autocorrelation by using equation

$$C(\tau, \delta) = \frac{E[(N_{t+\tau} - N_t)(N_{t+2\tau+\delta} - N_{t+\tau+\delta})] - E[(N_{t+\tau} - N_t)]E[(N_{t+2\tau+\delta} - N_{t+\tau+\delta})]}{\sqrt{var(N_{t+\tau} - N_t)var(N_{t+2\tau+\delta} - N_{t+\tau+\delta})}}$$

where τ is the length of the time interval, δ is the time lag, i.e., the length between two consecutive time intervals. In the last equation, if τ is fixed, then

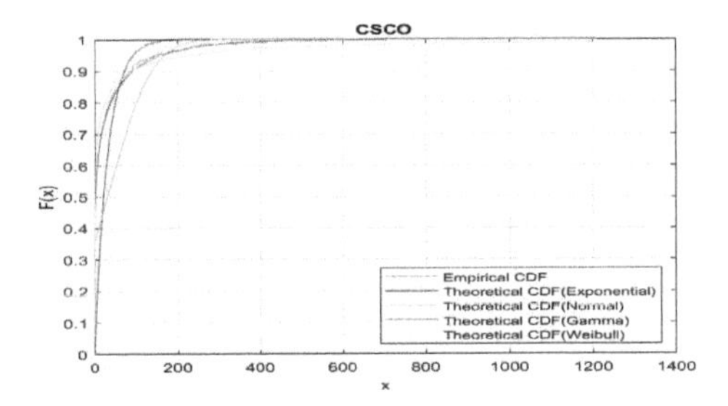

FIGURE 7.12
CSCO-20141107.

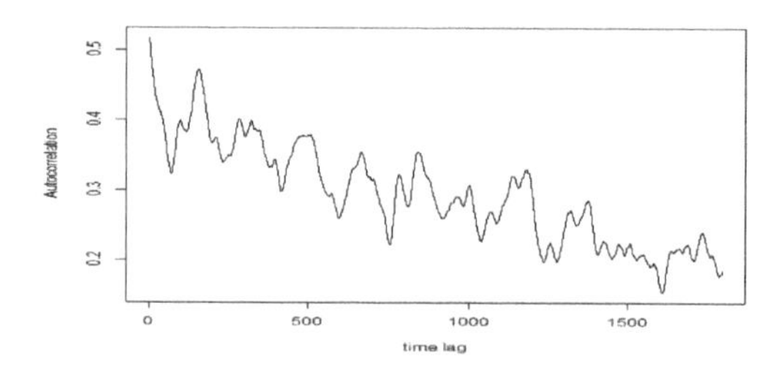

FIGURE 7.13
tao=20.

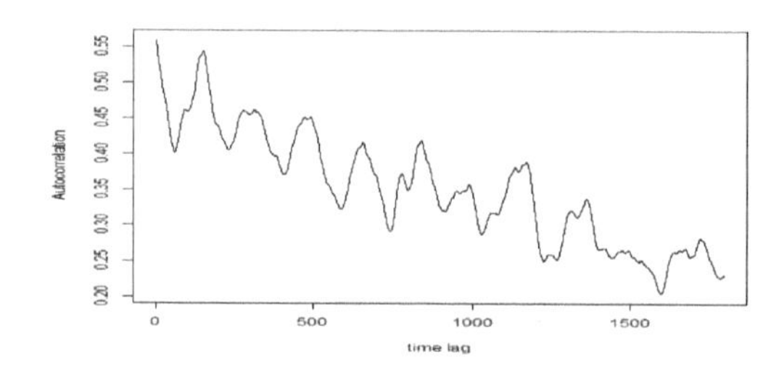

FIGURE 7.14
tao=30.

the autocorrelation function $C(\tau, \delta)$ will become a function that is respective to δ, i.e., $C(\delta)$. Take FB20141110 data as example. In the four plots shown in Figures 7.13–7.16, the length of the time interval is changed, (a) $\tau = 20$ seconds, (b) $\tau = 30$ seconds, (c) $\tau = 60$ seconds, (d) $\tau = 90$ seconds. In each graph, the X-coordinate represents the time lag δ which is changed from 1 second to 30 minutes by a step of 1 second.

From Figures 7.13–7.16, it is obvious to see that the shape of the autocorrelation function are nearly identical even if the plot is noisier when τ decreases.

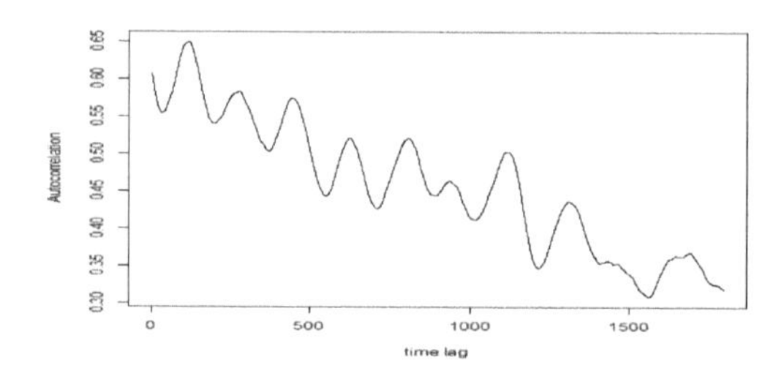

FIGURE 7.15
tao=60.

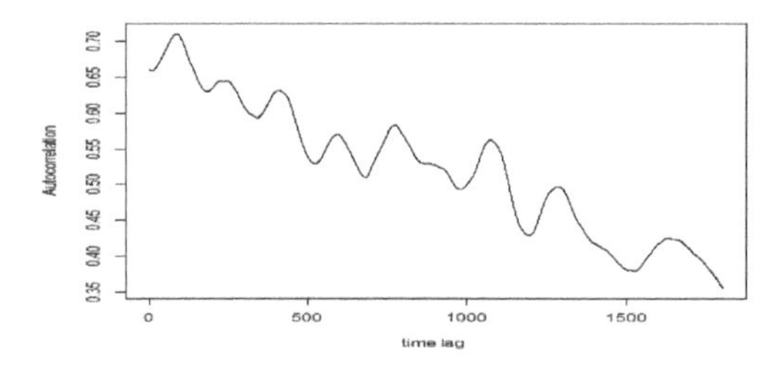

FIGURE 7.16
tao=90.

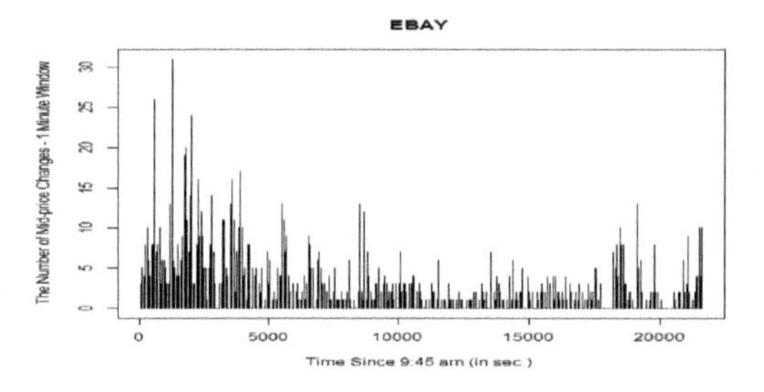

FIGURE 7.17
EBAY-clustering

Moreover, something more interesting could also be discovered from Figures 7.13–7.16. Even if $\delta = 30$ minutes, the value of the correlation function in each plot is still above 0.2 which is not neglectable. This means that suppose there is an arrival of mid price change at current time t, and then the impact of this arrival will last more than 30 minutes in the trading system, which is much longer than we can imagine.

7.3.3.3 Clustering Feature

At the end of the Descriptive Data Analysis, we plot the number of mid price changes occurring in every minute during the trading day. The clustering feature is obvious to be discovered in Figures 7.17–7.22.

From Sections 7.3.1–7.3.3, the main conclusion that we can make is trading activities are not a completely random and memoryless process.

7.4 Hawkes Process and Models Calibrations

In this section we calibrate Hawkes process' parameters λ, α, β (Section 7.1) and our mid-price S_t model's parameters $a(1), ..., a(n)$ (Section 4.2) for different GCHP (see Subsections 7.2.1–7.2.3) to find the price's volatility

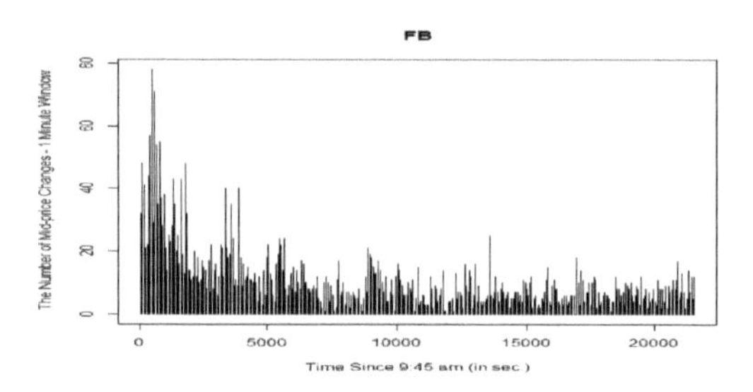

FIGURE 7.18
FB(clustering).

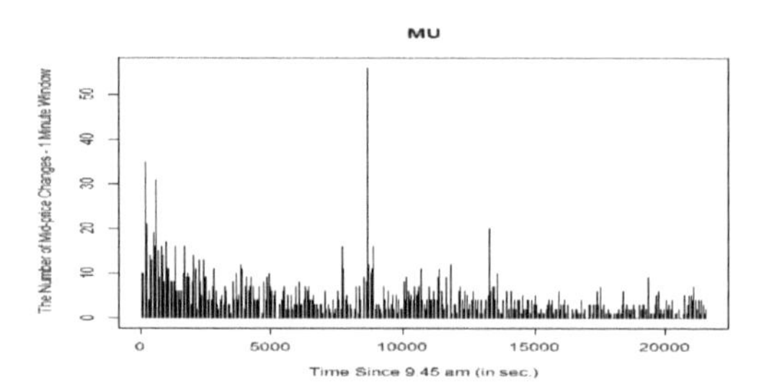

FIGURE 7.19
MU-clustering.

$\sigma^* \sqrt{\lambda/(1-\alpha/\beta)}$, obtained in Subsection 7.2.2.2. To estimate the models fitting accuracy we define in Section 7.6 an error rate for each model, and set the threshold value as 15%. The model with error rate less than the threshold is considered as well fitted. Thus, we define which of the model is the best fit for which real data.

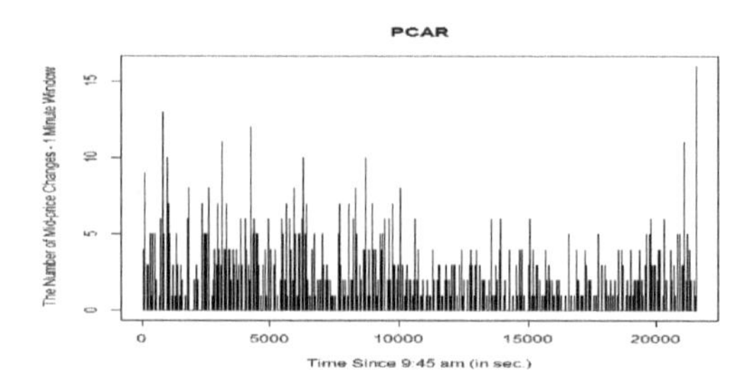

FIGURE 7.20
PCAR-clustering.

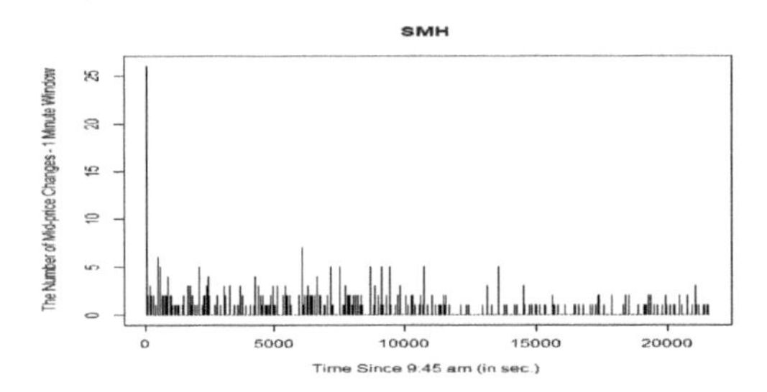

FIGURE 7.21
SMH-clustering.

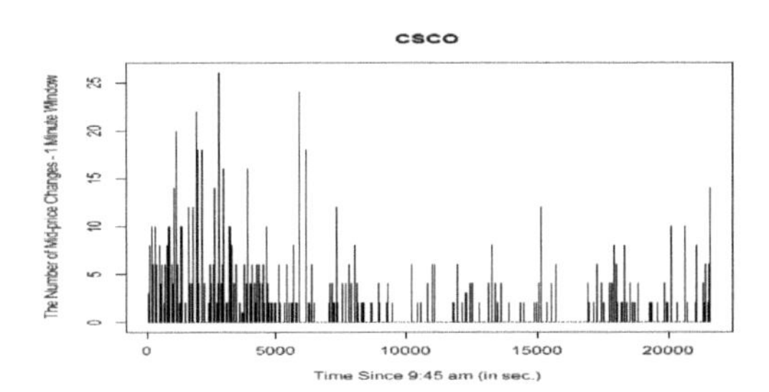

FIGURE 7.22
CSCO-20141107-clustering.

7.4.1 Hawkes Process' Parameters Calibration

Particle Swarm Optimization (PSO) method is used to solve the log-likelihood function in equation (4) and find the global optimization of the parameters $\hat{\lambda}$, $\hat{\alpha}$, $\hat{\beta}$. If the calibration result is appropriate, $E[N([0,1])] \to \frac{\hat{\lambda}}{1-\frac{\hat{\alpha}}{\hat{\beta}}}$ (see [1]), where $E[N([0,1])]$ is the empirical expectation of the number of mid price changes arriving in a unit time interval.

From Table 7.1, it is obvious to see that the values in the last two columns are super close, which means that our parameter estimation is appropriate.

7.4.2 Mid Price Modelling and Calibration

After Section 7.4, we can successfully model $N(t)$, the cumulative number of mid price changes up to current time t. Then, in this section, we will move forward to the mid price modelling by General Compound Hawkes Process. We will start from the most typical GCHPDO model and then move forward to the most general GCHPnSDO model. Under each model, the most important thing to figure out is he value of $a(X_k)$.

7.4.2.1 GCHPDO

In the case of GCHPDO model, the value of $a(X_k)$ can only take the fixed number. Suppose that $X_k \in \{-\delta, \delta\}$ and $a(x) = x$, i.e., $a(X_k) = -\delta$ or $a(X_k) = \delta$, where δ is the tick size of mid price in the trading system. In our case, suppose the tick size in the trading system is 1 cent. Whenever the ask/bid price goes up/down by 1 cent, the mid price will go up/down by 0.5 cent. Therefore, the tick size of the mid price movement is 0.5 cent.

Rearrange the equation in diffusive limit,

$$S_{nt} - N(nt)\hat{a}^* \overset{n\to\infty}{\longrightarrow} \hat{\sigma}^* \sqrt{n} \sqrt{\frac{\hat{\lambda}}{1 - \frac{\hat{\alpha}}{\hat{\beta}}}} W(t)$$

From 9:45am to 15:45pm, cut this 6 trading hours into some disjoint windows of size n, with $t = 1$, i.e., the disjoint time intervals are $[in, (i+1)n]$. Then, the left hand side of equation (14) can be discretized into

$$S_i^* = (S_{(i+1)n} - S_{in}) - (N((i+1)n) - N(in))\hat{a}^*$$

Combining last two equations, a converging formula between the standard deviation and its theoretical counterpart can be received,

$$std\{S_i^*\} \overset{n\to\infty}{\longrightarrow} \hat{\sigma}^* \sqrt{nt} \sqrt{\frac{\hat{\lambda}}{1 - \frac{\hat{\alpha}}{\hat{\beta}}}}, \quad t = 1$$

According to last equation, we can use this to estimate the correctness of the model fitting. If our model fitting is perfect, then

$$std\{S_i^*\} \approx \hat{\sigma}^* \sqrt{n} \sqrt{\frac{\hat{\lambda}}{1 - \frac{\hat{\alpha}}{\hat{\beta}}}}$$

Below we present GCHPDO model fitting results.

TABLE 7.1
Calibration result of Hawkes process.

Stock	$\hat{\lambda}$	$\hat{\alpha}$	$\hat{\beta}$	$E[N([0,1])]$	$\frac{\hat{\lambda}}{1-\frac{\hat{\alpha}}{\hat{\beta}}}$
EBAY	0.05142964	24.16091427	208.361411	0.058169396	0.05817548
FB	0.1460757	7.0368564	33.6354758	0.184708815	0.184721079
MU	0.06685776	8.49267842	47.38655865	0.081446361	0.081456495
PCAR	0.04105567	28.42153021	234.1878409	0.046720101	0.046726496
SMH	0.0157736	10.6972559	102.6908889	0.017602398	0.017607795
CSCO	0.0223269	642.6391460	1311.8684573	0.043657407	0.043766646

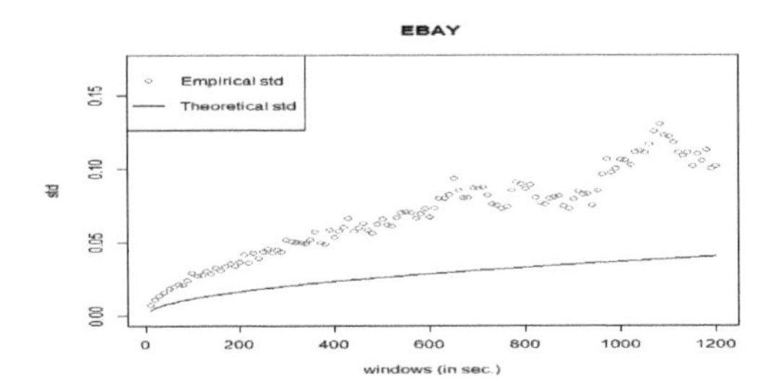

FIGURE 7.23
EBAY.

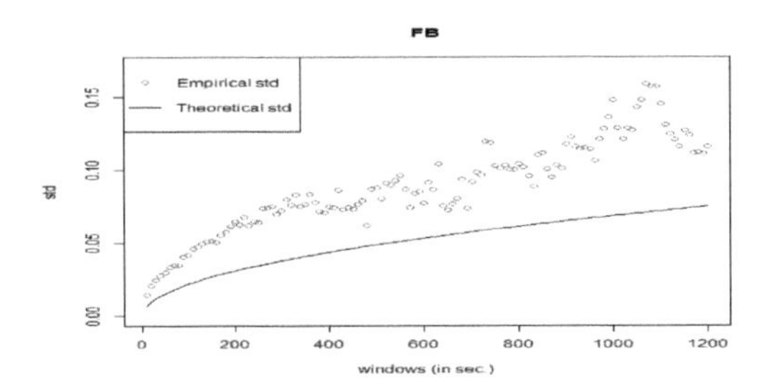

FIGURE 7.24
FB.

Figures 7.23–7.28 shows the empirical standard deviation against its theoretical counterpart for various window sizes, starting from 10 seconds to 20 minutes by the step of 10 seconds. Except for CSCO result, it is obvious to see that the differences between the empirical standard deviation and its theoretical counterpart are kind of large in the rest of five stocks. The reason for these unsatisfying results is that GCHPDO model assumes at each change, the mid

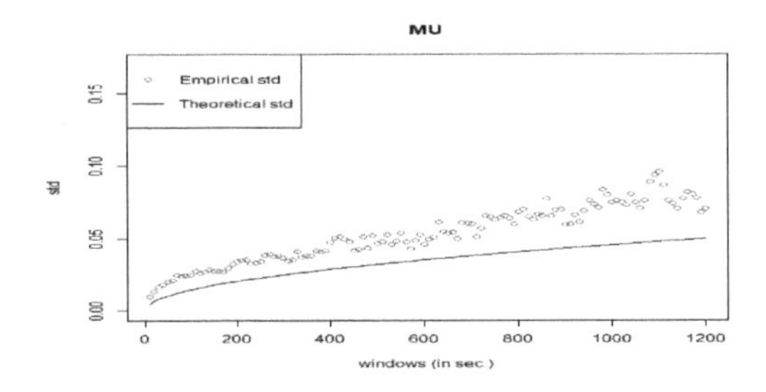

FIGURE 7.25
MU.

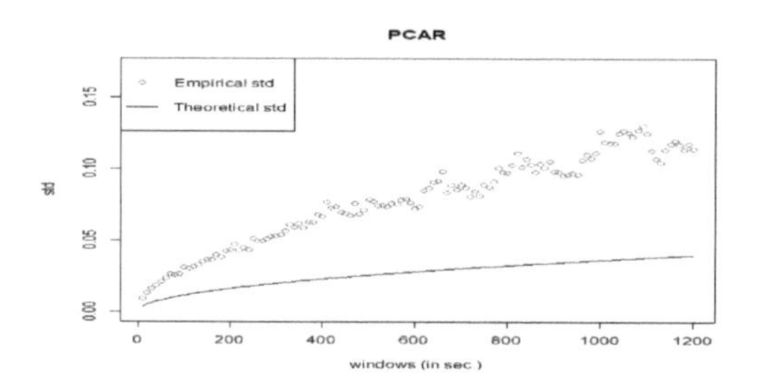

FIGURE 7.26
PCAR.

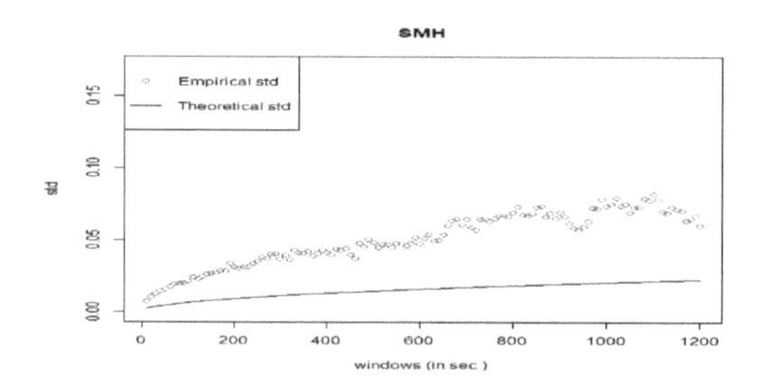

FIGURE 7.27
SMH.

price can only go up/down by tick size = 0.05 cent. However, in reality, only a small percentage of mid price movements change in this way (see Table 7.2).

7.4.2.2 GCHP2SDO

According to Section 7.5.1, GCHPDO is too special to model the mid price movements of stocks EBAY, FB, MU, PCAR, and SMH, and thus we need

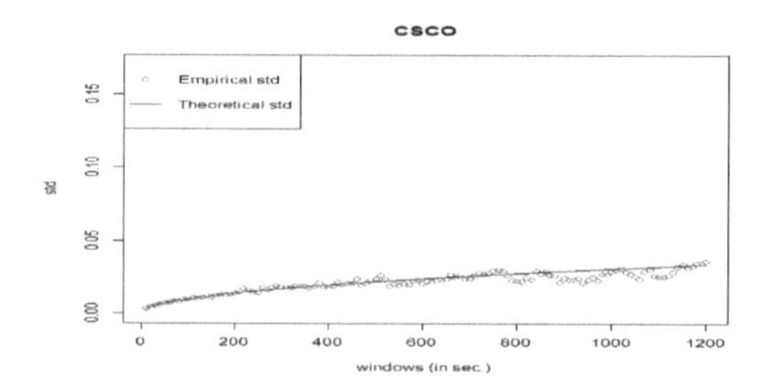

FIGURE 7.28
CSCO-20141107.

TABLE 7.2
Tick-size rate

Stock	Tick-size Rate
EBAY	21.35%
FB	18.18%
MU	9.85%
PCAR	22.22%
SMH	12.96%
CSCO	100%

TABLE 7.3
$a(X_k)$ for GCHP2SDO

Stock	a(1)	a(2)
EBAY	0.0091730474	−0.0092358804
FB	0.0095449949	−0.0093280239
MU	0.0098359729	−0.0096846330
PCAR	0.0116390041	−0.0108269962
SMH	0.0152972973	−0.0134196891

to consider some more general models. In GCHP2SDO model, the way we determine the value of $a(X_k)$ depends on the real data rather than just on the trading system. Take the mean of the upward and downward mid price movements and then assign them to $a(1)$ and $a(2)$. The result in Table 7.3 shows that in GCHP2SDO model, the mid price does not go up/down by exactly 0.5 cent at each change, which leads the model fitting become much better than GCHPDO model (see Figures 7.29–7.33).

Below we present GCHP2SDO Model Fitting Results.

7.4.2.3 GCHPnSDO

For upward or downward mid price movement, the above GCHPDO and GCHP2SDO model only consider 1 possible price interval since X_k is only

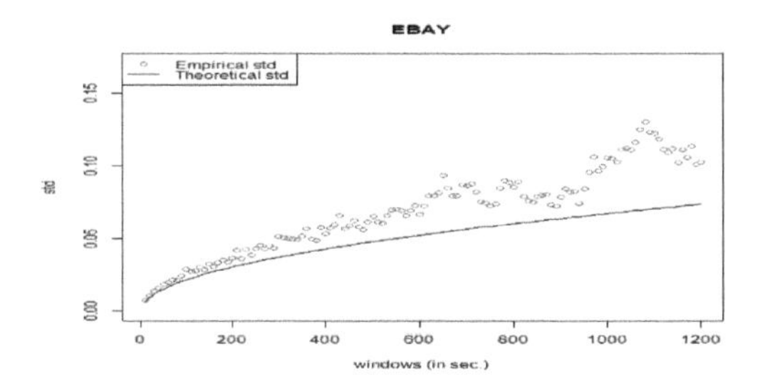

FIGURE 7.29
EBAY.

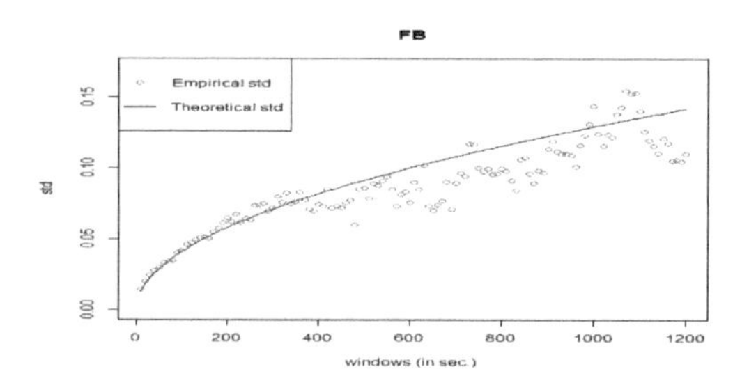

FIGURE 7.30
FB.

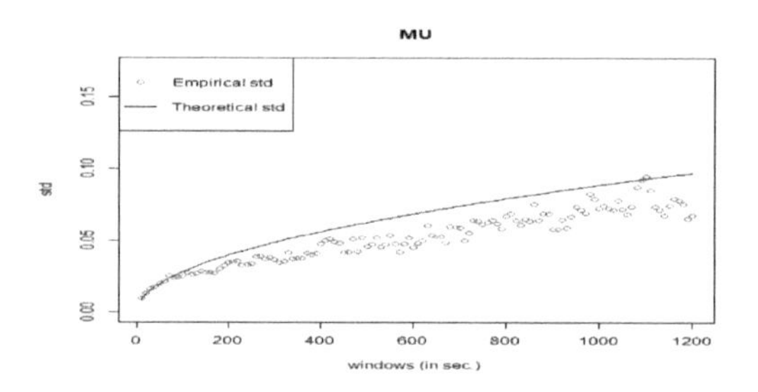

FIGURE 7.31
MU.

a 2-state Markov Chain. In this section, we are trying to figure out whether there is more than 1 possible price interval that could better fit the mid price movements than GCHP2SDO model.

Firstly, we need to figure out a way to determine the value of $a(X_k)$. **Step1:** After calculating the mid price changes, we separate the data into upward and downward movements. **Step 2:** we calculate the (evenly/unevenly) distributed quantiles for both data sets. Depending on the data, several quantiles may be identical and then we reject any duplicates. At the end of this

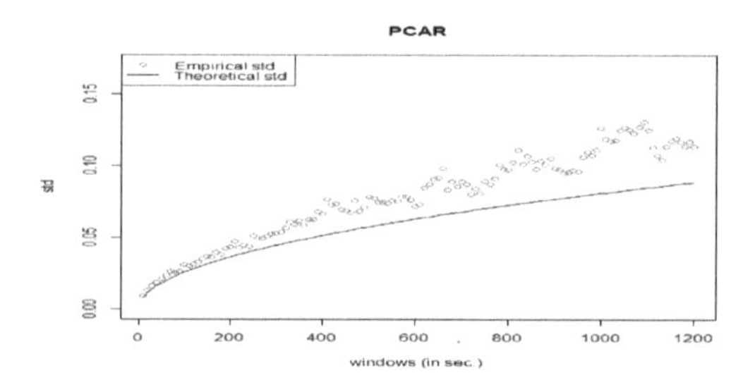

FIGURE 7.32
PCAR.

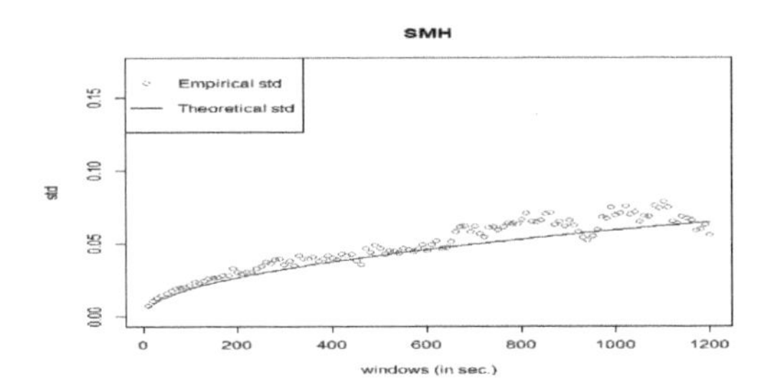

FIGURE 7.33
SMH.

step, we have gotten a list of bounds B_k, which we will use for determine the value of $a(X_k)$. **Step 3:** $a(X_k)$ = the average of all mid-price changes located between two neighbouring boundary values, i.e. $[B_k, B_{k+1})$. **Step 4:** There is an exception for the largest upper boundary, i.e., after step 3, we won't get the value for $a(X_n)$. And thus, we need to calculate the mean of the rest data manually and assign it to $a(X_n)$.

Secondly, we need to determine the best n for GCHPnSDO model. For different stocks, the Mean Square Errors (MSE) between the empirical standard deviation and its theoretical counterpart under varied number of states are shown in Table 7.3. We tried limited n, and then pick up the best n under the smallest MSE. For EBAY, the best GCHPnSDO model is GCHP8SDO model; for FB, the best GCHPnSDO model is GCHP4SDO model; for MU, the best GCHPnSDO model is GCHP6SDO model; for PCAR, the best GCHPnSDO model is GCHP15SDO model; for SMH, the best GCHPnSDO model is GCHP9SDO model.

In the end, compare the MSE of the best GCHPnSDO models with that of the previous GCHPDO and GCHP2SDO model. Again, pick up the best model for mid price movements under the smallest MSE (see Tables 7.4–7.7). Under the best model, the model fitting results can be seen in Figures 7.34–7.39.

Below we present Best Model Fitting Results

TABLE 7.4
n Selection for EBAY

n	4	5	6	7	8
MSE	0.00046701	0.000418703	0.000399311	0.000397898	0.000393683

TABLE 7.5
n Selection for FB

n	4	5	6	7	8
MSE	0.000317633	0.000330058	0.000344656	0.000349087	0.000361214

TABLE 7.6
n Selection for MU

n	4	6	8
MSE	0.000263588	0.000272237	0.000275675

TABLE 7.7
n Selection for PCAR

n	6	9	12	15
MSE	0.000238185	0.000132224	0.000129238	0.000124427

7.5 Error Measurement

Since the value of MSE is always very small under GCHPDO, GCHP2SDO, and GCHPnSDO model, it makes no sense to measure the correctness of model fitting under the best model only by MSE. Therefore, in this paper, we proposed a new way to measure the correctness. If we take the square of both

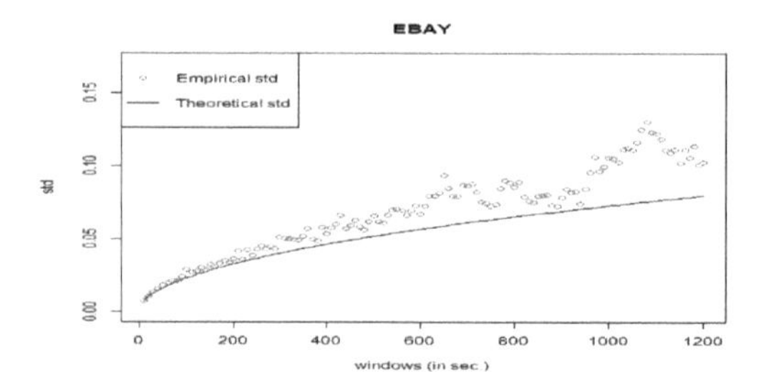

FIGURE 7.34
EBAY.

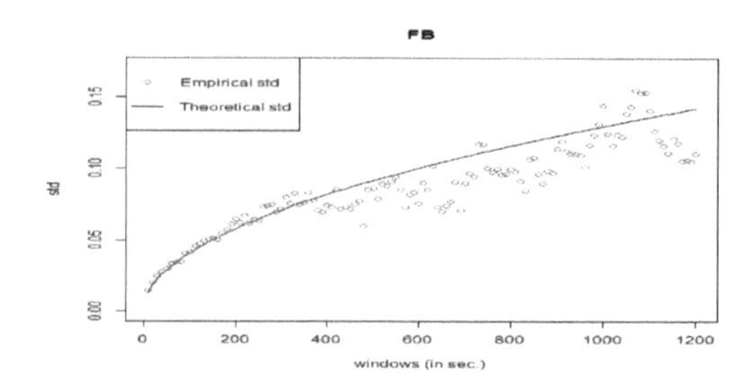

FIGURE 7.35
FB.

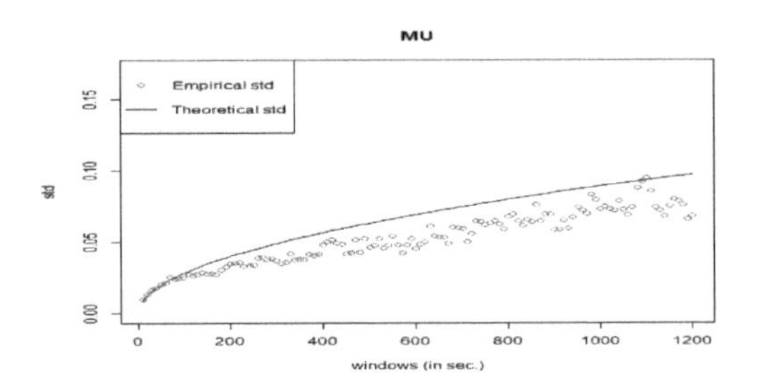

FIGURE 7.36
MU.

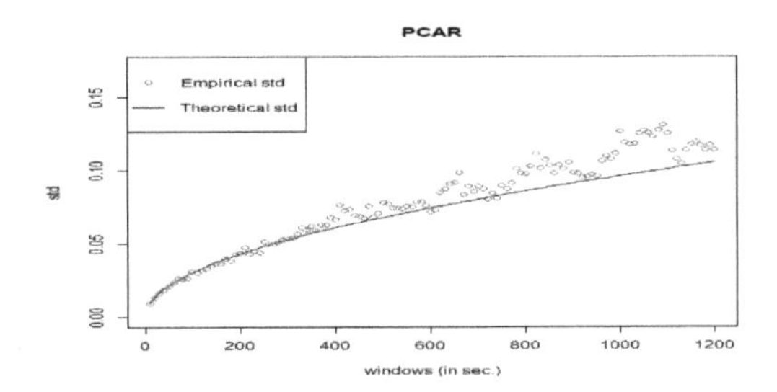

FIGURE 7.37
PCAR.

sides in equation (16), we can get

$$(std\{S_i^*\})^2 \overset{n\to\infty}{\longrightarrow} ((\hat{\sigma}^*)^2 \frac{\hat{\lambda}}{1-\frac{\hat{\alpha}}{\hat{\beta}}})nt,\ t=1$$

Last expression means that when n goes to infinity, the square of $std\{S_i^*\}$ converges to a linear regression function $L_1 = ((\hat{\sigma}^*)^2 \frac{\hat{\lambda}}{1-\frac{\hat{\alpha}}{\hat{\beta}}})n$ with respective

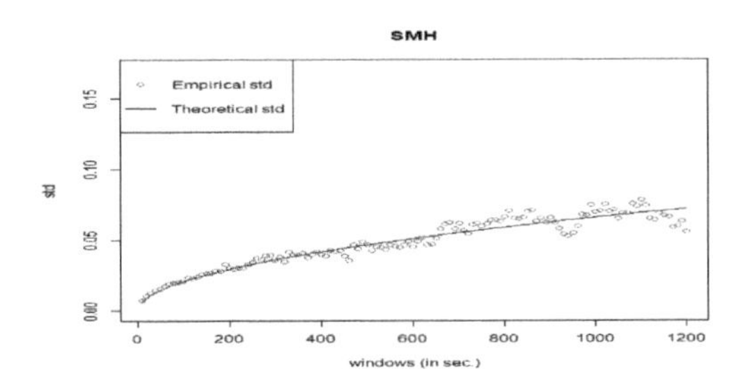

FIGURE 7.38
SMH.

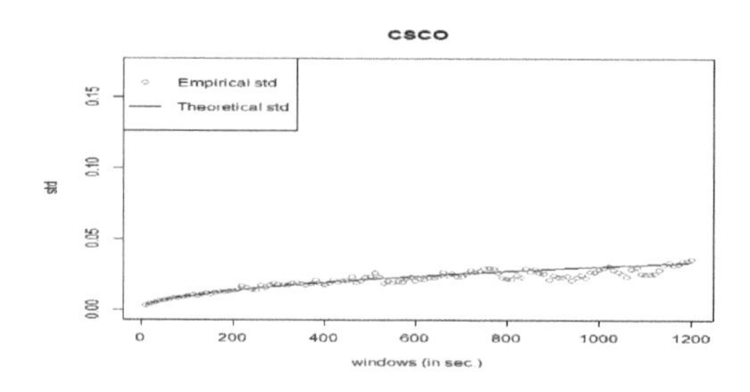

FIGURE 7.39
CSCO-20141107.

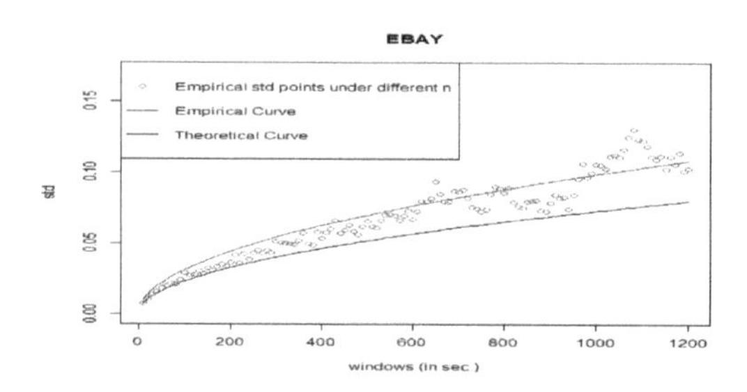

FIGURE 7.40
EBAY.

to n. Therefore, based on the real points $(std\{S_i^*\})^2$ under different n, we can estimate the linear regression function $L_1 = cn$ by least square method, where c is the estimated slope. Then, we take the square root of estimated function L, we can get our estimated curve of $std\{S_i^*\}$. The comparison between the empirical curve and the theoretical curve is shown in Figures 7.40–7.45.

TABLE 7.8
n Selection for PCAR

n	6	7	9	10
MSE	0.00003853463	0.00003436168	0.00002395992	0.00002475711

TABLE 7.9
Model comparison

stock	GCHPDO (MSE)	GCHP2SDO (MSE)	GCHPnSDO (MSE)	Best Model
EBAY	0.002200996	0.000566305	0.000393683	GCHP8SDO
FB	0.00173283	0.000241354	0.000317633	GCHP2SDO
MU	0.000450242	0.000227604	0.0002722367	GCHP2SDO
PCAR	0.003129217	0.000468104	0.000124427	GCHP15SDO
SMH	0.001477308	0.000053474	0.00002395992	GCHP9SDO

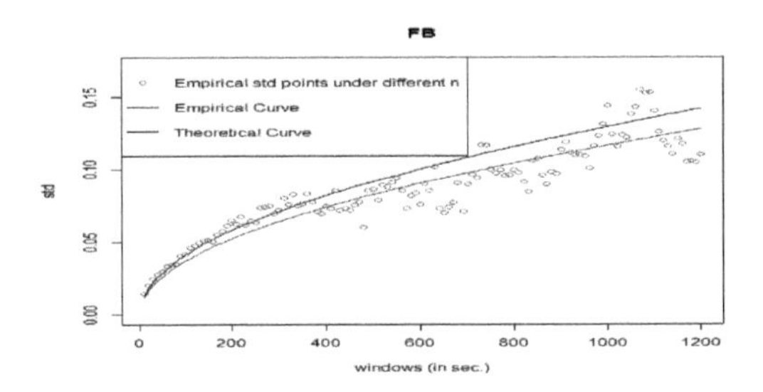

FIGURE 7.41
FB.

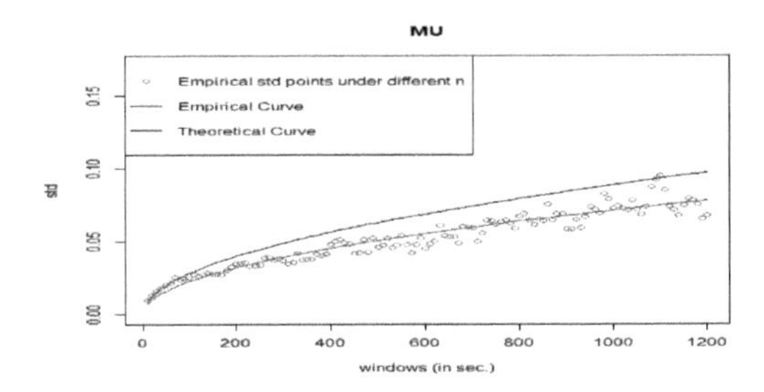

FIGURE 7.42
MU.

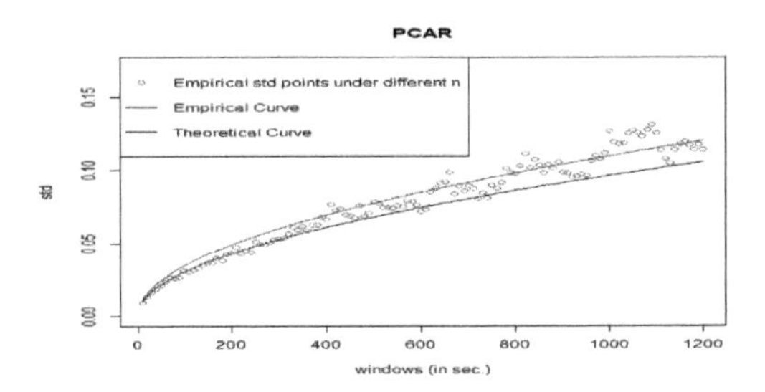

FIGURE 7.43
PCAR.

The error measurements are presented in the figures below.
We define the error rate as

$$\left| \frac{\sqrt{c} - \hat{\sigma}^* \sqrt{\frac{\hat{\lambda}}{1-\frac{\hat{\alpha}}{\hat{\beta}}}}}{\sqrt{c}} \right| \times 100\% \tag{7.2}$$

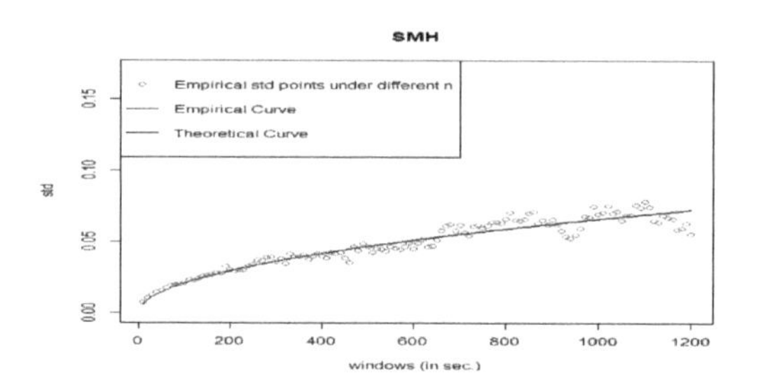

FIGURE 7.44
SMH.

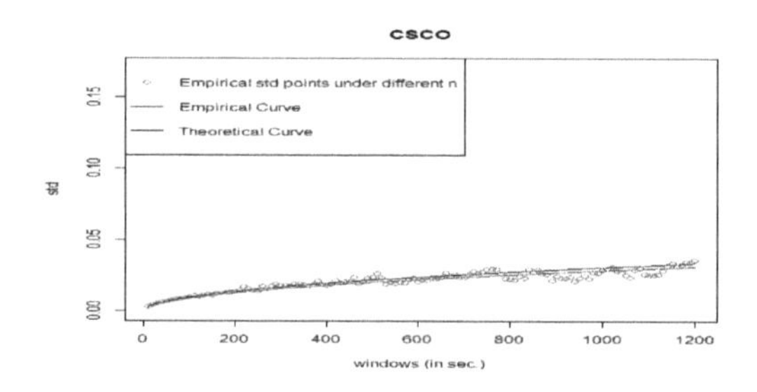

FIGURE 7.45
CSCO-20141107.

TABLE 7.10
Error rate under the best model

Stock	Error Rate
EBAY	26.05%
FB	10.95%
MU	24.42%
PCAR	12.18%
SMH	0.67%
CSCO	7.99%

Under the best model, the error rate of each stock is listed in Table 7.10. If we set the threshold value as 15%, the model with error rate less than threshold can be considered as well fitted. Table 7.10 shows that our General Compound Hawkes Process correctly model the mid price movements of 4 stocks among 6.

7.6 Conclusion

The main contribution and novelty of the chapter consists of: (i) solving the problem of quantitative and comparative analysis of modelling big data arising in high-frequency and algorithmic trading using real data; (ii) introducing of different new types of General Compound Hawkes Processes (GCHPDO, GCHP2SDO, GCHPnSDO) and their diffusive limits to model the mid price movements of 6 stocks, EBAY, FB, MU, PCAR,SMH, CSCO, (provided by Cartea et al. (2015), [13], book)[14]), and (iii) defining the error rates to estimate the models fitting accuracy. Maximum Likelihood Estimation (MLE) and Particle Swarm Optimization (PSO) have been used for Hawkes processes and parameters calibrations. Error measurements and comparative analysis were provided for those 6 stocks.

Bibliography

[1] Laub, P., Taimre, T. and Pollett, P. (2015): Hawkes Processes. arXiv: 1507.02822v1[math.PR]10 Jul 2015.

[2] Lorenzen, A. (2012) Analysis of order clustering using high frequency data: A point process approach. *Ph.D. thesis*, Swiss Federal Institute of Technology Zurich (ETH Zurich).

[3] Swishchuk, A. and Huffman, A. (2020): General Compound Hawkes Process in Limit Order Books. *Risks* 2020, 8(1), 28; https://doi.org/10.3390/risks8010028

[4] Gould, M. et al. (2013): Limit order books. *Quantitative Finance*, 13(11), 1709–1742.

[5] Swishchuk, A. (2017): General Compound Hawkes Processes in Limit Order Books. *Working Paper*, U of Calgary, 32 pages, June 2017. Available on arXiv: http://arxiv.org/abs/1706.07459.

[6] Swishchuk, A. and Vadori, N. (2017): Semi-Markov modelling of limit order books. *SIAM J. Finan. Math.*, 8, 240–273.

[7] Swishchuk, A., Cera, K., Hofmeister, T. and Schmidt, J. (2017): General semi-Markov model for limit order books. *Inern. J. Theor. Appl. Finance*, 20, 1–21.

[8] Swishchuk, A., Elliott, R., Remillard, B and Chavez-Casillas, J. (2019): Compound Hawkes processes in limit order books. *Financial*

Mathematics, Volatility and Covariance Modelling, Volume 2, 1st Ed., Eds: Chevallier J et al.

[9] Chavez-Casillas, J., Elliott, R., Remillard, B. and Swishchuk, A. (2019): *Methodology and Computing in Applied Probability*, 1–21; https://doi.org/10.1007/s11009-019-09715-7

[10] Cont, R. and de Larrard, A. (2013): Price dynamics in a Markovian limit order markets. *Quant. Finance*, 4, 1–25.

[11] Bacry, E., Mastromatteo, I. and Muzy, J.-F. (2015): Hawkes processes in finance. arXiv: 1502.04592v2

[12] Da Fonseca, J. and Zaatour, R. (2013): Hawkes Processes: Fast calibration, application to trade clustering, and diffusive limit. *J. Futures Markets,* 34(6), 548-579.

[13] Hawkes, A. G. (1971): Spectra of some self-exciting and mutually exciting point processes. *Biometrika*, 58, 83–90.

[14] Cartea, A., Jaimungal, S. and Penalva, J. (2015): *Algorithmic and High-Frequency Trading.* Cambridge University Press.

[15] Chavez-Cassilas, J., Elliott, R., Remillard, B and Swishchuk, A. (2019): A level-1 limit order book with time dependent arrival rates. *Mathod. Comput. Appl. Probab.*, 21, 699–719.

[16] He, Q. and Swishchuk, A. (2019): Quantitative and Comparative Analyses of Limit Order Books with General Compound Hawkes Processes. *Risks* 7(4), 110; https://doi.org/10.3390/risks7040110.

Part III

Multivariate Modelling of Big Data in Finance

8

Multivariate General Compound Hawkes Processes in BDF

This chapter focuses on various new multivariate Hawkes processes. We constructed multivariate general compound Hawkes processes (MGCHP) and investigate their properties in limit order books. For the MGCHP, we proved Law of Large Number (LLN) and two Functional Central Limit Theorems (FCLT), the latter provide insights into the link between price volatilities and order flows in limit order books with several assets. Numerical examples with Intel and Microsoft trading data are also provided in this chapter.

8.1 Introduction

In the multivariate case, we consider a vector of Hawkes process. Comparing to the one-dimensional Hawkes process, each element in this vector is not only a self-excited point process but also a mutually-excited point process, namely the evolution of each element depends on the history of itself and other elements. As for the widely applications, multivariate Hawkes processes were applied in mathematical finance, P. Embrechts et al. fitted a multivariate Hawkes process to daily stock market index data in 2011 [7]; In 2012, E. Bacrya et al. applied 1-dimensional and 2-dimensional Hawkes processes to analyze high frequency financial price data [2]. In social science area, D. Luo et al. modelling event sequences where there exists multiple triggering patterns within sequences and structures across sequences in 2015 [13]; K. Zhou et al. discovered the hidden network of social influence by modelling the recurrent events at different individuals as multi-dimensional Hawkes processes in 2013 [24]; L. Li et al. recovered the hidden diffusion network and predicting the occurrence time of social events in 2014 [12].

In 2013, E. Bacrya et al. conducted research of the asymptotic behaviour of the multivariate Hawkes process. They proved the Law of Large Numbers (LLN) and Function Central Limit Theorem (FCLT) for the multi-dimensional Hawkes process and applied them to reproduce the Epps effect and Lead-lag effect in Economics [3]. This work reveals the link between the macroscopic trace and the microscopic noise.

DOI: 10.1201/9781003265986-8

Compound Hawkes process is a new type of Hawkes process which was proposed by Stabile and Torrisi. They replaced Poisson process by a simple Hawkes process in studying the classical problem of the probability of ruin in 2010 [20]. The general version of compound Hawkes process was first introduced by Anatoliy Swishchuk in 2017 to model the risk process in insurance and studied in detailed in [22]. The LLN and FCLT of one-dimensional compound Hawkes process have been proved in [21] in 2017.

In this work, we mainly focus on the multivariate general compound Hawkes process which is a multi-dimensional version of the compound Hawkes process. In the previous research, the one-dimensional compound Hawkes process can be used to model one mid-price of the limit order books. However, in the high-dimensional case, we applied this kind of model to describe several stocks. We study the asymptotic behaviour of the multivariate general compound Hawkes process and proved two FCLTs with stochastic and deterministic centralization for this kind of process. A corresponding LLN was also provided. For applications, we applied 2-dimensional high-frequency trading data [1] to estimate parameters and conduct empirical analysis. Maximum likelihood estimation (MLE) method was applied for calibration and particle swarm optimization (PSO) was used for solving MLE numerically. This work provides a link between price volatilities and order flows for several price processes in the limit order books.

8.2 Hawkes Processes and Limit Theorems

In this section, we give a brief introduction about the univariate and multivariate Hawkes processes and present the law of large numbers (LLN) and the functional central limit theorem (FCLT) for the multivariate Hawkes process. This can help us to construct the multi-dimensional compound Hawkes process and consider corresponding limit theorems.

8.2.1 One-dimensional Hawkes Process

Definition 8.1 ***Counting Process***(*see, eg., [6]*): *We called a stochastic process $\{N(t), t \geq 0\}$ counting process if it satisfies*

- $N(t) \geq 0$, $N(0) = 0$.
- $N(t+s) \geq N(t)$, *for all* $t, s \geq 0$.
- $N(t)$ *is an integer.*

$N(t+s) - N(t)$ is the number of events occurred during the time interval $[t, t+s]$. It is a right continuous step function.

Definition 8.2 ***Point Process****(see, eg., [6]): Let $(T_1, T_2, T_3, \cdots)$ be a sequence of non-negative random variables with $P(0 \le T_1 \le T_2 \le T_3 \le \cdots) = 1$, and the number of points in a bounded region is almost surely finite, then $(T_1, T_2, T_3, \cdots)$ is called a point process.*

The point process was characterized by the conditional intensity function $\lambda(t)$ in the form of

$$\lambda(t) = \lim_{h \to 0} \frac{E[N(t+h) - N(t)|\mathcal{F}^N(t)]}{h}, \tag{8.1}$$

where $\lambda(t)$ is a non-negative function and $\mathcal{F}^N(t)$, $t > 0$ is the corresponding natural filtration. The conditional intensity function describes the possibility of a jump occurs.

Definition 8.3 ***One-dimensional Hawkes Process*** *[10]: We say N_t is a one-dimensional Hawkes process when the intensity function is in the form of*

$$\lambda(t) = \lambda + \int_0^t \mu(t-s)dN_s, \tag{8.2}$$

where λ is a positive constant called background intensity and $\mu(t)$ is the excitation function from $\mathbb{R}_+$ to $\mathbb{R}_+$ with $\int_0^{+\infty} \mu(s)ds < 1$.

From (8.2), we could find that one event occurs will cause the intensity function $\lambda(t)$ to increase which means it's more likely to have a new event to happen. Because of this property, the Hawkes process is also called self-exciting process.

Exponential function is one of the most common choices for the excitation function $\mu(t)$, namely

$$\mu(t) = \alpha \mathrm{e}^{-\beta t}, \text{with parameters } \alpha, \beta > 0.$$

Let $\{t_k\}_{k=1,2,\ldots}$ be a realization of one-dimensional Hawkes process with exponential decay on the interval $[0, T]$. Then, the intensity function can be written as

$$\begin{aligned}\lambda(t) &= \lambda + \int_0^t \alpha \mathrm{e}^{-\beta(t-s)} \mathrm{d}N(s) \\ &= \lambda + \sum_{t_k < t} \alpha \mathrm{e}^{-\beta(t-t_k)}.\end{aligned}$$

Another frequently-used excitation function is the power law function. It was also widely applied into mathematical finance [9] and social science area [14].

8.2.2 Multivariate Hawkes Process (MHP) and Limit Theorems

Next, we introduce the multivariate Hawkes process (MHP) and its limit theorems. E. Bacry et al. proved the corresponding law of large numbers (LLN) and functional central limit theorem (FCLT) in 2013 [3]. The LLN and the FCLT provide a relation between the macroscopic diffusion and microscopic randomness. There are also some relevant limit theorems for the other types of Hawkes processes such as limit theorems for nearly unstable Hawkes processes [11] and central limit theorem for nonlinear Hawkes processes [25]. In this section, we just demonstrate the LLN and the FCLT, details of proofs can be found in E. Bacry's paper.

Definition 8.4 *We say $\vec{N}_t = (N_{1,t}, N_{2,t}, \cdots, N_{d,t})$ is a d-dimensional Hawkes process when the intensity function for each N_i is in the form of*

$$\lambda_i(t) = \lambda_i + \int_{(0,t)} \sum_{j=1}^{d} \mu_{ij}(t-s)dN_{j,s}, \tag{8.3}$$

where $\lambda_i \in \mathbb{R}_+$ and the intensity $\mu_{ij}(t)$ is a function from $\mathbb{R}_+$ to $\mathbb{R}_+$.

From the (8.3), we can find the difference between the multivariate case and one-dimensional Hawkes process. In the multivariate case, the ith element $N_{i,t}$ not only depends on the history of itself but also depends on other processes' histories. In this way, if N_i has one jump happened, it will stimulate other $d-1$ processes and also stimulate itself. This is controlled by excitation functions μ_{ij}.

For the stability, we impose some assumptions for the intensity kernel μ_{ij} and intensity back ground λ_i. Consider a multivariate Hawkes process $\vec{N} = (\vec{N}_t)_{t\geq 0}$ on a rich enough probability space $(\Omega, \mathcal{F}, \mathcal{P})$ with $\vec{\lambda} = (\lambda_1, \lambda_2, \cdots, \lambda_d)^T$ and a $d \times d$-matrix valued function $\boldsymbol{\mu} = (\mu_{ij})_{1 \leq i,j \leq d}$, we have following assumptions:

- $\int_0^\infty \mu_{ij}(t)dt < \infty$ for all i and j.
- Let $\mathbf{K} = \int_0^\infty \boldsymbol{\mu}(t)dt$, then $\rho(\mathbf{K})$ is the spectral radius of $\mathbf{K}$ with $\rho(\mathbf{K}) < 1$.

Remark 8.1 *The second assumption implies that*

$$||\mathbf{K}^n|| \to 0 \ as \ n \to \infty,$$

where $||\cdot||$ denotes the Euclidean norm.

Theorem 8.5 ***LLN for MHP***
Assume previous assumptions holds. Then $\vec{N}_t \in L^2(P)$ for all $t \geq 0$, we have

$$sup_{v\in[0,1]} \left|\left| T^{-1}\vec{N}_{Tv} - v(\mathbf{I} - \mathbf{K})^{-1}\vec{\lambda} \right|\right| \to 0 \tag{8.4}$$

as $T \to \infty$ almost-surely and in $L^2(P)$. Here $\mathbf{I}$ is a d-dimensional identity matrix.

Let $\boldsymbol{\mu}_n$ be the nth convolution of $\boldsymbol{\mu}$ defined on $\mathbb{R}_+$ in the form of

$$\boldsymbol{\mu}_1 = \boldsymbol{\mu},\ \boldsymbol{\mu}_{n+1}(t) = \int_0^t \boldsymbol{\mu}(t-s)\boldsymbol{\mu}_n(s)ds.$$

With assumptions before, we have $\int_0^\infty \boldsymbol{\mu}_n(t)dt = \mathbf{K}^n$ and $\sum_{n\geq 1}\boldsymbol{\mu}_n$ coverges in L^1. Set $\boldsymbol{\psi} = \sum_{n\geq 1}\boldsymbol{\mu}_n$, , we have $\int_0^\infty \boldsymbol{\psi}(t)dt = (\mathbf{I}-\mathbf{K})^{-1} - \mathbf{I}$. Note that $\boldsymbol{\psi}$, $\mathbf{K}$, and $\boldsymbol{\mu}$ are d-by-d matrices with non-negative entries.

Lemma 8.1 *Let $(\vec{M}_t)_{t\geq 0}$ be a d-dimensional martingale and $\vec{M}_t = \vec{N}_t - \int_0^t \vec{\lambda}_s ds$. For all $t \geq 0$, we have*

$$E(\vec{N}_t) = t\vec{\lambda} + \left(\int_0^t \boldsymbol{\psi}(t-s)sds\right)\vec{\lambda},$$

and

$$\vec{N}_t - E(\vec{N}_t) = \vec{M}_t + \int_0^t \boldsymbol{\psi}(t-s)\vec{M}_s ds.$$

Theorem 8.6 ***FCLT for MHP***
With assumptions before and Lemma 8.1, we have

$$\frac{1}{\sqrt{T}}(\vec{N}_{Tv} - E(\vec{N}_{Tv})),\ v \in [0,1]$$

converge in law of the Skorohod topology to $(\mathbf{I}-\mathbf{K})^{-1}\boldsymbol{\Sigma}^{1/2}\vec{W}_v$ as $T \to \infty$, where $\vec{W}_v$ is a standard d-dimensional Brownian motion and $\boldsymbol{\Sigma}$ is a diagonal matrix such that $\boldsymbol{\Sigma}_{ii} = ((\mathbf{I}-\mathbf{K})^{-1}\vec{\lambda})_i$.

Next, we consider an additional restriction on intensity kernel $\boldsymbol{\mu}$:

$$\int_0^\infty \boldsymbol{\mu}(t)t^{1/2}dt < \infty.$$

Then, we can obtain a corollary of the FCLT:

Corollary 8.1

$$\sqrt{T}\left(\frac{1}{T}\vec{N}_{Tv} - v(\mathbf{I}-\mathbf{K})^{-1}\vec{\lambda}\right),\quad \textit{for all } v \in [0,1]$$

converge in law for the Skorokhod topology to $(\mathbf{I}-\mathbf{K})^{-1}\boldsymbol{\Sigma}^{1/2}\vec{W}_v$ as $T \to \infty$.

8.3 Multivariate General Compound Hawkes Processes (MGCHP) and Limit Theorems

In this section, we introduce a new type of MHP, namely the multivariate general compound Hawkes process (MGCHP). The compound Hawkes process was first proposed by Stabile and Torrisi in 2010 [20]. They replaced Poisson process by a simple Hawkes process in studying the classical problem of the probability of ruin. In 2017, A. Swishchuk et al. applied the general compound Hawkes process to the limit order books for modelling the mid-price [21].

The general one-dimensional compound Hawkes process is defined as

$$S_t = S_0 + \sum_{k=1}^{N(t)} a(X_k),$$

where $N(t)$ is an one-dimensional Hawkes process, X_n is an ergodic continuous-time finite state (or countable state) Markov chain, independent of N_t, with space state X, and $a(x)$ is a bounded continuous function on X.

Let $\vec{N}_t = (N_{1,t}, N_{2,t}, \ldots, N_{d,t},)$ be a d-dimensional Hawkes process which defined above. The MGCHP $\vec{S}_t = (S_{1,t}, S_{2,t}, \cdots, S_{d,t},)$ is defined as

$$S_{i,t} = S_{i,0} + \sum_{k=1}^{N_{i,t}} a(X_{i,k}),$$

where $X_{i,k}$ are independent ergodic continuous-time Markov chains.

Theorem 8.7 ***FCLT I for MGCHP: Stochastic Centralization***

Let $X_{i,k}$, $i = 1, 2, \ldots, d$ be independent ergodic Markov chains with n states $\{1, 2, \cdots, n\}$ and with ergodic probabilities $\left(\pi^*_{i,1}, \pi^*_{i,2}, \ldots, \pi^*_{i,n}\right)$. Let $\vec{S}_{nt}$ be d-dimensional compound Hawkes process, we have

$$\frac{\vec{S}_{nt} - \tilde{a}^* \vec{N}_{nt}}{\sqrt{n}} \longrightarrow \tilde{\sigma}^* \mathbf{\Sigma}^{1/2} \vec{W}(t), \; for\, all\, t > 0 \tag{8.5}$$

as $n \to \infty$, where $\vec{W}(t)$ is a standard d-dimensional Brownian motion, $\mathbf{\Sigma}$ is a diagonal matrix such that $\mathbf{\Sigma}_{ii} = ((\mathbf{I} - \mathbf{K})^{-1} \vec{\lambda})_i$, $\vec{N}_{nt}$ is a d-dimensional vector, $\tilde{a}^*$ and $\tilde{\sigma}^*$ are diagonal matrices:

$$\tilde{a}^* = \begin{bmatrix} a_1^* & \cdots & 0 \\ \vdots & \ddots & \vdots \\ 0 & \cdots & a_d^* \end{bmatrix}, \vec{N}_{nt} = \begin{bmatrix} N_{1,nt} \\ \vdots \\ N_{d,nt} \end{bmatrix}, \tilde{\sigma}^* = \begin{bmatrix} \sigma_1^* & \cdots & 0 \\ \vdots & \ddots & \vdots \\ 0 & \cdots & \sigma_d^* \end{bmatrix}.$$

Here, $a_i^* = \sum_{k \in X_i} \pi^*_{i,k} a\left(X_{i,k}\right)$, and $\left(\sigma_i^*\right)^2 := \sum_{k \in X_i} \pi^*_{i,k} v_i(k)$ with

$$v_i(k) = b_i(k)^2 + \sum_{j \in X_i} (g_i(j) - g_i(k))^2 P_i(k,j) - 2b_i(k) \times$$

$$\sum_{j \in X_i} (g_i(j) - g_i(k)) P_i(k,j),$$

$$\begin{aligned} b_i &= (b_i(1), b_i(2), \ldots, b_i(n))', \\ b_i(k) :&= a_i(k) - a_i^*, \\ g_i :&= \left(P_i + \Pi_i^* - I\right)^{-1} b_i, \end{aligned}$$

where P_i is the transition probability matrix for the Markov chain X_i, Π_i^* is the matrix of stationary distributions of P_i, and $g_i(j)$ is the jth entry of g_i.

Proof *From (8.5) it follows that*

$$S_{i,nt} = S_{i,0} + \sum_{k=1}^{N_{i,nt}} a(X_{i,k}),$$

and

$$S_{i,nt} = S_{i,0} + \sum_{k=1}^{N_{i,nt}} (a(X_{i,k}) - a_i^*) + a_i^* N_{i,nt},$$

where $a_i^* := \pi_{i,1}^* a(1) + \pi_{i,2}^* a(2) + \ldots + \pi_{i,n}^* a(n)$. In this way, we have

$$\frac{S_{i,nt} - a_i^* N_{i,nt}}{\sqrt{n}} = \frac{S_{i,0} + \sum_{i=1}^{N_{i,nt}} \left(a\left(X_{i,k}\right) - a_i^*\right)}{\sqrt{n}}.$$

When $n \to +\infty$, $\frac{S_{i,0}}{\sqrt{n}} \to_{n\to+\infty} 0$. So, we just need to consider the limit for

$$\frac{\sum_{i=1}^{N_{i,nt}} \left(a\left(X_{i,k}\right) - a_i^*\right)}{\sqrt{n}}.$$

Consider the following sums

$$R_{i,n}^* := \sum_{k=1}^{n} \left(a\left(X_{i,k}\right) - a_i^*\right)$$

and

$$U_{i,n}^*(t) := n^{-1/2}\left[(1 - (nt - \lfloor nt \rfloor))R_{i,\lfloor nt \rfloor}^* + (nt - \lfloor nt \rfloor)R_{i,\lfloor nt \rfloor+1}^*\right],$$

where $\lfloor \cdot \rfloor$ is the floor function. Following the martingale method from [23] and [26], we have the following weak convergence in the Skorokhod topology [19]

$$U_{i,n}^*(t) \to_{n\to+\infty} \sigma_i^* W_i(t). \tag{8.6}$$

Recall the LLN of multivariate Hawkes process, for each element $N_i(t)$ in the vector $\vec{N}(t)$, we have

$$\frac{N_i(t)}{t} \to_{t\to+\infty} ((\mathbf{I} - \mathbf{K})^{-1}\vec{\lambda})_i$$

and also

$$\frac{N_i(nt)}{n} \to_{n\to+\infty} t((\mathbf{I} - \mathbf{K})^{-1}\vec{\lambda})_i$$

Using change of time in (8.6) and let $t \to N_i(nt)/n$, we have

$$U_{i,n}^*(N_i(nt)/n) \to_{n\to+\infty} \sigma_i^* \sqrt{((\mathbf{I} - \mathbf{K})^{-1}\boldsymbol{\lambda})_i} W_i(t),$$

and the result follows.

Theorem 8.8 ***LLN for MGCHP***

Let $\vec{S}_{nt}$ be a multivariate general compound Hawkes process defined before, we have

$$\frac{\vec{S}_{nt}}{n} \to \tilde{a}^* \Sigma t$$

in the law of Skorohod topology as $n \to +\infty$, where $\tilde{a}^*$ and Σ are defined before.

Proof *From (8.5) we have*

$$\frac{S_{i,nt}}{n} = \frac{S_0}{n} + \sum_{k=1}^{N_{i,nt}} \frac{a(X_{i,k})}{n}.$$

When $n \to +\infty$, $S_{i,0}/n$ approaches to 0, so we just need to consider the second term. From the strong LLN of Markov chain (see, eg,. [15]), we have

$$\frac{1}{n}\sum_{k=1}^{n} a(X_{i,k}) \to_{n\to+\infty} a_i^*,$$

where a_i^* is defined before. Recall the LLN for multivariate Hawkes process:

$$\frac{N_i(nt)}{n} \to_{n\to+\infty} t((\mathbf{I}-\mathbf{K})^{-1}\vec{\lambda})_i,$$

we can obtain

$$\frac{1}{n}\sum_{k=1}^{N_{i,nt}} a(X_{i,k}) = \frac{N_{i,nt}}{n}\frac{1}{N_{i,nt}}\sum_{k=1}^{N_{i,nt}} a(X_{i,k}) \to_{n\to+\infty} a_i^* t((\mathbf{I}-\mathbf{K})^{-1}\vec{\lambda})_i,$$

and the result follows.

Next, we consider a special case of the MGCHP. Let the mid-price process $S_{i,t}$ be

$$S_{i,t} = S_{i,0} + \sum_{k=1}^{N_{i,t}} X_{i,k},$$

where $X_{i,k}$ are independent ergodic Markov chains with two states $(+\delta, -\delta)$ and ergodic probabilities $(\pi_i^*, 1-\pi_i^*)$. In the limit order book (LOB), we can assume that δ is the fixed tick size, and $N_{i,t}$ is the number of price changes up to moment t, described by d-dimensional Hawkes process.

Corollary 8.2 ***FCLT I for two-state MGCHP: Stochastic Centralization***

$$\frac{\vec{S}_{nt} - \tilde{a}^*\vec{N}_{nt}}{\sqrt{n}} \longrightarrow \tilde{\sigma}^*\Sigma^{1/2}\vec{W}(t),\ for\, all\, t > 0$$

as $n \to \infty$, where $\vec{W}(t)$ is a standard d-dimensional Brownian motion, $\vec{N}_{nt}$ is a d-dimensional vector, $\tilde{a}^*$ and $\tilde{\sigma}^*$ are diagonal matrices

$$\tilde{a}^* = \begin{bmatrix} a_1^* & \cdots & 0 \\ \vdots & \ddots & \vdots \\ 0 & \cdots & a_d^* \end{bmatrix}, \vec{N}_{nt} = \begin{bmatrix} N_{1,nt} \\ \vdots \\ N_{d,nt} \end{bmatrix}, \tilde{\sigma}^* = \begin{bmatrix} \sigma_1^* & \cdots & 0 \\ \vdots & \ddots & \vdots \\ 0 & \cdots & \sigma_d^* \end{bmatrix}.$$

Here $a_i^* := \delta(2\pi_i^* - 1)$, and let (p_i, p_i') be transition probabilities

$$\sigma_i^{*2} := 4\delta^2\left(\frac{1 - p' + \pi^*(p' - p)}{(p + p' - 2)^2} - \pi^*(1 - \pi^*)\right).$$

Corollary 8.3 ***LLN for two-state MGCHP***
Let $\vec{S}_{nt}$ be a MGCHP defined before, we have

$$\frac{\vec{S}_{nt}}{n} \to \tilde{a}^*\Sigma t$$

in the law of Skorohod topology as $n \to +\infty$, where $\tilde{a}^*$ and Σ are defined in (8.7).

Proof *Corollary 8.3 and 8.2 can be proved directly from the Theorem 8.7 and Theorem 8.8 by setting the Markov chain $X_{i,k}$ with two states $(+\delta, -\delta)$ and function $a(x) = x$.*

Remark 8.2 *By the FCLT I for MGCHP, we can obtain*

$$\vec{S}_{nt} \sim \tilde{\sigma}^*\mathbf{\Sigma}^{1/2}\vec{W}(t)\sqrt{n} + \tilde{a}^*\vec{N}_{nt}, \text{ for some large enough } n. \tag{8.7}$$

(8.7) can be applied to approximate the price process when we choose a large window size n. However, in the real estimation problem, we cannot have the order flow $\vec{N}_{nt}$ in advance. So, (8.7) is not a practical approximation for the price process $\vec{S}_{nt}$. This motivate us to consider a further diffusion limit for the MGCHP in next section.

8.4 FCLT II for MGCHP: Deterministic Centralization

In this section, we consider a further diffusion limit theorem of the MGCHP, namely the FCLT II for MGCHP: Deterministic Centralization.

Theorem 8.9 ***FCLT II for MGCHP: Deterministic Centralization***
Let $X_{i,k}$, $i = 1, 2, \cdots, d$ be independent ergodic Markov chains with n states

$\{1, 2, \cdots, n\}$ and with ergodic probabilities $(\pi^*_{i,1}, \pi^*_{i,2}, \ldots, \pi^*_{i,n})$. Let $\vec{S}_{nt}$ be d-dimensional compound Hawkes process, we have

$$\frac{\vec{S}_{nt} - \tilde{a}^* nt(\mathbf{I} - \mathbf{K})^{-1}\vec{\lambda}}{\sqrt{n}} \longrightarrow \tilde{\sigma}^* \mathbf{\Sigma}^{1/2} \vec{W}_1(t) + \tilde{a}^* (\mathbf{I} - \mathbf{K})^{-1} \mathbf{\Sigma}^{1/2} \vec{W}_2(t), \; for\, all\, t > 0$$

as $n \to \infty$, where $\vec{W}_1(t)$ and $\vec{W}_2(t)$ are independent standard d-dimensional Brownian motions. Other parameters $\tilde{a}^*$, $\tilde{\sigma}^*$, $\mathbf{\Sigma}^{1/2}$, $\vec{\lambda}$, and $\mathbf{K}$ are defined before.

Proof *Recall the corollary of the FCLT for MHP, we have*

$$\sqrt{n}\left(\frac{1}{n}\vec{N}_{nt} - t(\mathbf{I} - \mathbf{K})^{-1}\vec{\lambda}\right) \longrightarrow (\mathbf{I} - \mathbf{K})^{-1} \mathbf{\Sigma}^{1/2} \vec{W}(t)$$

in law for the Skorokhod topology, as $n \to \infty$. And from theorem 8.7, we have the FCLT I for MGCHP

$$\frac{\vec{S}_{nt} - \tilde{a}^* \vec{N}_{nt}}{\sqrt{n}} \longrightarrow \tilde{\sigma}^* \Sigma^{1/2} \vec{W}(t),$$

as $n \to \infty$. Note that we assume two Brownian motions in above are independent. In order to distinguish them, we call them $\vec{W}_2(t)$ and $\vec{W}_1(t)$. Next, consider

$$\frac{\vec{S}_{nt}}{\sqrt{n}} - \tilde{a}^* \sqrt{n} t(\mathbf{I} - \mathbf{K})^{-1}\vec{\lambda} = \frac{\vec{S}_{nt} - \tilde{a}^* \vec{N}_{nt}}{\sqrt{n}} + \tilde{a}^* \left(\frac{1}{\sqrt{n}}\vec{N}_{nt} - t\sqrt{n}(\mathbf{I} - \mathbf{K})^{-1}\vec{\lambda}\right).$$

With (8.5) and Corollary 8.1 we can derive

$$\frac{\vec{S}_{nt} - \tilde{a}^* \vec{N}_{nt}}{\sqrt{n}} + \tilde{a}^* \left(\frac{1}{\sqrt{n}}\vec{N}_{nt} - t\sqrt{n}(\mathbf{I} - \mathbf{K})^{-1}\vec{\lambda}\right) \longrightarrow \tilde{\sigma}^* \mathbf{\Sigma}^{1/2} \vec{W}_1(t) + \tilde{a}^* (\mathbf{I} - \mathbf{K})^{-1} \mathbf{\Sigma}^{1/2} \vec{W}_2(t)$$

as $n \to \infty$ which proves the result.

Remark 8.3 *For some large n and all $t > 0$, we have the following approximation:*

$$\vec{S}_{nt} \sim \sqrt{n}\tilde{\sigma}^* \mathbf{\Sigma}^{1/2} \vec{W}_1(t) + \sqrt{n}\tilde{a}^* (\mathbf{I} - \mathbf{K})^{-1} \mathbf{\Sigma}^{1/2} \vec{W}_2(t) + \tilde{a}^* nt(\mathbf{I} - \mathbf{K})^{-1}\vec{\lambda},$$

This can be derived directly from the Theorem 8.9. It provides an approximation for multivariate stock prices. Note that although we require n to be some large enough numbers, we still can control the time scale by choosing an appropriate t.

TABLE 8.1
Data Description and stock liquidity of Microsoft and Intel for June 21, 2012

Ticker	#Ord./Day	Avg#Ord./Sec.	#Pr.Ch./Day	Avg#Pr.Ch./Sec.
INTC	404986	17.3071	3218	0.1375
MSFT	411409	5.0640	4016	0.1716

8.5 Numerical Example

In this section, we conducted numerical simulation to test our multivariate model with LOBSTER data. We first presented our data by proving that it cannot be modelled by Poisson process and it has clustering effect property which means the frequency of this data has a similar distribution as the Hawkes process. Next, to calibrate our model, we applied the maximum likelihood estimation (MLE) method and solved it numerically by applying the particle swarm optimization (PSO) method. At last, we used the corollary 8.2 and the theorem 8.9 for the empirical analysis.

8.5.1 Data Description

We applied level one limit order book data of Intel and Microsoft on June 21, 2012 [1]. In this data, time is measured in milliseconds and the tick size is one cent which means the corresponding $\delta = 0.005$. Basic data description and liquidity checking can be found from the Table 8.1.

We used a set of *Q*-*Q* plots to show that the inter-arrival time does not follow an exponential distribution. As we can see in Figures 8.1(a) and 8.1(b), none of these data sticks to the straight line, so we can conclude that the inter-arrival time is non-exponentially distributed. What's more, we counted the number of price changes for each one minute window. Graphs for each day can be found in Figures 8.2(a) and 8.2(b). The arrival time is distributed similarly as the Hawkes process, it has same shape as the intensity function of Hawkes process if we choose an exponential decay.

Remark 8.4 *Here, we assume INTC and MSFT data are mutually independent thus we can model the arrival time by the 2-dimensional compound Hawkes process.*

8.5.2 Maximum Likelihood Estimation (MLE)

In the following numerical example, we applied the exponential decay as the excitation function. Compare to the one-dimensional excitation function, we can write down the exponential decay for the multi-dimensional case, namely

$$u_{ij}(t) = \alpha_{ij} e^{-\beta_{ij} t}, \text{with parameters } \alpha_{ij},\ \beta_{ij} > 0.$$

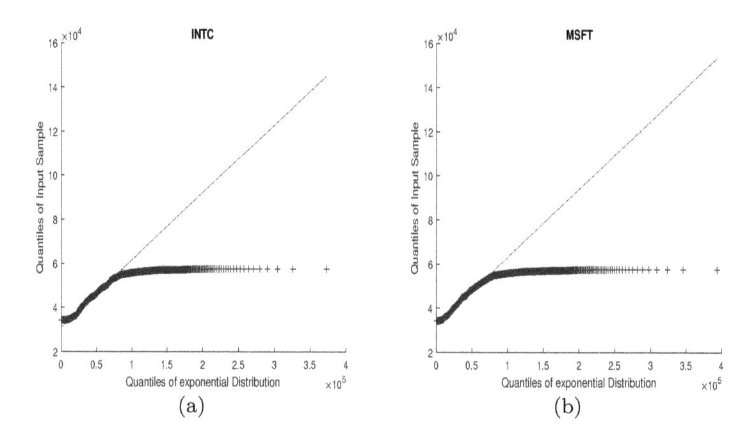

FIGURE 8.1
Non-exponential inter-arrival times' distributions for INTC and MSFT.

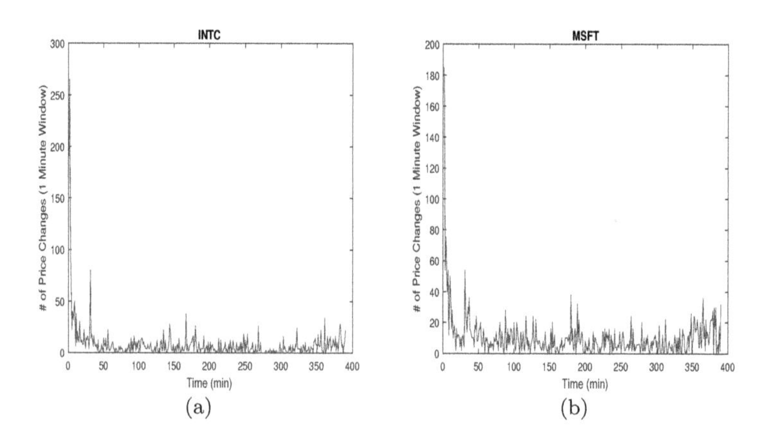

FIGURE 8.2
Clustering effects for INTC and MSFT (1 minute window size).

Let $\{t_k^i\}_{k=1,2,\ldots}$ be a realization of d-dimensional Hawkes process with exponential decay on the interval $[0, T]$. Then, the intensity function can be written as

$$\begin{aligned}\lambda_i(t) &= \lambda_i + \sum_{j=1}^{d} \int_0^t \alpha_{ij} \mathrm{e}^{-\beta_{ij}(t-s)} \mathrm{d}N_j(s) \\ &= \lambda_i + \sum_{j=1}^{d} \sum_{t_k^j < t} \alpha_{ij} \mathrm{e}^{-\beta_{ij}(t-t_k^j)}.\end{aligned}$$

To estimate parameters λ_i, α_{ij}, and β_{ij}, we used the MLE method. The log-likelihood function with exponential decay excitation was given by proposition 1.

Proposition 1 ***Likelihood function for MHP[5]***

Given a data set ω which contains values of each dimension m $(m = 1, 2, ..., d)$ at each time t_k $(k = 1, 2, ..., t_K), t_k \in [0, T]$. The log-likelihood function for a d-variate Hawkes process is:

$$\ln L^m \left(\theta | \{t_k^n\}_{n=1,\dots,d}\right) = -\lambda_m T - \sum_{n=1}^{d} \frac{\alpha_{mn}}{\beta_{mn}} \sum_{\{k: t_k^n < T\}} \left[1 - e^{-\beta_{mn}(T - t_k^n)}\right]$$

$$+ \sum_{\{k: t_k^m < T\}} \ln \left[\lambda_m + \sum_{n=1}^{d} \alpha_{mn} R_{mn}(k)\right],$$

where

$$R_{mn}(k) = e^{-\beta_{mn}(t_k^m - t_{k-1}^m)} R_{mn}(k-1) + \sum_{\{i: t_{k-1}^m \le t_i^n < t_k^m\}} e^{-\beta_{mn}(t_k^m - t_i^n)}.$$

Once we have the log-likelihood function, we wish to have an optimal estimations for parameters. We consider the particle swarm optimization (PSO) method for finding the maximum value of log-likelihood function numerically.

Remark 8.5 *Here, 2 stocks were applied for the simulation example. So, the dimension d is* 2 *and we have* 10 *parameters to estimate, namely λ_i, α_{ij}, β_{ij}, $i, j = 1, 2$.*

8.5.3 Calibration and Empirical Analysis by FCLT I for MGCHP with Stochastic Centralization

We consider the model in Section 8.3, namely 2-dimensional compound Hawkes process with two states $(+\delta, -\delta)$. The transition matrix P of Markov chain X_k is denoted as

$$P = \begin{bmatrix} p_{uu} & 1 - p_{uu} \\ 1 - p_{dd} & p_{dd} \end{bmatrix}.$$

Then we calculate frequency in our data to estimated the p_{uu} and p_{dd} in P by

$$p_{uu} = \frac{t_{uu}}{t_{uu} + t_{ud}},$$
$$p_{dd} = \frac{t_{dd}}{t_{dd} + t_{du}},$$

where t_{uu} is the number of price goes up twice in a row, t_{dd} is the number of price goes down twice in a row, t_{ud} is number that the price goes up and then down and t_{du} is number that the price goes down and then up. And the result is in Table 8.2.

TABLE 8.2
Transition matrix and constant parameters, a^* and σ^*.

Ticker	P_{uu}	P_{dd}	σ^*	a^*
INTC	0.5373	0.5814	0.0057	-2.5023e-04
MSFT	0.5711	0.6044	0.0060	-2.0145e-04

TABLE 8.3
Calibration result for 2-dimensional Hawkes process

Ticker	α_{i1}	α_{i2}	β_{i1}	β_{i2}	λ
INTC	115.7317	0.4492	280.9249	2.9611	0.0545
MSFT	0.0218	123.2703	0.0669	307.2993	0.0593

In the next step, we estimated 10 parameters in 2-dimensional Hawkes process by applying the PSO for the log-likelihood function of multivariate Hawkes process. Calibration results can be found from Table 8.3.

To test the calibration results, we compare expected empirical arrivals on 1s unit interval with expected value calculated by estimated parameters in Table 8.3. From Table 8.4, we can find that estimation results by MLE method are very close to the empirical expectations which means estimated results are good enough for the next step simulation.

Furthermore, to test whether the calibration result fits data well or not, we compare the standard deviation for the left hand side and right hand side in the FCLT I for MGCHP with stochastic centralization:

$$\frac{\vec{S}_{nt} - \tilde{N}_{nt}\vec{a^*}}{\sqrt{n}} \longrightarrow_{n\to\infty} \tilde{\sigma}^*\Sigma^{1/2}\vec{W}(t).$$

That is to say, we first cut our data into disjoint windows of size n, specifically $[in, (i+1)n]$ with $t = 0.001$ and by setting the left bound as our starting time we can calculate:

$$\vec{S_i^*} = \vec{S}_{(i+1)nt} - \vec{S}_{int} - (\tilde{N}((i+1)nt) - \tilde{N}(int))\vec{a^*},$$

and the equation for standard deviation is given by

$$\text{std}\{S_i^*\} \approx \sqrt{n}\tilde{\sigma}^*\Sigma^{1/2}\sqrt{t}.$$

TABLE 8.4
Expected arrivals on 1s unit interval and expected value calculated by MLE parameters

Ticker	Empirical $E[N[0,1]]$	Expected arrivals by MLE
INTC	0.1363	0.1375
MSFT	0.1722	0.1739

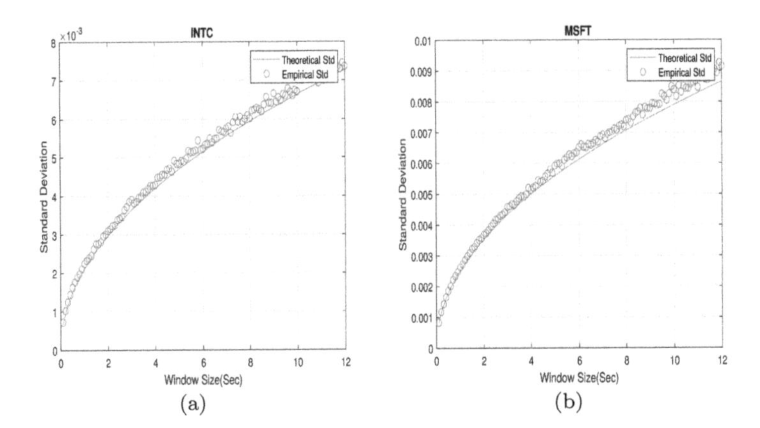

FIGURE 8.3
Standard deviation comparisons for 2 stocks by FCLT I for MGCHP.

Figures 8.3(a) and 8.3(b) gives a standard deviation comparison for 2 stocks in different window sizes from 0.1 second to 12 seconds in steps of 0.1 second. It shows that the parameters we that estimated makes the standard deviation of LHS very similar to the RHS for each stocks when n is large and so we can say our model fits the data well.

8.5.4 Empirical Analysis by FCLT II for MGCHP with Deterministic Centralization

In this part, we tested the FCLT II for MGCHP with deterministic centralization:

$$\frac{\vec{S}_{nt}-\tilde{a}^* nt(\mathbf{I}-\mathbf{K})^{-1}\vec{\lambda}}{\sqrt{n}} \longrightarrow_{n\to\infty} \tilde{\sigma}^*\mathbf{\Sigma^{1/2}}\vec{W}_1(t) + \tilde{a}^*(\mathbf{I}-\mathbf{K})^{-1}\mathbf{\Sigma^{1/2}}\vec{W}_2(t),\ for\, all\, t > 0. \tag{8.8}$$

Following the same idea as the FCLT I for MGCHP with stochastic centralization, we also cut data into mutually disjoint windows of size n with $t = 0.001$ and compare the standard deviation for right-hand side and left-hand side of (8.7), namely

$$\mathrm{std}\left\{\vec{S}_{(i+1)nt} - \vec{S}_{int}\right\} \approx \sqrt{(\tilde{\sigma}^*\mathbf{\Sigma^{1/2}})^{\circ 2} n\vec{t} + (\tilde{a}^*(\mathbf{I}-\mathbf{K})^{-1}\mathbf{\Sigma^{1/2}})^{\circ 2} n\vec{t}},$$

where $\circ 2$ is the entrywise product, namely if $B = A^{\circ 2}$, then $B_{ij} = A_{ij}^2$.

Comparison results of two stocks can be found in Figures 8.4(a) and 8.4(b). Window sizes also start from 0.1 second and increase to 12 seconds by time step 0.1.

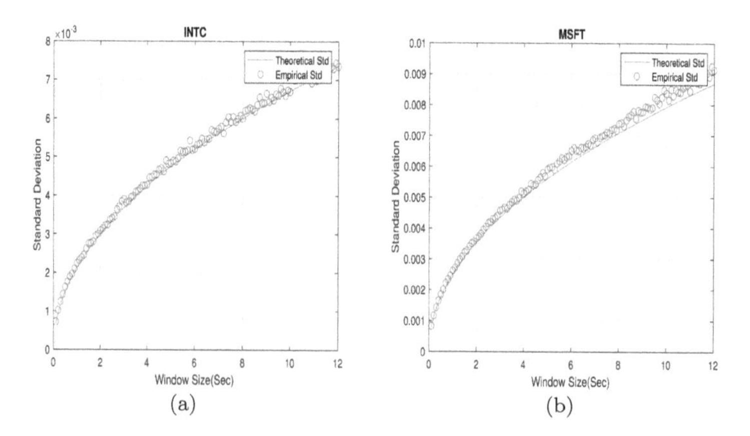

FIGURE 8.4
Standard deviation comparison for 2 stocks by FCLT II for MGCHP.

We can find that results of the FCLT II for MGCHP with deterministic centralization are as good as the diffusion limit theorem with stochastic centralization in figure below. The empirical standard deviations are very close to theoretical curves. So, we can applied this further limit theorem to model mid-price and conduction forecast simulation. We present one of trajectories of simulated mid-price and real mid-price in figure below.

Remark 8.6 *As can be found in both Figure 8.5(a) and Figure 8.5(b), when the window size increases, the spread of data increases. This is because the number of windows decreases as the window size increases.* 1 *second window size yields* 23,400 *windows while we just have 1,950 windows if we set window size to* 12 *seconds.*

Remark 8.7 *In this chapter, we just applied MGCHP model to* 2 *stocks data for the numerical simulation. Theoretically, we may consider more stocks and actually there are 3 more stocks data from [1]. However, more stocks data requires significant increase in the number of parameters, which will lead to a much more time consuming optimization program. For example, if we have a* 5*-dimensional compound Hawkes process with exponential decay excitation function, we need to estimate* 55 *parameters. The PSO program will not be able to generate a global maximum after 72 hours even for a* 3*-dimensional case (*21 *parameters). So, in the future, we need to consider more efficient global optimization algorithms for the high-dimensional compound Hawkes process.*

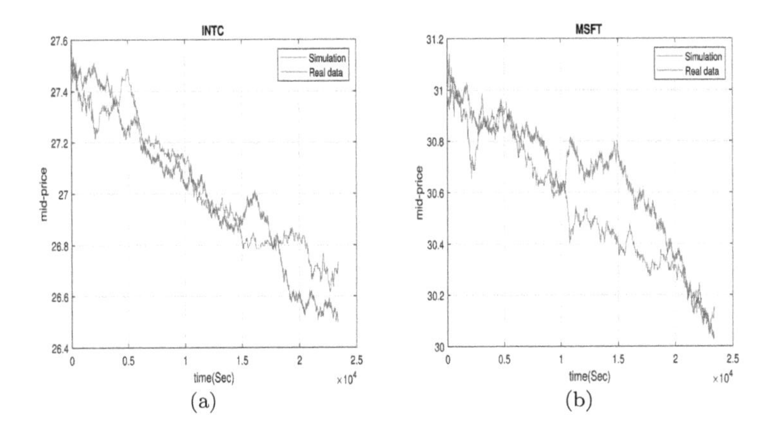

FIGURE 8.5
One simulated mid-price trajectory and real mid-price of INTC and MSFT.

8.6 Conclusion

This chapter focused on various new multivariate Hawkes processes. We constructed multivariate general compound Hawkes processes (MGCHP) and investigate their properties in limit order books. For the MGCHP, we proved Law of Large Number (LLN) and two Functional Central Limit Theorems (FCLT), the latter provide insights into the link between price volatilities and order flows in limit order books with several assets. Numerical examples with Intel and Microsoft trading data have been also provided in this chapter.

Bibliography

[1] LOBSTER data. n.d. Retrieved from https://lobsterdata.com/info/DataSamples.php.

[2] Bacry, E., Dayri, K., and Muzy, J. F. 2012. Non-parametric kernel estimation for symmetric Hawkes processes. Application to high frequency financial data. *The European Physical Journal B*, 85(5), 157.

[3] Bacry, E., Delattre, S., Hoffmann, M., and Muzy, J. F. 2013. Some limit theorems for Hawkes processes and application to financial statistics. *Stochastic Processes and their Applications*, 123(7), 2475–2499.

[4] Chavez-Demoulin, V., and McGill, J. A. 2012. High-frequency financial data modelling using Hawkes processes. *Journal of Banking and Finance*, 36(12), 3415–3426.

[5] Yuanda, C. 2016. Likelihood function for multivariate hawkes processes. Retrieved from https://www.math.fsu.edu/ychen/research/HawkesLikelihood.pdf.

[6] Daley, D. J., and Vere-Jones, D. 2007. *An Introduction to the Theory of Point Processes: Volume II: General Theory and Structure*. New York: Springer Science and Business Media.

[7] Embrechts, P., Liniger, T., and Lin, L. 2011. Multivariate Hawkes processes: an application to financial data. *Journal of Applied Probability*, 48(A), 367–378.

[8] Guo, Q., and Swishchuk, A. (2020): Multivariate general compound Hawkes processes and their applications in limit order books. *Wilmott*, 107: 42–51.

[9] Hardiman, S. J., Bercot, N., and Bouchaud, J. P. 2013. Critical reflexivity in financial markets: a Hawkes process analysis. *The European Physical Journal B*, 86(10), 442.

[10] Hawkes, A. G. 1971. Spectra of some self-exciting and mutually exciting point processes. *Biometrika*, 58(1), 83–90.

[11] Jaisson, T., and Rosenbaum, M. 2015. Limit theorems for nearly unstable Hawkes processes. *The Annals of Applied Probability*, 25(2), 600–631.

[12] Li, L., and Zha, H. 2014, June. Learning parametric models for social infectivity in multi-dimensional hawkes processes. In *Twenty-Eighth AAAI Conference on Artificial Intelligence*.

[13] Luo, D., Xu, H., Zhen, Y., Ning, X., Zha, H., Yang, X., and Zhang, W. 2015, June. Multi-task multi-dimensional hawkes processes for modelling event sequences. In *Twenty-Fourth International Joint Conference on Artificial Intelligence*.

[14] Mitchell, L., and Cates, M. E. 2009. Hawkes process as a model of social interactions: a view on video dynamics. *Journal of Physics A: Mathematical and Theoretical*, 43(4), 045101.

[15] Norris, J. R., and Norris, J. R. 1998. *Markov Chains*. Cambridge: Cambridge University Press.

[16] Ogata, Y. 1999. Seismicity analysis through point-process modelling: A review. In *Seismicity patterns, their statistical significance and physical meaning* (pp. 471–507). Birkhäuser, Basel.

[17] Reynaud-Bouret, P., Rivoirard, V., and Tuleau-Malot, C. 2013. December. Inference of functional connectivity in neurosciences via Hawkes processes. In *2013 IEEE Global Conference on Signal and Information Processing* (pp. 317–320). IEEE.

[18] Reynaud-Bouret, P., and Schbath, S. 2010. Adaptive estimation for Hawkes processes; application to genome analysis. *The Annals of Statistics*, 38(5), 2781–2822.

[19] Skorokhod, A. V. 1982. *Studies in the Theory of Random Processes* (Vol. 7021). New York: Dover Publications.

[20] Stabile, G., and Torrisi, G. L. 2010. Risk processes with non-stationary Hawkes claims arrivals. *Methodology and Computing in Applied Probability*, 12(3), 415–429.

[21] Swishchuk, A. 2017. General compound Hawkes processes in limit order books. Available on *arXiv*: Available at: arxiv.org/submit/1929048.

[22] Swishchuk, A. 2018. Risk model based on compound Hawkes process. *Wilmott*, (94), 50–57.

[23] Swishchuk, A., Hofmeister, T., Cera, K., and Schmidt, J. 2017. General semi-Markov model for limit order books. *International Journal of Theoretical and Applied Finance*, 20(03), 1750019.

[24] Zhou, K., Zha, H., and Song, L. 2013, February. Learning triggering kernels for multi-dimensional hawkes processes. In *International Conference on Machine Learning* (pp. 1301–1309).

[25] Zhu, L. 2013. Central limit theorem for nonlinear Hawkes processes. *Journal of Applied Probability*, 50(3), 760–771.

[26] Vadori, N., and Swishchuk, A. 2015. Strong law of large numbers and central limit theorems for functionals of inhomogeneous Semi-Markov processes. *Stochastic Analysis and Applications*, 33(2), 213–243.

9

Multivariate General Compound Point Processes in BDF

This chapter focuses on a new generalization of multivariate general compound Hawkes process (MGCHP), which we referred to as the multivariate general compound point process (MGCPP). Namely, we applied a multivariate point process to model the order flow instead of the Hawkes process. The law of large numbers (LLN) and two functional central limit theorems (FCLTs) for the MGCPP were proved in this chapter. Applications of the MGCPP in the limit order market were also considered. We provided numerical simulations and comparisons for the MGCPP and MGCHP by applying Google, Apple, Microsoft, Amazon, and Intel trading data.

9.1 Introduction

In this chapter, we introduced a new class of stochastic models, which can be considered as a generalization of the multivariate general compound Hawkes process (MGCHP) in [11]. We called this model the multivariate general compound point processes (MGCPP). A Law of Large Numbers (LLN) and two Functional Central Limit Theorems (FCLT) for MGCPP were proved. FCLTs of the MGCPP can be viewed as a link between price volatility and the order flow. Thus, we applied this asymptotic method to study the mid-price modelling in the limit order book (LOB).

Hawkes process was applied to financial modelling for the first time in 2007 [2]. In [1] a LLN and FCLT were proved for multivariate Hawkes process and applied them to study some economic phenomenons in 2013. Volatilities between five stocks were estimated by a 5-dimensional Hawkes process in [5] in 2009. Other types of Hawkes processes have been studied widely as well. The nonlinear Hawkes process was considered by [3] and the corresponding FCLT was proved in [26]. Some applications of multivariate Hawkes process to financial data are given in [10]. The regime-switching Hawkes process was considered by [22] to describe the dynamics dependency on the bid–ask spread in limit order book. In [19], a semi-Markov process based on a renewal process was applied to the mid-price modelling in LOB. [20] also considered the

DOI: 10.1201/9781003265986-9

general case of the semi-Markovian models in 2017. A good textbook for algorithmic and High-Frequency trading methods was written by [7] in 2015. [25] introduced a multivariate point process describing the dynamics of the Bid and Ask price of a financial asset. The point process is similar to a Hawkes process, with additional constraints on its intensity corresponding to the natural ordering of the best Bid and Ask prices. [8] developed a new approach for investigating the properties of the Hawkes process without the restriction to mutual excitation or linear link functions. They employed a thinning process representation and a coupling construction to bound the dependence coefficient of the Hawkes process. Using recent developments on weakly dependent sequences, a concentration inequality for second-order statistics of the Hawkes process was established. This concentration inequality was applied to cross-covariance analysis in the high-dimensional regime, and it was verified the theoretical claims with simulation studies. A framework for fitting multivariate Hawkes process for large-scale problems (long history and a wide variety of events) was proposed in [13]. Liniger thesis addresses theoretical and practical questions arising in connection with multivariate, marked, linear Hawkes process [14]. [24] developed a nonparametric and online learning algorithm that estimates the triggering functions of a multivariate Hawkes process. An introduction to point processes from a martingale point of view may be found in Bjork's lecture notes [4].

Paper [11] constructed a multivariate general compound Hawkes process (MGCHP) which is an extended model from [9] and [18]. In [11], they applied the multivariate Hawkes process to model the order flow of several stocks in limit order market and proved limit theorems for the MGCHP. In this chapter, we proposed a new mid-price model which is a generalization of the MGCHP and we called it the *multivariate general compound point process* (MGCPP). For the MGCPP, we applied a multi-dimensional simple point process to represent the order flow in LOB instead of the Hawkes process. We also proved the corresponding LLN and FCLTs for the MGCPP. One of the reasons why we considered the generalized model is parameters for simple point process are much easier to estimate than Hawkes process. So, we provided the numerical comparisons of the MGCPP and MGCHP by real high-frequency trading data and we found that results of the new generalized model are as good as the MGCHP.

This chapter is organized as follows. Definitions and assumptions of the multivariate general compound point process (MGCPP) can be found in Section 9.2. Functional central limit theorem (FCLT) 1 and law of large numbers were proved in Section 9.3. We also provided numerical examples simulated by real data for the FCLT 1 in Section 9.3. In Section 9.4, we considered a FCLT 2 for the MGCPP and applied it in the mid-price prediction.

9.2 Definition of Multivariate General Compound Point Process (MGCPP)

In this section, we proposed a multivariate stochastic model for the mid-price in the limit order book. This is a generalization for models in [9], [11], and [18]. Here, we assume the order flow was described by a multivariate simple point process with some good asymptotic properties.

Definition 9.1 (Counting Process. (see, e.g., [4])) *We called a stochastic process $\{N(t), t \geq 0\}$ counting process if it satisfies: the trajectories of N_t are right continues and piecewise constant with probability one, $N(0) = 0$, and $\Delta N_t = N_t - N_{t-} = 0$ or 1 with probability one.*

Counting process is the simplest type of point process. In the following discussion of the chapter, we adopt above Definition as the definition of a point process. The point process can be determined by the conditional intensity function $\lambda(t)$ in the form of

$$\lambda(t) = \lim_{h \to 0} \frac{E[N(t+h) - N(t) | \mathcal{F}^N(t)]}{h},$$

where $\lambda(t) \geq 0$ and $\mathcal{F}^N(t)$ is the corresponding natural filtration.

9.2.1 Assumptions for Multivariate Point Processes

Let $\vec{N}_t = (N_{1,t}, N_{2,t}, \cdots, N_{d,t},)$ be d-dimensional point process with following assumptions:

Assumption 9.1 *We assume there's a law of large numbers (LLN) of the $\vec{N}_t$ in the form of:*

$$\frac{\vec{N}(nt)}{n} \to \vec{\bar{\lambda}} t$$

as $n \to +\infty$ almost surely, where $\vec{\bar{\lambda}} = (\bar{\lambda}_1, \bar{\lambda}_2, \bar{\lambda}_3, \cdots, \bar{\lambda}_d)$.

Assumption 9.2 *We also assume there's a Functional Central Limit Theorem (FCLT) of the $\vec{N}_t$ in the form of:*

$$\frac{1}{\sqrt{n}}(\vec{N}_{nt} - E(\vec{N}_{nt})) \xrightarrow{n \to \infty} \Sigma^{1/2} \vec{W}_t, \ t \in [0, 1]$$

in law of the Skorokhod topology [17], where $\vec{W}_t$ is a standard d-dimensional Brownian motion and Σ is a d-by-d covariance matrix.

Here, $\vec{N}_t$ denotes the order flow in the limit order market for d stocks. Liquidity for the high-frequency trading data guarantees there are enough price changes in one day or even a small window size nt. So, it is reasonable to consider those two limit assumptions in the limit order book modelling.

Remark 9.1 *For a simple example, if we consider the point process as a multivariate homogeneous Poisson process with independent coordinates, then two assumptions above are LLN and FCLT for the multi-dimensional Poisson process. Let $\vec{P}_t$ be a d-dimensional Poisson process with intensity $\vec{\lambda}$. Here, we used notation $\vec{P}_t$ to distinguish the general case and the Poisson example. Then, we have the LLN in the form of*

$$sup_{t\in[0,1]}||n^{-1}\vec{P}_{nt} - t\vec{\lambda}|| \to 0$$

as $n \to \infty$ almost surely. Further, the FCLT in the form of

$$\sqrt{n}\left(\frac{1}{n}\vec{P}_{nt} - t\vec{\lambda}\right)$$

converge in law for the Skorokhod topology to $\vec{W}_t \circ \vec{\lambda}^{1/2}$ as $n \to \infty$, where $\circ$ is the element-wise product.

Remark 9.2 *Another interesting example is limit theorems for the multivariate Hawkes process (MHP) in [1]. Let $\vec{H}_t = (H_{1,t}, H_{2,t}, \cdots, H_{d,t})$ be a d-dimensional Hawkes process. The intensity function for each H_i is in the form of*

$$\lambda_i(t) = \lambda_i + \int_{(0,t)} \sum_{j=1}^{d} \mu_{ij}(t-s)dH_{j,s},$$

Let $\boldsymbol{\mu} = (\mu_{ij})_{1\leq i,j\leq d}$, $\vec{\lambda} = (\lambda_1, \lambda_2, \cdots, \lambda_d)^T$, and $\mathbf{K} = \int_0^\infty \boldsymbol{\mu}(t)dt$, then the LLN for MHP is in the form of

$$sup_{t\in[0,1]}||n^{-1}\vec{H}_{nt} - t(\mathbf{I} - \mathbf{K})^{-1}\vec{\lambda}|| \to 0$$

as $n \to \infty$ almost surely, where $\mathbf{I}$ is a d-by-d identity matrix. We can also have the FCLT for MHP:

$$\frac{1}{\sqrt{n}}(\vec{H}_{nt} - E(\vec{H}_{nt})) \xrightarrow{n\to\infty} (\mathbf{I} - \mathbf{K})^{-1}\mathbf{D}^{1/2}\vec{W}_t,\ t \in [0,1]$$

in law of the Skorokhod topology [17], where $\vec{W}_t$ is a standard d-dimensional Brownian motion and $\mathbf{D}$ is a diagonal matrix determined by $\mathbf{D}_{ii} = ((\mathbf{I} - \mathbf{K})^{-1}\vec{\lambda})_i$. Details about the LLN and FCLT of MHP can be found in [1].

9.2.2 Definition for MGCPP

Next, we consider a price process $\vec{S}_t$ in the form $\vec{S}_t = (S_{1,t}, S_{2,t}, \cdots, S_{d,t},)$ as:

$$S_{i,t} = S_{i,0} + \sum_{k=1}^{N_{i,t}} a_i(X_{i,k}),$$

where $X_{i,k}$ are independent ergodic continuous-time Markov chains, independent of $\vec{N}_t$. The state space of $X_{i,k}$ is denoted by $X^{\{i\}} = \{1, 2, \cdots, \mathcal{N}_i\}$. $a_i(\cdot)$ are bounded continuous functions. We refer $\vec{S}_t$ as multivariate general compound point processes (MGCPP).

Remark 9.3 *If we consider the one-dimensional case, let N_t be a Poisson process, $a_i(x) = (-\delta) \vee (x \wedge \delta)$, and X_k is a sequence of independent random variables such that $P(X_1 = \delta) = P(X_1 = -\delta) = 1/2$, then S_t is a stochastic model for the dynamics of a limit order book discussed in [9].*

Remark 9.4 *When $\vec{N}_t$ is a multivariate Hawkes process, then $\vec{S}_t$ is a multivariate general compound Hawkes processes (MGCHP), which was proposed in [11].*

9.3 LLNs and Diffusion Limits for MGCPP

In this section, we considered the diffusion limit theorems for the MGCPP. It provides us a link between the order flow $\vec{N}_t$ and the price process $\vec{S}_t$. The functional central limit theorem (FCLT) and law of large numbers (LLN) for the MGCPP are generalizations for the diffusion limit theorems of the MGCHP in [11].

9.3.1 LLN for MGCPP

Theorem 9.2 (LLN for MGCPP) *Let $\vec{S}_{nt} = (S_{1,nt}, S_{2,nt}, S_{3,nt}, \cdots, S_{d,nt})$ be a d-dimensional MGCPP defined before, we have*

$$\frac{\vec{S}_{nt}}{n} \to \tilde{a^*} \vec{\lambda} t$$

as $n \to \infty$ almost surely.

Proof *From the definition of MGCPP we have*

$$\frac{S_{i,nt}}{n} = \frac{S_{i,0}}{n} + \sum_{k=1}^{N_{i,nt}} \frac{a_i\left(X_{i,k}\right)}{n}.$$

Since $S_{i,0}$ is a constant, we have

$$\lim_{n \to \infty} \left(\frac{S_{i,t}}{n}\right) = \lim_{n \to \infty} \left(\frac{S_{i,0}}{n}\right) + \lim_{n \to \infty} \frac{\sum_{k=1}^{N_{i,nt}} a_i(X_{i,k})}{n}$$
$$= 0 + \lim_{n \to \infty} \frac{\sum_{k=1}^{N_{i,nt}} a_i(X_{i,k})}{n}.$$

Recall the strong LLN of Markov chain (see, e.g., [15]), we have

$$\frac{1}{n}\sum_{k=1}^{n} a_i\left(X_{i,k}\right) \xrightarrow{n\to+\infty} a_i^*, \quad a.s.,$$

where a_i^* *is defined by* $a_i^* = \sum_{k\in X^{\{i\}}} \pi_{i,k}^* a_i\left(X_{i,k}\right)$. *Consider the LLN of MPP in Assumption 1, we have*

$$\frac{N_{i,nt}}{n} \to \bar{\lambda}_i t$$

as $n\to\infty$ *almost surely, we obtain*

$$\frac{1}{n}\sum_{k=1}^{N_{i,nt}} a_i\left(X_{i,k}\right) = \frac{N_{i,nt}}{n}\frac{1}{N_{i,nt}}\sum_{k=1}^{N_{i,nt}} a_i\left(X_{i,k}\right) \xrightarrow{n\to+\infty} a_i^*\bar{\lambda}_i t, \quad a.s.$$

Rewriting the last expression in the multivariate case, we derive the LLN for the MGCPP.

9.3.2 Diffusion Limits for MGCPP: Stochastic Centralization

Theorem 9.3 (FCLT 1: Stochastic Centralization) *Let* $X_{i,k}$, $i = 1,2,\ldots,d$ *be independent ergodic Markov chains defined before.* $X^{\{i\}} = \{1,2,\cdots,\mathcal{N}_i\}$ *is the state space and the ergodic probabilities is given by* $\left(\pi_{i,1}^*, \pi_{i,2}^*, \ldots, \pi_{i,n}^*\right)$. *We assume* $X_{i,k}$ *is independent of* $\vec{N}_t$. *Let* $\vec{S}_{nt}$ *be d-dimensional MGCPP, we have*

$$\frac{\vec{S}_{nt} - \tilde{a^*}\vec{N}_{nt}}{\sqrt{n}} \xrightarrow{n\to\infty} \tilde{\sigma}^*\Lambda^{1/2}\vec{W}(t), \; for\, all\, t>0,$$

where $\vec{W}(t)$ *is a standard d-dimensional Brownian motion.* Λ *is a d-by-d diagonal matrix in the form of* $\Lambda = diag(\bar{\lambda}_1, \bar{\lambda}_2, \bar{\lambda}_3, \cdots, \bar{\lambda}_d)$. $\vec{N}_{nt}$, $\tilde{a^*}$ *and* $\tilde{\sigma}^*$ *are given by*

$$\tilde{a^*} = \begin{bmatrix} a_1^* & \cdots & 0 \\ \vdots & \ddots & \vdots \\ 0 & \cdots & a_d^* \end{bmatrix}, \vec{N}_{nt} = \begin{bmatrix} N_{1,nt} \\ \vdots \\ N_{d,nt} \end{bmatrix}, \tilde{\sigma}^* = \begin{bmatrix} \sigma_1^* & \cdots & 0 \\ \vdots & \ddots & \vdots \\ 0 & \cdots & \sigma_d^* \end{bmatrix}.$$

Here, $a_i^* = \sum_{k\in X^{\{i\}}} \pi_{i,k}^* a_i\left(X_{i,k}\right)$, and $\left(\sigma_i^*\right)^2 := \sum_{k\in X^{\{i\}}} \pi_{i,k}^* v_i(k)$ with

$$v_i(k) = b_i(k)^2 + \sum_{j\in X^{\{i\}}} (g_i(j)-g_i(k))^2 P_i(k,j) - 2b_i(k)\times$$

$$\sum_{j\in X^{\{i\}}} (g_i(j)-g_i(k)) P_i(k,j),$$

$$\begin{aligned} b_i &= (b_i(1), b_i(2), \ldots, b_i(n))', \\ b_i(k) &: = a_i(k) - a_i^*, \\ g_i &: = \left(P_i + \Pi_i^* - I\right)^{-1} b_i, \end{aligned}$$

where P_i is the transition probability matrix for X_i, Π_i^* is the matrix of stationary distributions of P_i, and $g_i(j)$ is the jth entry of g_i.

Proof *From the definition of MGCPP, we have*

$$S_{i,nt} = S_{i,0} + \sum_{k=1}^{N_{i,nt}} a_i(X_{i,k}),$$

and

$$S_{i,t} = S_{i,0} + \sum_{k=1}^{N_{i,nt}} (a_i(X_{i,k}) - a_i^*) + a_i^* N_{i,nt},$$

here the a_i^* is defined by $a_i^* = \sum_{k \in X\{i\}} \pi_{i,k}^* a_i(X_{i,k})$. Then, for some n, we have

$$\frac{S_{i,t} - a_i^* N_{i,nt}}{\sqrt{n}} = \frac{S_{i,0} + \sum_{k=1}^{N_{i,nt}} (a_i(X_{i,k}) - a_i^*)}{\sqrt{n}}.$$

Since $S_{i,0}$ is a constant, when $n \to \infty$, we have

$$\begin{aligned} \lim_{n\to\infty} \left(\frac{S_{i,t} - a_i^* N_{i,nt}}{\sqrt{n}} \right) &= \lim_{n\to\infty} \left(\frac{S_{i,0}}{\sqrt{n}} \right) + \lim_{n\to\infty} \left(\frac{\sum_{k=1}^{N_{i,nt}} (a_i(X_{i,k}) - a_i^*)}{\sqrt{n}} \right) \\ &= 0 + \lim_{n\to\infty} \left(\frac{\sum_{k=1}^{N_{i,nt}} (a_i(X_{i,k}) - a_i^*)}{\sqrt{n}} \right). \end{aligned}$$

Consider the following sums:

$$R_{i,n}^* := \sum_{k=1}^{n} \left(a_i \left(X_{i,k} \right) - a_i^* \right),$$

and

$$U_{i,n}^*(t) := n^{-1/2} \left[(1 - (nt - \lfloor nt \rfloor)) R_{i,\lfloor nt \rfloor}^* + (nt - \lfloor nt \rfloor) R_{i,\lfloor nt \rfloor + 1}^* \right],$$

where $\lfloor \cdot \rfloor$ is the floor function. By applying the martingale method in [19], we have

$$U_{i,n}^*(t) \xrightarrow{n \to +\infty} \sigma_i^* W_i(t)$$

converge weakly in Skorokhod topology [17]. From the Assumption 1, we have the LLN for the MPP in the form of

$$\frac{N_i(nt)}{n} \xrightarrow{n \to \infty} \bar{\lambda}_i t.$$

Using change of time in pre-last expression and let $t \to N_i(nt)/n$, we have

$$U_{i,n}^*(N_i(nt)/n) \xrightarrow{n \to +\infty} \sigma_i^* \sqrt{\bar{\lambda}_i} W_i(t).$$

Rewriting $\frac{S_{i,t}-a_i^* N_{i,nt}}{\sqrt{n}}$ in the multivariate form, we derive the weak convergence for MGCPP:

$$\frac{\vec{S}_{nt} - \tilde{a^*}\vec{N}_{nt}}{\sqrt{n}} \xrightarrow{n\to\infty} \tilde{\sigma}^* \Lambda^{1/2} \vec{W}(t), \; for\, all\, t > 0.$$

Next, we considered a simple special case of the MGCPP. Let $X_{i,k}$ be a Markov chain with two dependent states $(+\delta, -\delta)$ and the ergodic probabilities $(\pi_i^*, 1-\pi_i^*)$. In the limit order market, the δ is the fixed tick size and the d-dimensional point process $\vec{N}_{nt}$ represents the order flow for d stocks. Thus, we set $a_i(x) = (-\delta)\vee(x\wedge\delta)$ in our setting. Then, we can derive the corresponding limit theorems for this kind of special case.

Corollary 9.1 (FCLT 1 two-state MGCPP: Stochastic Centralization)

$$\frac{\vec{S}_{nt} - \tilde{a^*}\vec{N}_{nt}}{\sqrt{n}} \longrightarrow_{n\to\infty} \tilde{\sigma}^* \Lambda^{1/2} \vec{W}(t), \; for\, all\, t > 0.$$

$\tilde{a^*}$ and $\tilde{\sigma}^*$ are given by

$$\tilde{a^*} = \begin{bmatrix} a_1^* & \cdots & 0 \\ \vdots & \ddots & \vdots \\ 0 & \cdots & a_d^* \end{bmatrix}, \vec{N}_{nt} = \begin{bmatrix} N_{1,nt} \\ \vdots \\ N_{d,nt} \end{bmatrix}, \tilde{\sigma}^* = \begin{bmatrix} \sigma_1^* & \cdots & 0 \\ \vdots & \ddots & \vdots \\ 0 & \cdots & \sigma_d^* \end{bmatrix},$$

where $a_i^* = \delta(2\pi_i^* - 1)$, and

$$\sigma_i^{*2} := 4\delta^2 \left(\frac{1 - p_i' + \pi_i^* (p_i' - p_i)}{(p_i + p_i' - 2)^2} - \pi_i^* (1 - \pi_i^*) \right)$$

(p_i, p_i') are transition probabilities of the Markov chain $X_{i,k}$.

Corollary 9.2 (LLN for two-state MGCPP) *Let $\vec{S}_{nt}$ be d-dimensional general compound point process with two-state Markov chain $X_{i,k}$, we have*

$$\frac{\vec{S}_{nt}}{n} \to \tilde{a^*}\vec{\lambda} t, \quad a.s.$$

Here, $\tilde{a^*}$ and $\vec{\lambda}$ are constants defined in Corollary 9.1.

Proof of Corollaries 9.1 and 9.2 . *Set Markov chain $X_{i,k}$ with two states $(+\delta, -\delta)$ and $a_i(x) = (-\delta) \vee (x \wedge \delta)$ in previous two FCLT Theorems, we can derive Corollaries 9.1 and 9.2 directly.*

Remark 9.5 *From the FCLT 1 of MGCPP, we can derive an approximation for the mid-price $\vec{S}_{nt}$:*

$$\vec{S}_{nt} \sim \tilde{\sigma}^* \Lambda^{1/2} \vec{W}(t)\sqrt{n} + \tilde{a^*}\vec{N}_{nt},$$

for all $t > 0$ and some lagre enough n. Since $\vec{S}_{nt}$ is the price process in high-frequency trading, the time is always measured in a very short period (e.g., milliseconds). So, even if the window size $nt = 10$ s with $t = 0.001$, the n will equal to 10,000 which is a very large number. In this way, it is reasonable to consider this kind of approximation in the LOB.

Remark 9.6 *When $\vec{N}_t$ is a multivariate Hawkes process in Remark 9.2, then the $\vec{S}_{nt}$ is a MGCHP model, corresponding FCLTs and LLNs were considered in [11]. To distinguish with the general case, we also applied the $\vec{H}_{nt}$ to denote the multivariate Hawkes process and $\vec{S}_{Hawkes}(nt)$ to denote the price process by MGCHP. Then we have the FCLT for MGCHP in the form of*

$$\frac{\vec{S}_{Hawkes}(nt) - \tilde{a}^* \vec{H}_{nt}}{\sqrt{n}} \xrightarrow{n\to\infty} \tilde{\sigma}^* \mathbf{D}^{1/2} \vec{W}(t), \text{ for all } t > 0,$$

where $\vec{W}(t)$ is a multivariate standard Brownian motion and $\mathbf{D}$ is defined as $\mathbf{D}_{ii} = ((\mathbf{I} - \mathbf{K})^{-1}\vec{\lambda})_i$. We can find clearly that the limit theorem for MGCPP is a generalization of the Hawkes case. Also, when we consider an one-dimensional case, if N_t is a renewal process, the corresponding limit theorems for the semi-Markovian model S_t were discussed in [19] and [20].

9.3.3 Numerical Examples for FCLT: Stochastic Centralization

In this section, we tested the FCLT 1 of MGCPP model with the LOBSTER data and compared our results with the simulation results by MGCHP in [11]. In their paper, they applied two stocks in the LOBSTER data set, namely the mid-price of Microsoft and Intel. As for the Markov chain part, they used the two-state Markov chain $(+\delta, -\delta)$.

In order to make our results comparable with the MGCHP, we first applied the same data set (Microsoft and Intel) and same two-state Markov chain $(+\delta, -\delta)$ for the MGCPP model. Next, we explore more simulation examples (by Apple, Amazon, and Google data) which were mentioned in [11]. For those three stocks, we applied the MGCPP model with both two-state Markov chain and $\mathcal{N}$-state Markov chain.

9.3.3.1 Data Description and Parameter Estimations

The level one LOBSTER data was considered in this paper. The LOBSTER data set contained the stock prices and order flows of Apple, Amazon, Google, Microsoft, and Intel on 21 June 2012. The tick size is one cent ($\delta = 0.005$) and time was measured in milliseconds (0.001 s). We can find the basic data description and check the liquidity from Table 9.1. Notation # is the number sign.

Next, we estimate $\vec{\bar{\lambda}} = (\bar{\lambda}_1, \bar{\lambda}_2, \bar{\lambda}_3, \cdots, \bar{\lambda}_d)$ via the LLN assumption of $\vec{N}_t$. From Assumption 1, when n is large enough, we can derive the approximation:

TABLE 9.1
Basic data description and the liquidity

Ticker	#of Ord. (1 Day)	Avg # of Ord.	#of Pri. Ch. (1 Day)	Avg # of Pr. Ch.
INTC	404,986	17.3071	3218	0.1375
MSFT	411,409	5.0640	4016	0.1716
AAPL	118,497	5.0640	64,351	2.7500
AMZN	57,515	2.4579	27,558	1.1777
GOOG	49,482	2.1146	24,085	1.0293

TABLE 9.2
Estimated parameters $\vec{\bar{\lambda}}$ for 5 stocks

Ticker	$\bar{\lambda}$
INTC	0.1366
MSFT	0.1729
AAPL	2.2938
AMZN	1.0374
GOOG	0.8178

$$\frac{\vec{N}(nt)}{nt} \sim \vec{\bar{\lambda}}, \;\; t \in [0,1]. \tag{9.1}$$

Take the expectation for (9.1), we have

$$\frac{\mathrm{E}(\vec{N}(nt))}{nt} \sim \vec{\bar{\lambda}}, \;\; t \in [0,1]. \tag{9.2}$$

In this way, we derived the estimated parameters $\vec{\bar{\lambda}}$ for 5 stocks in Table 9.2.

In the definition of the MGCPP, we assumed Markov chains $X_{i,k}$ are independent. So, we checked correlations of the price increments between 5 stocks in Table 9.3. As can be seen from this table, correlations are relatively weak (around 0.3). So, it is reasonable to consider Markov chains $X_{i,k}$ here are independent.

TABLE 9.3
Correlations of the price increments between 5 stocks

Ticker	INTC	MSFT	AAPL	AMZN	GOOG
INTC	1.0000	0.3870	0.2948	0.2932	0.2389
MSFT	0.3870	1.0000	0.4373	0.3984	0.3474
AAPL	0.2948	0.4373	1.0000	0.3697	0.3322
AMZN	0.2932	0.3984	0.3697	1.0000	0.3251
GOOG	0.2389	0.3474	0.3322	0.3251	1.0000

TABLE 9.4
Estimated parameters for the Markov chain

Ticker	p_{uu}	p_{dd}	σ^*	a^*
INTC	0.5373	0.5814	0.0057	-2.5023×10^{-4}
MSFT	0.5711	0.6044	0.0060	-2.0145×10^{-4}
AAPL	0.4954	0.4955	0.0050	-2.1529×10^{-7}
AMZN	0.4511	0.4590	0.0046	-3.6077×10^{-5}
GOOG	0.4536	0.4886	0.0047	-1.6584×10^{-4}

Next, we estimated parameters for the Markov chain by applying the two-state MGCPP model in Corollary 9.2. The transition matrix P of two dependent state Markov chain X_k is denoted as

$$P = \begin{bmatrix} p_{uu} & 1 - p_{uu} \\ 1 - p_{dd} & p_{dd} \end{bmatrix}.$$

We calculated frequency in our data to estimate the p_{uu} and p_{dd} in P by

$$p_{uu} = \frac{q_{uu}}{q_{uu} + q_{ud}},$$
$$p_{dd} = \frac{q_{dd}}{q_{dd} + q_{du}},$$

where q_{uu}, q_{dd}, q_{ud}, and q_{du} are the number of price goes up twice, goes down twice, goes up and then down, goes down and then up, respectively. The result is in Table 9.4.

9.3.3.2 Comparison with MGCHP with Two Dependent Orders

In this section, we compared the simulation results of MGCPP with the multivariate general compound Hawkes process (MGCHP) model to show that the simple generalized model can also reach a good accuracy as the MGCHP who has a sophisticated intensity function. In [11], they simulated the MGCHP with two dependent states for Microsoft and Intel's data. So here we also conduct simulations for Microsoft and Intel's data with the two-state MGCPP.

We tested the MGCPP model by comparing the standard deviation for the left hand side and right hand side in the FCLT:

$$\frac{\vec{S}_{nt} - \tilde{N}_{nt}\vec{a^*}}{\sqrt{n}} \xrightarrow{n\to\infty} \tilde{\sigma}^* \Lambda^{1/2} \vec{W}(t).$$

We separated the data set into disjoint windows $[int, (i+1)nt]$. Since the time was measured in milliseconds, we set $t = 0.001$. Then we can calculate:

$$\vec{S}_i^* = \vec{S}_{(i+1)nt} - \vec{S}_{int} - (\vec{N}((i+1)nt) - \vec{N}(int))\vec{a^*},$$

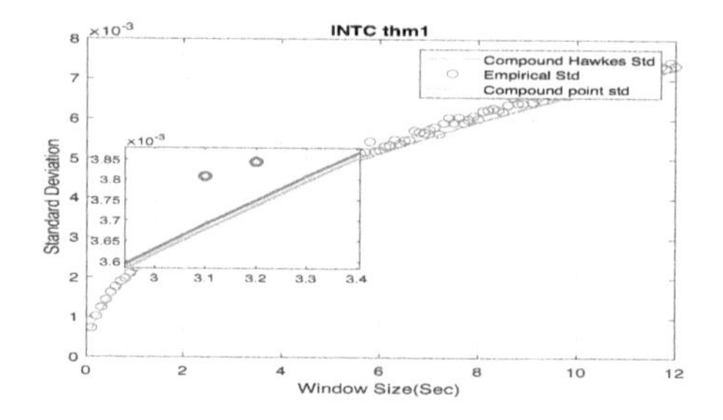

FIGURE 9.1
Standard Deviation Comparison of MGCPP, MGCHP, and the raw data for INTC stock.

and the standard deviation is in the form of

$$\text{std}\left\{\vec{S}^*\right\} \approx \sqrt{n}\tilde{\sigma}^*\Lambda^{1/2}\sqrt{\vec{t}}. \tag{9.3}$$

Figures 9.1 and 9.2 gives a standard deviation comparisons of MGCPP, MGCHP, and the raw data for 2 stocks in different window sizes from 0.1 s to 12 s in steps of 0.1 s. First, we could find the MGCPP parameters make the standard deviation of LHS very similar to the RHS for each stocks when n is large. So, generally speaking, we can say our MGCPP model fits the data well. Second, the MGCPP curve is very close to the MGCHP curve or we could say the simulation results via Intel and Microsoft stocks data are nearly same. It shows that even we do not have a sophisticated intensity function as the Hawkes process, we still can reach a relative good result with a simple point process model. This can help us deal with the computing efficiency problem when using the MGCHP model. We'll give more quantitative error analysis later.

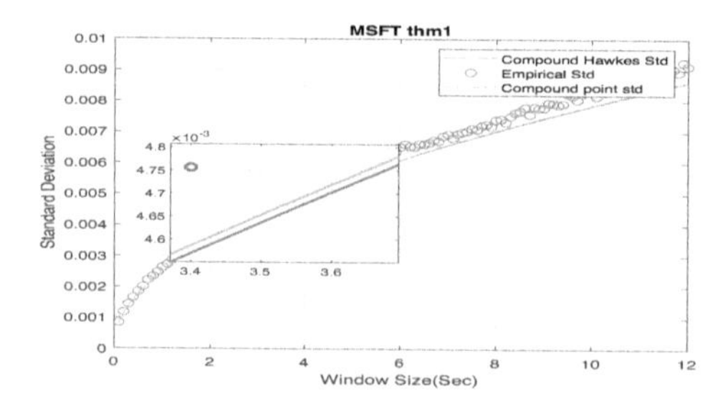

FIGURE 9.2
Standard Deviation Comparison of MGCPP, MGCHP, and the raw data for MSFT stock.

TABLE 9.5
Mean square error (MSE) of the real standard deviation and theoretical standard deviations

Ticker	MGCHP MSE	MGCPP MSE
INTC	3.4039×10^{-8}	3.9858×10^{-8}
MSFT	9.6454×10^{-8}	8.6189×10^{-8}

TABLE 9.6
Percentage error of both stochastic models

Ticker	MGCHP Coeff.	MGCPP Coeff.	Bench. Coeff.	MGCHP Er.	MGCPP Er.
INTC	0.002086	0.002089	0.002162	3.515%	3.377%
MSFT	0.002494	0.002487	0.002609	4.408%	4.676%

Remark 9.7 *Since the number of windows decreases as the window size nt increases, we can find that the spread of data increases when the window size increases in Figure 9.1. For example, when we consider $nt = 0.1$ s, the number of windows is 234,000. However, a 12-s window size yields 1950 windows which will lead the standard deviation increases.*

Intuitively, Figure 9.1 shows that the standard deviation of MGCHP and MGCPP are very close and both of them fit the real standard deviation very well. Next, we analyze MGCHP and MGCPP models quantitatively.

We computed the mean square error (MSE) of the real standard deviation and theoretical standard deviations in Table 9.5. As can be seen from Table 9.5, MGCHP model performs better than the MGCPP model with both Intel and Microsoft data. For Intel stock data, the MSE of MGCHP is 17% better than MGCPP and nearly 10% better than MGCPP model with the Microsoft stock data. However, when we compare the order of magnitude of the MSE (-8) with the real standard deviation (-2 and -3), we still can conclude that MGCPP is good enough for the mid-price modelling task.

Recall Equation (9.3), we can find the standard deviation and the square root of time step have a linear relationship. So, we can fit the real standard deviation data with the square root curve by using the least-square regression. Then, we can set the regression curve as a benchmark and compare the benchmark coefficients with two stochastic models.

From Table 9.6, we can find that the percentage error of both two stochastic models are all smaller than 5% and there is no significant difference between the MGCPP coefficient and the MGCHP coefficient.

Based on the previous analysis, we can conclude that the empirical results of MGCHP and MGCPP are very close and all of them have a very good performance in the mid-price modelling. However, as for the MGCHP, we need to estimate many parameters. As the [11] mentioned, if we consider a two-dimensional MGCHP (two stocks), we have to estimate 5 parameters for the Hawkes process part and the number of parameters increases dramatically to 55 when we consider a 5-dimensional case (5 stocks). The parameter estimation procedure is also quite time consuming for the MGCHP because of the complicated likelihood function of multivariate Hawkes process. For example, it takes a dozen hours to estimate parameters for a 3-dimensional Hawkes process (21 parameters) with LOBSTER data set by using the maximum likelihood estimation (MLE) and the particle swarm optimization (PSO) method in [11]. On the contrary, the number of parameters for MGCPP is much smaller than the MGCHP. In the two-dimensional case, we have 2 parameters to be estimated in the simple point process part and this increases to 5 parameters in the 5-dimensional case, which is much smaller than 55. The parameter estimation procedure is also quite simple and fast (in several seconds with the same data set) because we do not have to deal with the likelihood function. In this way, from the numerical perspective, the generalized model MGCPP is better than the MGCHP because of the fast and simple estimation procedure.

Remark 9.8 *Note that the numbers of parameters we mentioned before are all parameters of the order flow $\vec{N}_t$. Parameters of Markov chains for MGCHP and MGCPP are same.*

In general, we showed that the results of the new generalized model are as good as the MGCHP and this kind of generalization has better numerical properties. In the following parts, we will explore the MGCPP model more.

9.3.3.3 MGCPP with $\mathcal{N}$-State Dependent Orders

We give more simulation examples by using the Google, Apple, and Amazon data with the MGCPP model with $\mathcal{N}$-state dependent orders in this section. Thanks to [21], we can conclude that the accuracy of the general compound Hawkes process model increases when the number of states increases. For Google, Apple, and Amazon in the LOBSTER data set, the best number of states is 4 to 7. In the previous section, we also showed that simulation results of MGCPP are nearly same as the MGCHP. So, it is reasonable to consider a MGCPP model with 7-state Markov chain here.

We applied the method in [21] to calculate the state values $a(X_{i,k})$ for each stock. First, we compute the changes of mid-price and separate the data into two sets by positive increments or negative increments. Next, we calculate the quantiles for both data sets and split the data set according to the quantiles. If there are identical quantiles, we merge them into one. Then, we set the state values $a(X_{i,k})$ as the average of mid-price changes located in each quantile (or merged quantile).

Figures 9.3–9.4 give standard deviation comparisons for MGCPP with 2-state Markov chain and 7-state Markov chain simulated by different tickers' data. Since the 2-state simulation results here are not as good as the results simulated by Intel's and Microsoft's data, we take bigger time steps and window sizes (from 10 s to 20 min with 10 s time step) to capture more dynamics.

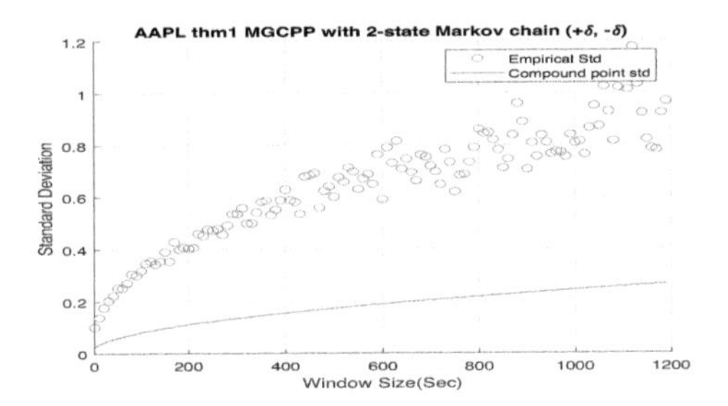

FIGURE 9.3
Standard deviation comparisons for MGCPP with 2-state Markov chain simulated by Apple's stock data.

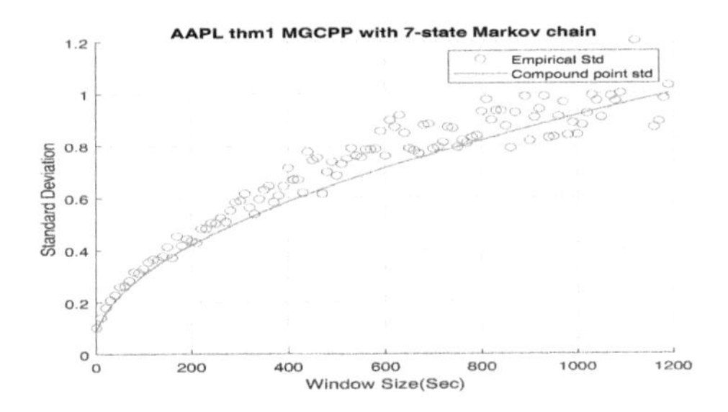

FIGURE 9.4
Standard deviation comparisons for MGCPP 7-state Markov chain simulated by Apple's stock data.

From figures we can find that the 7-state model has a significant improvement than the 2-state model. Seven-state curves for AAPL and GOOG are very close to the real standard deviation, although the theoretical curve of AMZN is underestimated even with the 7-state model.

Table 9.7 lists the MSE and coefficients of the 2-state and 7-state models with different tickers. We can find the improvement of 7-state model quantitatively from the table. The results of AAPL and GOOG are good enough for the mid-price modelling. As for AMZN, although we derive a remarkable improvement from 2-state model (74.60% error) to 7-state model (28.29% error), we cannot make the error smaller than 5% or 10%. This is to say, MGCPP model may not be able to capture the full dynamics for AMZN data, but it still can be a strong candidate for modelling the mid-price, which is consistent with the conclusion of compound Hawkes model in [21].

Remark 9.9 *The MGCPP is not only a generalization of MGCHP, but also a generalization for all multivariate compound models whose point processes $\vec{N}_t$ satisfy the Assumptions 1and 2. The reason we use Hawkes process for comparison is we want to take the advantage of numerical examples in references.*

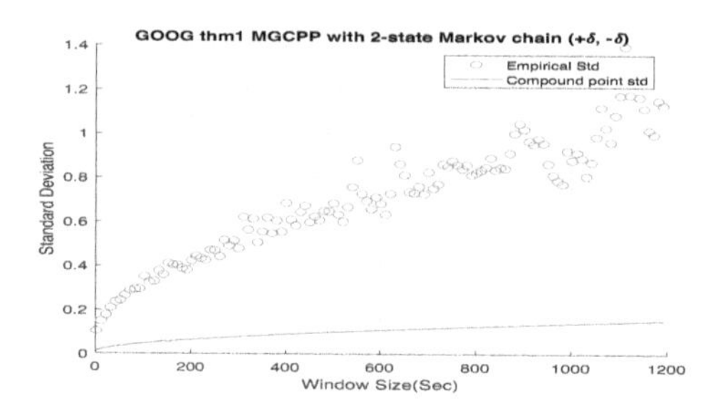

FIGURE 9.5
Standard deviation comparisons for MGCPP with 2-state Markov chain simulated by Google's stock data.

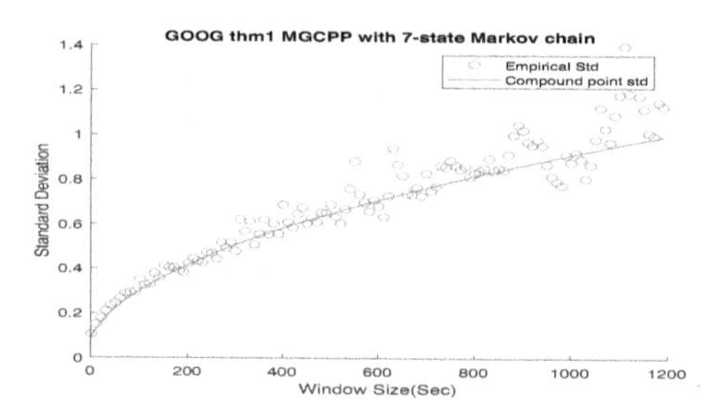

FIGURE 9.6
Standard deviation comparisons for MGCPP 7-state Markov chain simulated by Google's stock data.

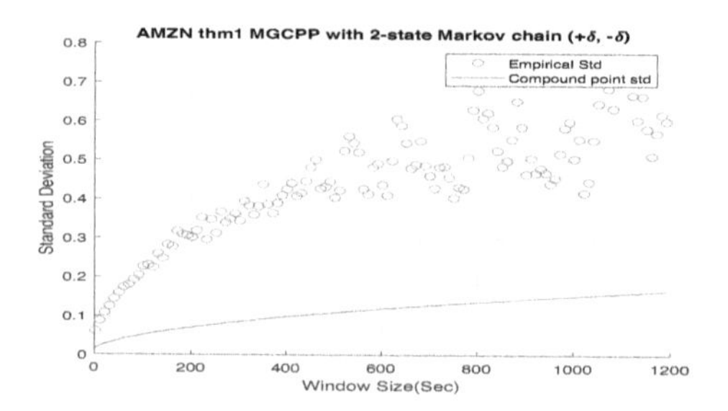

FIGURE 9.7
Standard deviation comparisons for MGCPP with 2-state Markov chain simulated by Amazon's stock data.

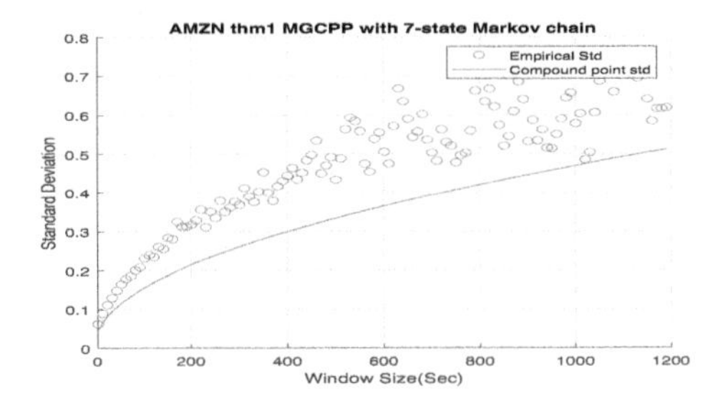

FIGURE 9.8
Standard deviation comparisons for MGCPP 7-state Markov chain simulated by Amazon's stock data.

TABLE 9.7
MSE and coefficients of the 2-state and 7-state models with different tickers

Ticker	MSE	Regr. Ceoff.	MGCPP Coeff.	Perc. Error
AAPL 2-state	0.2467	0.0278	0.0076	72.66%
AAPL 7-state	0.0064	0.0311	0.0288	7.40%
GOOG 2-state	0.4161	0.0307	0.0044	85.67%
GOOG 7-state	0.0081	0.0307	0.0287	6.51%
AMZN 2-state	0.1233	0.0189	0.0048	74.60%
AMZN 7-state	0.0225	0.0205	0.0147	28.29%

9.4 Diffusion Limit for the MGCPP: Deterministic Centralization

We proved a LLN and FCLT for the MGCPP in the previous section. Limit theorems provide us an approximation for the mid-price modelling in the LOB. Recall the approximation in Remark 9.5, we have

$$\vec{S}_{nt} \sim \tilde{\sigma}^* \Lambda^{1/2} \vec{W}(t)\sqrt{n} + \tilde{a^*} \vec{N}_{nt}.$$

where the $\vec{S}_{nt}$ is the price process and $\vec{N}_{nt}$ is the order flow. However, in the real-world problems, last equation cannot help us with the forecasting task directly because we cannot observe the future order flow $\vec{N}_{nt}$ in advance. This is the motivation for us to consider a FCLT 2 for the MGCPP model.

9.4.1 FCLT for MGCPP: Deterministic Centralization

Theorem 9.4 *(FCLT 2: Deterministic Centralization). Let $X_{i,k}$, $i = 1, 2, \cdots, d$ be independent ergodic Markov chains defined before. $X^{\{i\}} = \{1, 2, \cdots, \mathcal{N}_i\}$ is the state space and the ergodic probabilities is given by*

$(\pi^*_{i,1}, \pi^*_{i,2}, \ldots, \pi^*_{i,n})$. *Assume* $X_{i,k}$ *is independent of* $\vec{N}_t$. *Let* $\vec{S}_{nt}$ *be d-dimensional MGCPP, we have*

$$\frac{\vec{S}_{nt} - \tilde{a^*} E(\vec{N}_{nt})}{\sqrt{n}} \xrightarrow{n\to\infty} \tilde{\sigma}^* \Lambda^{1/2} \vec{W}_1(t) + \tilde{a^*} \Sigma^{1/2} \vec{W}_2(t), \; for\, all\, t > 0,$$

where $\vec{W}_1(t)$ *and* $\vec{W}_2(t)$ *are independent d-dimensional Brownian motions. Parameters* $\tilde{\sigma}^*$, $\tilde{a^*}$, Λ, *and* Σ *are defined in Theorem 9.3.*

Proof *Recall the FCLT for MPP (Assumption 1), we have*

$$\left(\frac{1}{\sqrt{n}}\vec{N}_{nt} - \frac{1}{\sqrt{n}}E(\vec{N}_{nt})\right) \xrightarrow{n\to\infty} \mathbf{\Sigma}^{1/2}\vec{W}_t$$

in law for the Skorokhod topology [17]. From FCLT 1 for MGCPP, we have

$$\frac{\vec{S}_{nt} - \tilde{a^*}\vec{N}_{nt}}{\sqrt{n}} \xrightarrow{n\to\infty} \tilde{\sigma}^* \Lambda^{1/2} \vec{W}_t, \; for\, all\, t > 0$$

in the weak law of Skorokhod topology [17]. Here, we assume two multivariate Brownian motions in two last expressions are mutually independent and we denote them as $\vec{W}_2(t)$ *and* $\vec{W}_1(t)$. *Let* $\mathcal{G}_t$ *be the* σ*-algebra generated by* $N_i(s)$, $s \le t$, $1 \le i \le d$. *Since* $\vec{N}_t$ *and the Markov chain* $a(X_{i,k})$ *are independent,* $\vec{S}_t$ *is only determined by* $\vec{N}_t$ *and* $a(X_{i,k})$, *we can have processes*

$$\left(\frac{1}{\sqrt{n}}\vec{N}_{nt} - \frac{1}{\sqrt{n}}E(\vec{N}_{nt})\right),$$

and

$$\frac{\vec{S}_{nt} - \tilde{a^*}\vec{N}_{nt}}{\sqrt{n}}$$

are $\mathcal{G}_t$*-conditional independent. Similar to the central limit theorem in [16], we consider the convergence of conditional on* $\mathcal{G}_t$ *expectations for corresponding processes. Then with the characteristic functions for both limiting processes, we have the joint convergence*

$$\left(\frac{1}{\sqrt{n}}\vec{N}_{nt} - \frac{1}{\sqrt{n}}E\left(\vec{N}_{nt}\right), \frac{\vec{S}_{nt} - \tilde{a}^*\vec{N}_{nt}}{\sqrt{n}}\right) \xrightarrow{\text{conditional on } \mathcal{G}_t} \left(\Sigma^{1/2}\vec{W}_2(t), \tilde{\sigma}^*\Lambda^{1/2}\vec{W}_1(t)\right)$$

as $n \to \infty$. *Next, consider*

$$\frac{\vec{S}_{nt}}{\sqrt{n}} - \frac{\tilde{a}^* E(\vec{N}_{nt})}{\sqrt{n}} = \frac{\vec{S}_{nt} - \tilde{a}^*\vec{N}_{nt}}{\sqrt{n}} + \tilde{a}^*\left(\frac{1}{\sqrt{n}}\vec{N}_{nt} - \frac{1}{\sqrt{n}}E(\vec{N}_{nt})\right).$$

Thus, we can derive

$$\frac{\vec{S}_{nt} - \tilde{a}^*\vec{N}_{nt}}{\sqrt{n}} + \tilde{a}^*\left(\frac{1}{\sqrt{n}}\vec{N}_{nt} - \frac{1}{\sqrt{n}}E(\vec{N}_{nt})\right) \longrightarrow \tilde{\sigma}^*\Lambda^{1/2}\vec{W}_1(t) + \tilde{a^*}\Sigma^{1/2}\vec{W}_2(t)$$

as $n \to \infty$ which proves the result.

Remark 9.10 *We can also consider a special case as the FCLT 1. Let $X_{i,k}$ be a Markov chain with two dependent states $(+\delta, -\delta)$ and the ergodic probabilities are $(\pi_i^*, 1-\pi_i^*)$. Set $a_i(x) = (-\delta) \vee (x \wedge \delta)$. Then, we can derive a similar result for FCLT 2. Parameters $\tilde{a}^*$ and $\tilde{\sigma}^*$ can be computed i the similar way.*

Remark 9.11 *For the FCLT 2, we can also consider a similar approximation as the FCLT 1. For some large enough n, we have*

$$\vec{S}_{nt} \sim \sqrt{n}\tilde{\sigma}^* \Lambda^{1/2}\vec{W}_1(t) + \sqrt{n}\tilde{a^*}\Sigma^{1/2}\vec{W}_2(t) + \tilde{a^*}E(\vec{N}_{nt}), \textit{ for all } t > 0.$$

To deal with the $E(\vec{N}_{nt})$ term, we consider the approximation derived from Assumption 1 in Equation:

$$E(\vec{N}(nt)) \sim nt\vec{\bar{\lambda}}.$$

In tis way, we have the following new approximation

$$\vec{S}_{nt} \sim \sqrt{n}\tilde{\sigma}^* \Lambda^{1/2}\vec{W}_1(t) + \sqrt{n}\tilde{a^*}\Sigma^{1/2}\vec{W}_2(t) + \tilde{a^*}nt\vec{\bar{\lambda}}.$$

9.4.2 Numerical Examples for FCLT: Deterministic Centralization

In this section, we applied the LOBSTER data to test the FCLT 2. According to the numerical examples of FCLT 1, we consider the standard deviation of the approximation, namely

$$\text{std}\left\{\vec{S}_{(i+1)nt} - \vec{S}_{int}\right\} \approx \sqrt{(\tilde{\sigma}^*)^2\Lambda n\vec{t} + (\tilde{a^*})^2\Sigma n\vec{t}.}$$

First, we estimated the covariance matrix Σ by applying the Assumption 2. When n is large enough, have the approximation:

$$\frac{1}{\sqrt{n}}(\vec{N}_{nt} - E(\vec{N}_{nt})) \sim \Sigma^{1/2}\vec{W}_t, \quad t \in [0,1].$$

Take the covariance for both side of last expression, we have

$$\frac{1}{nt}(\text{Cov}(N_i(nt), N_j(nt))) \sim \Sigma_{i,j}, \quad t \in [0,1], \quad i,j = 1,2,\cdots,d.$$

Then, we can derive the estimated Σ for 5 stocks in Table 9.8.

Comparisons of real standard deviation and theoretical standard deviation can be found in Figures 9.9–9.12. Since results of INTC and MSFT are good enough with the 2-state Markov chain $(+\delta, -\delta)$ in FCLT 1, we also applied 2-state Markov chain for INTC and MSFT here. As for AAPL, GOOG, and AMZN, we used the MGCPP model with 7-state Markov chain. Window sizes here start from 1 s and increase to 20 min in time steps of 10 s. As can be seen in Figures 9.9–9.13, the results for FCLT 2 are as good as the FCLT

TABLE 9.8
Estimated Σ for 5 stocks

Ticker	INTC	MSFT	AAPL	AMZN	GOOG
INTC	0.4844	0.1719	1.2393	0.5317	0.5312
MSFT	0.1719	0.5634	1.8361	0.7834	0.7162
AAPL	1.2393	1.8361	62.3800	6.7811	6.4331
AMZN	0.5317	0.7834	6.7811	19.2883	1.9617
GOOG	0.5312	0.7162	6.4331	1.9617	22.7980

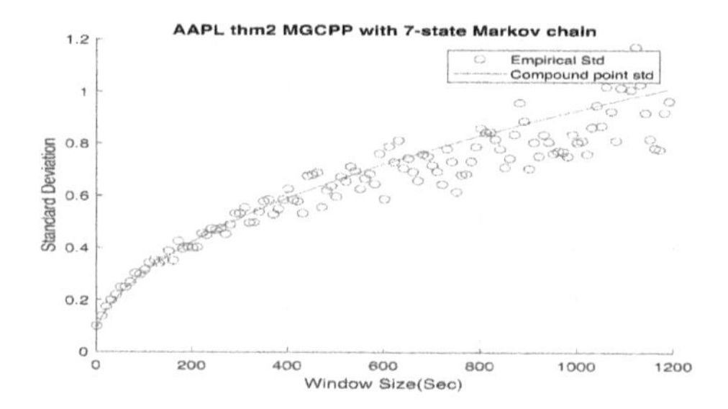

FIGURE 9.9
Standard deviation comparisons for 5 stocks by FCLT II for the MGCPP. AAPL is using 7-state Markov chain.

1 results in Figures 9.3–9.6. We also computed the MSE and coefficients in Table 9.9. Benchmark coefficients are from the least-square regression curves which are similar as benchmarks in Table 9.9.

We see that the percentage errors of MSFT and AAPL are very small (less than 5%) and the results of INTC and GOOG are also good (less than 10%). The percentage error of AMZN is large, but it is still smaller than the error derived from FCLT 1 in Table 9.7. In general, the simulation results of FCLT 2 is as good as the FCLT 1 and we can apply this FCLT 2 to model a mid-price.

9.4.3 Rolling Cross-Validation

In this section, we tested the forecast ability of the MGCPP model. Since we did not assume the multivariate point process $\vec{N}_t$ is stationary, we cannot apply the K-fold cross-validation directly. Here, we used the rolling K-fold cross-validation method which proposed in [6]. We divided the last 50 mins' data into 5 disjoint 10-min windows for each stock. For the fold 1, We take the first 280 mins' data as the training set to estimate parameters. Then, we applied the data in the next 10-min window to calculate the percentage error. Next, we merge the test set into the training set in fold 1 as the new

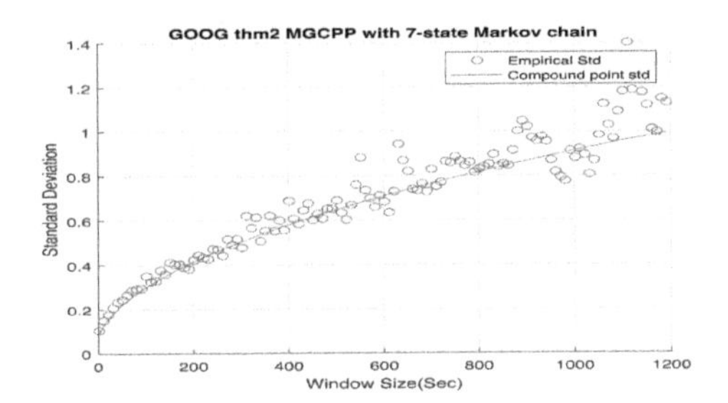

FIGURE 9.10
Standard deviation comparisons for 5 stocks by FCLT II for the MGCPP. GOOG is using 7-state Markov chain.

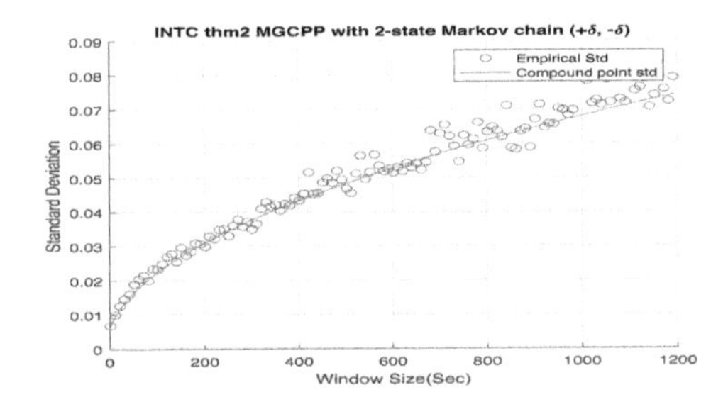

FIGURE 9.11
Standard deviation comparisons for 5 stocks by FCLT II for the MGCPP. INTC are simulated with 2-state Markov chain.

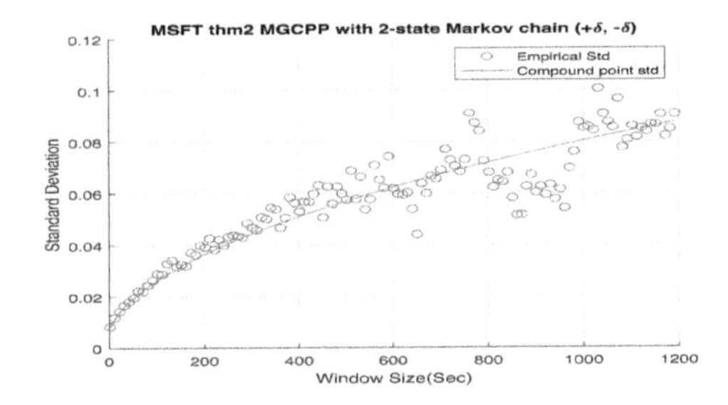

FIGURE 9.12
Standard deviation comparisons for 5 stocks by FCLT II for the MGCPP. MSFT are simulated with 2-state Markov chain .

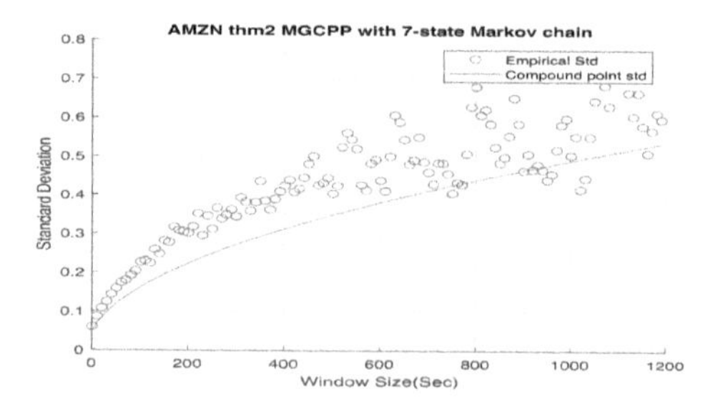

FIGURE 9.13
Standard deviation comparisons for 5 stocks by FCLT II for the MGCPP. AMZN is using 7-state Markov chain.

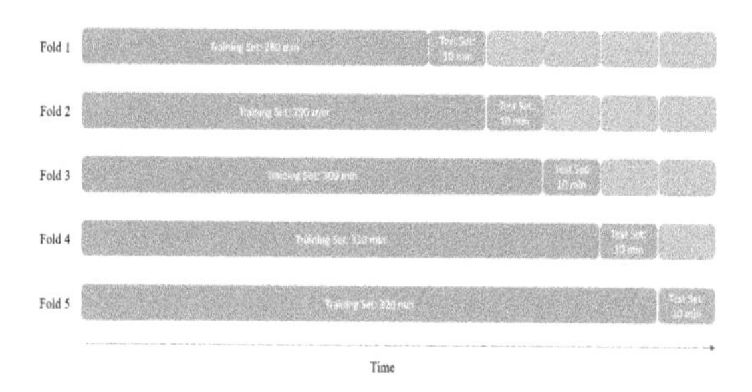

FIGURE 9.14
Rolling Cross-validation.

training set in fold 2 and apply the next 10-min window as a new test set. Repeating this procedure 5 times, we will get 5 percentage errors. The mean value of the 5 percentage errors will be the test error E for this stock. So, the overall test error for our multivariate model is the average of all test errors. Figure 9.14 gives an example diagram for the rolling cross-validation.

Table 9.10 lists test errors for different tickers and the overall test error for the MGCPP model. As can be seen from the table, the test error for each stock is relatively large and the overall test error (15.46%) is nearly double the overall percentage error (7.92%) in Table 9.9. That is because the results in Table 9.10 is a fitting error while the test errors in Table 9.9 is a kind of forecast error. We did not apply any future information when we conduct the forecast task. So, even the 15.46% overall test error is not as good as the fitting one, it is still a good prediction in the LOB and can provide lots of insights in the forecast task.

TABLE 9.9
MSE and coefficients

Ticker	MSE	Bench.Coeff.	MGCPPCoeff.	Perc.Error
INTC 2-state	1.4452×10^{-5}	2.2361×10^{-3}	2.0958×10^{-3}	6.27%
MSFT 2-state	6.6227×10^{-5}	2.5157×10^{-3}	2.4919×10^{-3}	0.94%
AAPL 7-state	6.1382×10^{-3}	2.7799×10^{-2}	2.8788×10^{-2}	3.56%
GOOG 7-state	8.0981×10^{-3}	3.0736×10^{-2}	2.8686×10^{-2}	6.67%
AMZN 7-state	1.1156×10^{-2}	1.8940×10^{-2}	1.4747×10^{-2}	22.14%
Overall Percentage Error		7.92%		

TABLE 9.10
Test errors for different tickers & overall test error for the MGCPP model

Ticker	Fold1	Fold2	Fold3	Fold4	Fold5	Mean Error
INTC	6.75%	0.39%	3.16%	14.32%	16.60%	8.24%
MSFT	20.33%	31.35%	16.96%	8.33%	22.61%	19.92%
AAPL	8.22%	0.51%	22.53%	21.34%	23.33%	15.01%
GOOG	19.60%	20.41%	16.41%	6.13%	12.51%	15.19%
AMZN	20.78%	4.87%	7.98%	18.81%	42.15%	18.92%
Overall				$E_{test} = 15.46\%$		
Test Error				$E_{test} = 15.46\%$		

9.5 Conclusion

This chapter focused on a new generalization of multivariate general compound Hawkes process (MGCHP), which we referred to as the multivariate general compound point process (MGCPP). Namely, we applied a multivariate point process to model the order flow instead of the Hawkes process. The law of large numbers (LLN) and two functional central limit theorems (FCLTs) for the MGCPP were proved in this chapter. Applications of the MGCPP in the limit order market were also considered. We provided numerical simulations and comparisons for the MGCPP and MGCHP by applying Google, Apple, Microsoft, Amazon, and Intel trading data.

Bibliography

[1] Bacry, Emmanuel, Sylvain Delattre, Marc Hoffmann, and Jean-François Muzy. 2013. Some limit theorems for Hawkes processes and application to financial statistics. *Stochastic Processes and their Applications* 123: 2475–2499.

[2] Bowsher, Clive G. 2007. Modelling security market events in continuous time: intensity based, multivariate point process models. *Journal of Econometrics* 141: 876–912.

[3] Brémaud, Pierre, and Laurent Massoulié. 1996. Stability of nonlinear Hawkes processes. *The Annals of Probability* 1: 1563–1588.

[4] Bjork, Tomas. 2011. *Introduction to Point Processes from a Martingale Point of View*; Stockholm: KTH.

[5] Bauwens, Luc, and Nikolaus Hautsch. 2009. *Modelling Financial High Frequency Data Using Point Processes*; Berlin/Heidelberg: Springer.

[6] Bergmeir, Christoph, Rob J. Hyndman, and Bonsoo Koo. 2018. A note on the validity of cross-validation for evaluating autoregressive time series prediction. *Computational Statistics and Data Analysis* 120: 70–83.

[7] Cartea, Álvaro, Sebastian Jaimungal, and José Penalva. 2015. *Algorithmic and High-Frequency Trading*; Cambridge: Cambridge University Press.

[8] Chen, Shizhe, Ali Shojaie, Eric Shea-Brown, and Daniela Witten. 2019. The Multivariate Hawkes Process in High Dimensions: Beyond Mutual Excitation. *arXiv*, arXiv:1707.04928v2.

[9] Cont, Rama, and Adrien De Larrard. 2013. Price dynamics in a Markovian limit order market. *SIAM Journal on Financial Mathematics* 4: 1–25.

[10] Embrechts, Paul, Thomas Liniger, and Lu Lin. 2011. Multivariate Hawkes processes: An application to financial data. *Journal of Applied Probability* 48: 367–378.

[11] Guo, Qi, and Anatoliy Swishchuk. 2020. Multivariate general compound Hawkes processes and their applications in limit order books. *Wilmott* 107: 42–51.

[12] Guo, Q., Remillard, B. nd Swishchuk, A. 2020. Multivariate General Compound Point Processes in Limit Order Books. *Risks*, 8, 98; doi:10.3390/risks8030098

[13] Lemonnier, Rémi, Kevin Scaman, and Argyris Kalogeratos. 2017. Multivariate Hawkes Processes for Large-Scale Inference. Paper presented at Thirty-First AAAI Conference on Artificial Intelligence, San Francisco, CA, USA, February 4–9.

[14] Liniger, Thomas Josef. 2009. Multivariate Hawkes Processes. Ph.D. dissertation, Swiss Federal Institute of Technology in Zurich, Zurich, Switzerland.

[15] Norris, James R. 1998. *Markov Chains*; Cambridge: Cambridge University Press.

[16] Rao, BLS Prakasa. 2009. Conditional independence, conditional mixing and conditional association. *Annals of the Institute of Statistical Mathematics* 61: 441–460.

[17] Skorokhod, A. 1965. *Studies in the Theory of Random Processes*; Courier Dover Publications, vol. 7021. (Reprinted by Addison-Wesley Publishing Company, Inc. 1982).

[18] Swishchuk, Anatoliy. 2017. Risk model based on compound Hawkes process. *Wilmott* 94: 50–57.

[19] Swishchuk, Anatoliy, and Nelson Vadori. 2017. A semi-Markovian modelling of limit order markets. *SIAM Journal on Financial Mathematics* 8: 240–273.

[20] Swishchuk, Anatoliy, Tyler Hofmeister, Katharina Cera, and Julia Schmidt. 2017. General semi-Markov model for limit order books. *International Journal of Theoretical and Applied Finance* 20: 1750019.

[21] Swishchuk, Anatoliy, and Aiden Huffman. 2020. General compound Hawkes processes in limit order books. *Risks* 8: 28.

[22] Vinkovskaya, Ekaterina. 2014. A Point Process Model for the Dynamics of LOB. Ph.D. dissertation, New York: Columbia University.

[23] Vadori, Nelson, and Anatoliy Swishchuk. 2015. Strong law of large numbers and central limit theorems for functionals of inhomogeneous Semi-Markov processes. *Stochastic Analysis and Applications* 33: 213–243.

[24] Yang, Yingxiang, Jalal Etesami, Niao He, and Negar Kiyavash. 2017. Online Learning for Multivariate Hawkes Processes. Paper Presented at 31st Conference on Neural Information Processing Systems, Long Beach, CA, USA, December 4–9.

[25] Zheng, Ban, François Roueff, and Frédéric Abergel. 2014. Ergodicity and scaling limit of a constrained multivariate Hawkes process. *SIAM Journal on Financial Mathematics* 5: 99–136.

[26] Zhu, Lingjiong. 2013. Central limit theorem for nonlinear Hawkes processes. *Journal of Applied Probability* 50: 760–771.

Part IV

Appendix: Basics in Stochastic Processes

A

Basics in Stochastic Processes

This Appendix contains all necessary basic definitions and facts in stochastic processes which we use in this book.

It includes discrete- and continuous-time Markov chains, Wiener, Poisson, compound Poisson, Hawkes, Markov, Markov renewal and semi-Markov processes, and martingales. We refer to [1]-[7] for more information and results.

A.1 Discrete-time Markov Chains

Let us consider a certain random process that retains *no memory* of the past. This means that only the current state of the process can influence where it goes next. Such a process is called a *Markov process.* In this section, we are concerned exclusively with the case where the process can assume only a finite or countable set of states, and such a process is usually called a *Markov chain.* We shall consider chains both in *discrete* time $n \in \mathbf{Z}^+ \equiv \{0, 1, 2, ...\}$ and *continuous time* $t \in \mathbf{R}_+ \equiv [0, +\infty)$. The letters n, m, k will always denote integers, whereas t and s will refer to real numbers. Thus we write $(y_n)_{n \in \mathbf{Z}^+}$ for a discrete-time process and $(y_t)_{t \in \mathbf{R}^+}$ or $y(t)$ for a continuous-time process.

Let us start with a **discrete-time Markov chain.** We now describe a discrete Markov chains by the following example where the *state space* Y contains three elements (states): $Y = \{1, 2, 3\}$.

In the process, the state is changed from state 1 to state 2 with probability 1; from state 3 either to 1 or to 2 with probability 2/3 or 1/3, respectively, and from state 2 to state 3 with probability 1/4, otherwise stay at 2. Therefore, we obtain a stochastic matrix

$$\mathbf{P} = \begin{pmatrix} p_{11} & p_{12} & p_{13} \\ p_{21} & p_{22} & p_{23} \\ p_{31} & p_{32} & p_{33} \end{pmatrix} = \begin{pmatrix} 0 & 1 & 0 \\ 0 & 3/4 & 1/4 \\ 2/3 & 1/3 & 0 \end{pmatrix}.$$

Now, we let Y be a countable set. Each $i \in Y$ is called a *state* and Y is called the *state space.* We say that a set of real numbers $p = (p_i; i \in Y)$ is a *measure* on Y if $0 \leq p_i < +\infty$ for all $i \in Y$. If, in addition, the *total mass* $\sum_{i \in Y} p_i = 1$, then we call p a *distribution.* In what follows, we fix a *probability*

DOI: 10.1201/9781003265986-A

space $(\Omega, \mathcal{F}, \mathcal{P})$. A *random variable* y with values in Y is a function $y : \Omega \to Y$. For a random variable y, if we define

$$p_i = \mathcal{P}\{y = i\} \equiv \mathcal{P}\{\omega : y(\omega) = i\},$$

then p defines a distribution, called the distribution of y. We therefore can think of y as modelling a random state which takes the value i with probability p_i.

We say that a matrix $\mathbf{P} = (p_{ij}; i, j \in Y)$ is *stochastic* if every row $(p_{ij}; j \in Y)$ is a distribution, namely, $p_{ij} \geq 0$ and $\sum_{j \in Y} p_{ij} = 1$.

A *semi-stochastic matrix* $\mathbf{P} = (p_{ij}; i, j \in Z_+)$ is one such that if every row $(p_{ij}; j \in Z_+)$ satisfies $0 \leq P_{ij} < +\infty$ and $\sum_{j \in Y} p_{ij} \leq 1$.

We say that $(y_n)_{n \in \mathbf{Z}^+}$ is a Markov chain with *initial distribution* p and *transition matrix* $\mathbf{P}$, if

(a) y_0 has distribution p;

(b) for $n \geq 0$, conditional on $y_n = i$, y_{n+1} has distribution $(p_{ij}; j \in Y)$ independent of $y_0, y_1, ..., y_{n-1}$, namely,

$$\mathcal{P}\{\omega : y_{n+1} = j | y_n = i\} = p_{ij}.$$

More explicitly, we have, for $n \geq 0$ and $i, j, i_0, i_1, ..., i_{n-1} \in Y$, that

(a)

$$\mathcal{P}\{y_0 = i\} = p_i;$$

(b)

$$\mathcal{P}\{y_{n+1} = j | y_0 = i_0, y_1 = i_1, ..., y_{n-1} = i_{n-1}, y_n = i\} = p_{ij}.$$

A Markov chain with a finite phase space is called a *finite Markov chain.* In general, the phase space is not required to be countable. We now describe a general case. Let $(Y, \mathcal{Y})$ be a measurable space, here $\mathcal{Y}$ is a σ-algebra of measurable sets on Y, which can be interpreted as a collection of observable subsets of states in the random environment. We assume that $\mathcal{Y}$ contains all one-point sets.

A function $P : Y \times \mathcal{Y} \to [0, +\infty)$ is called *a stochastic kernel* in the measurable phase space $(Y, \mathcal{Y})$ if it satisfies the following conditions:

(i) for fixed $y \in Y$, the function $P(y, A)$ is a probability distribution on A and $P(y, Y) = 1$;

(ii) for fixed $A \in \mathcal{Y}$, the function $P(y, A)$ is $\mathcal{Y}$-measurable with respect to $y \in Y$.

If (i) and (ii) are satisfied, except that $P(y, Y) = 1$ being replaced by $P(y, Y) \leq 1$ for $y \in Y$, then the kernel is said to be *semi-stochastic.*

In applications, the stochastic kernel $P(y, A)$ determines the probability of transitions of the random environment under consideration from the state y into the set of states A.

In a discrete phase space $Y = \{1, 2, \ldots, \}$, a stochastic kernel is given by a stochastic matrix $P = (p_{ij}, i, j \in Y)$ with non-negative elements as follows: $P(i, A) = \sum_{j \in A} p_{ij}$ for any $A \in \mathcal{Y}$.

For a Markov chain $(y_n)_{n \in Z^+}$, we can define the initial distribution $P(A) = \mathcal{P}\{y_0 \in A\}$ and the following stochastic kernels $P_n(y, A) = \mathcal{P}\{y_{n+1} \in A | y_n = y\}$, called the *probabilities of the (n-th step) one-step transitions of* $(y_n)_{n \in Z^+}$.

A Markov chain $(y_n)_{n \in Z^+}$ is *homogeneous* if the above probabilities of one-step transitions do not depend on the transition time n.

It is known that $(y_n)_{n \in Z^+}$ is a Markov chain if and only if the following *Markovian property* holds: the joint distributions of the states of this chain are determined only by the initial distribution and the probabilities of one-step transitions as follows

$$\begin{aligned} &\mathcal{P}\{y_n \in A_n, y_{n-1} \in A_{n-1}, \ldots, y_1 \in A_1, y_0 \in A_0\} \\ &= \int_{A_0} P(dy_0) \cdot \int_{A_1} P(y_0, dy_1) \cdot \int_{A_2} P(y_1, dy_2) \\ &\quad \cdots \int_{A_{n-1}} P(y_{n-2}, dy_{n-1}) P(y_{n-1}, A_n). \end{aligned}$$

According to the Markovian property, the *probability of the n-step transition* $P_n(y, A), n \geq 1$, satisfies the Chapman-Kolmogorov equation

$$P_{n+m}(y, A) = \int_Y P_n(y, dz) P_m(z, A),$$

which describes the Markovian property of the chain $(y_n)_{n \in Z^+}$ analytically: given a state of the chain at a fixed time, the probability law of the future evolution of the environment does not depend on the state of the system in the past.

A.1.1 Continuous-time Markov Chains.

We now turn to *continuous-time Markov chains.* Again, let Y be a countable space. A *continuous-time* random process $(y(t))_{t \in R^+}$ with values in Y is a family of random variables $y(t) : \Omega \to Y$. We shall abuse the notation and using both $y(t)$ and y_t if no confusion arises. There are some problems connected with the process to estimate, for example, $\mathcal{P}(y(t) = i)$. These problems are not presented for discrete-time cases since for a countable disjoint union $\mathcal{P}(\bigcup_n A_n) = \sum_n \mathcal{P}(A_n)$, but there is no such analogue for an uncountable union $\bigcup_{t \geq 0} A_t$. To avoid these problems, we shall restrict our attention to a process $(y(t))_{t \in R_+}$ which is *right-continuous* in the following sense: for all $\omega \in \Omega$ and $t \geq 0$ there exists $\epsilon > 0$ such that $y(s) = y(t)$ for $t \leq s < t + \epsilon$.

Every path $R^+ \ni t \to y(t, \omega)$ of a right-continuous process must remain constant for a while, so there are three cases:

(i) the path makes infinitely many jumps, but only finitely many in any finite interval $[0, t]$;

(ii) the path makes finitely many jumps and then becomes stuck in some state forever;

(iii) the process makes infinitely many jumps in a finite interval.

We are interested, throughout this book, in the first case. Such processes are called *regular processes.*

We need to introduce the notion of a Q-matrix in order to better describe continuous-time Markov chains. A *Q-matrix* on Y is a matrix $Q = (q_{ij}; i, j \in Y)$ satisfying the following conditions:

(i) $0 \leq -q_{ii} < \infty$ for all i;

(ii) $q_{ij} \geq 0$ for all $i \neq j$;

(iii) $\sum_{j \in Y} q_{ij} = 0$ for all i.

For example, the following matrix is Q-matrix:

$$Q = \begin{pmatrix} -2 & 1 & 1 \\ 1 & -1 & 0 \\ 2 & 1 & -3 \end{pmatrix}.$$

Intuitively, in a Q-matrix, the entry q_{ij} is the rate of changing from state i to state j, and $-q_{ii}$ is the rate of leaving state i.

Now we illustrate how to obtain a continuous-time Markov chain from a discrete-time Markov chain. We first assume the discrete parameter space $\{1, 2, ...\}$ is embedded in the continuous parameter space $[0, \infty)$. A natural way to interpolate the discrete sequence $\{p^n; n \in Z^+\}$ for $p \in (0, \infty)$ is by the function $(\exp\{tq\}; t \geq 0)$, where $q = \log p$. Consider a finite space Y and a matrix $P = (p_{ij}; i, j \in Y)$. Suppose that we can find a matrix Q with $\exp\{Q\} = P$. Then $\exp\{nQ\} = (\exp\{Q\})^n = P^n$. In this way, $(\exp\{tQ\}; t \geq 0)$ fills in the gaps in the discrete sequence. On the other hand, if Q is a given matrix on a finite space Y, and if $P(t) = \exp\{tQ\}$, then $(P(t))_{t \in R^+}$ has the following properties (Chapman-Kolmogorov equation):

(I1)

$$P(s) \cdot P(t) = P(t + s), \forall t, s \geq 0;$$

(I2)

$$\lim_{t \to 0} P(t) = \mathbf{I},$$

the identity matrix.

Conversely, if the properties (I1)–(I2) hold, then $P(t) = [P(t/n)]^n$ for all positive integer n, and hence

$$P(t) = \exp\{Qt\},$$

for all $t \geq 0$, where $Q = P^{'}(0)$ is the derivative of $P(t)$ at $t = 0$. It follows that if $P(t)$ is also a stochastic matrix for every $t \geq 0$, then

$$Q = (q_{ij})$$

satisfies:

$$\begin{cases} q_{ii} &= \lim_{t\to 0}(p_{ii}(t)-1)/t \leq 0 \qquad \text{for all } i \in Y, \\ q_{ij} &= \lim_{t\to 0} p_{ij}(t)/t \geq 0 \qquad \text{for all } i \neq j, \\ \sum_j q_{ij} &= 0 \qquad \text{for all } i \in Y. \end{cases}$$

Moreover, $P(t)$ satisfies the *backward* equation

$$\frac{dP(t)}{dt} = QP(t), P(0) = I,$$

and *forward* equation

$$\frac{dP(t)}{dt} = P(t)Q, P(0) = I.$$

Furthermore, for $k = 0, 1, 2, ...,$ we have

$$\frac{d^k P(t)}{dt}|_{t=0} = Q^k.$$

It is also easy to show that if Q is a Q-matrix, then for every $t \geq 0$, $P(t)$ is a stochastic matrix.

Therefore, $P(t) = \exp\{tQ\}$ is a stochastic matrix for every $t \geq 0$ if and only if Q is a Q-matrix.

We say $(y(t))_{t\in R^+}$ is a continuous-time Markov chain in a finite phase space Y if there exists a family of stochastic matrices $P(t) = (p_{ij}(t); i, j \in Y)$ satisfying $(I1)$ and $(I2)$ and such that

$$\mathcal{P}\{y(t_{n+1}) = i_{n+1}/y(t_0) = i_0, ..., y(t_n) = i_n) = p_{i_n i_{n+1}}(t_{n+1} - t_n)\}$$

for all $n \in Z^+$, all $0 \leq t_0 \leq ... \leq t_{n+1}$ and all $i_1, ..., i_{n+1} \in Y$.

In particular, $p_{ij}(t) = \mathcal{P}\{y(t) = j|y(0) = i\}$.

Since the process $(y(t))_{t\in R^+}$ is assumed to be right continuous, the system starts from some state y_1, and stays at this state for a duration of θ_1, then moves to a new state y_2 and stays there again for a duration θ_2, and then jumps to a new state y_3 where it stays for a duration θ_3 and this process continues. The resulted discrete-time process $(y_n)_{n\in Z^+}$ is a Markov chain with the transition probability

$$\mathcal{P}(y_{n+1} = j|\theta_1, ..., \theta_n; y_n = i) = q_{ij}/(-q_{ii}),$$

and $(\theta_n)_{n\in Z^+}$ is also a Markov chain with

$$\mathcal{P}(\theta_{n+1} > t|\theta_1, ..., \theta_n; y_{n+1} = j) = \exp(q_j t),$$

where $q_j := q_{jj}$. The transition probability of the two-component chain $((y_n, \theta_n))_{n\in Z_+}$ is given by, for $i \neq j$,

$$\mathcal{P}(y_{n+1} = j, \theta_{n+1} > t|y_n = i) = (q_{ij}/(-q_{ii})) \exp(q_i t).$$

Therefore, a continuous-time Markov chain $(y(t))_{t\in R^+}$ is characterized by an initial distribution p (the distribution of y_0) and a Q-matrix Q such that $P(t) = \exp(Qt)$ (the *generator matrix* of $(y(t))_{t\in R^+}$, which determines how the process involves from its initial state). More specifically, if the chain starts at i, then it stays there for an exponential time of parameter q_i and then jumps to a new state, choosing state j with probability $q_{ij}/(-q_{ii})$.

Consider a continuous Markov chain for which the system stays in state 0 for a random time with exponential distribution of parameter $\lambda \in (0, +\infty)$, then jumps to 1. Thus the distribution function of the waiting time T is given by $G_T(t) = 1-\exp\{-\lambda t\}$, $t \in R^+$. This is the so-called *exponential distribution of parameter* λ. This distribution plays a fundamental role in continuous-time Markov chains because of the *memoryless property:* $\mathcal{P}(T > t + s|T > s) = \mathcal{P}(T > t)$ for all $s, t \geq 0$. A random variable T has an exponential distribution if and only if it has this memoryless property.

In the second example, when the system reaches the state 1, it does not stop there but rather, after another independent exponential time of parameter λ jumps to state 2, and so on. The resulting process is called the *Poisson process of rate* λ. The associated matrix Q is given by:

$$\begin{pmatrix} -\lambda & \lambda & \dots & 0 \\ 0 & -\lambda & \lambda & \dots \\ 0 & 0 & -\lambda & \lambda \\ 0 & \dots & \dots & \dots \end{pmatrix}.$$

Here, the probability $p_{ij}(t)$ has the following form:

$$p_{ij}(t) = \exp(-\lambda t)((\lambda t)^{j-i}/(j-i)!).$$

Note that if $i = 0$, we obtain the *Poisson probabilities* of parameter λt.

A *birth process* is a generalization of the Poisson process in which λ is allowed to depend on the current state of the process. Thus a birth process is characterized by a sequence of *birth rates* $0 \leq q_j < \infty$ for $j = 0, 1, 2, ...$

The corresponding matrix Q is given by

$$\begin{pmatrix} -q_0 & q_0 & \dots & 0 \\ 0 & -q_1 & q_1 & \dots \\ 0 & 0 & -q_2 & q_2 \\ 0 & \dots & \dots & \dots \end{pmatrix}.$$

Finally, a *Birth and Death process* by $q_{ii} = -(\lambda_i + \mu_i)$, $q_{ij} = \lambda_j$ if $j = i+1$, $q_{ij} = \mu_j$ if $j = i - 1$, and $q_{ij} = 0$ for $j \neq j - 1, j + 1$,where λ_i are birth rates and μ_i are death rates.

A.1.2 Ergodicity and Reducibility of Markov Chains

The ergodic properties of homogeneous Markov chains are quite important for the averaging, merging and other limit theorems, and for the stability of populations of biological systems in random environment to be studied in later chapters.

We shall consider a homogeneous Markov chain $(y_n)_{n\in Z^+}$ defined in a measurable phase space of states $(Y, \mathcal{Y})$, and we assume a stochastic kernel $P(y, A), y\epsilon Y, A\epsilon\mathcal{Y}$, is defined by the chain, as explained in Section 10.1.

Let us start with the simple case where $Y = \{1, 2, \ldots, \ldots\}$, and let $P =$ with $(p_{ij}; i, j = 1, 2, \ldots)(p_{ij} \geq 0, \sum_{j=1}^{\infty} p_{ij} = 1,)$ be the transition probability matrix of the chain.

We write $j \to k$ to denote the fact that the state k *can be reached* from the state j (or j *leads* to k). More precisely, $j \to k$ if there is an integer $n \geq 1$ such that $p_{jk}^{(n)} > 0$, where $p_{jk}^{(n)}$ denotes the jkth element of matrix P^n (the conditional probability of obtaining the kth state from jth state at the nth step). We say j *communicates with* k if $j \to k$. We write $k \leftrightarrow j$, if $k \to j$ and $j \to k$. It is easy to see, that $\leftrightarrow$ is an equivalence relation on Y, and thus partitions Y into *communicating classes.*

We call a class of states C *closed* if $p_{jk} = 0$ whenever $j \in C$ and $k \notin C$, namely, no one-step transition is possible from any state belonging to C to any state outside C. Therefore, $p_{jk}^{(n)} = 0$ for all n, if $j \in C$ and $k \notin C$, and so once an environment reaches a state belonging to C it can never subsequently be in any state outside C.

In the special case where C consists of the single state j (that is, $p_{jj} = 1$), j is called an *absorbing state.* A Markov chain in which there are two or more closed sets is said to be *reducible.* The chain is called *irreducible* if there exists no closed set other than the set of all states.

Let $f_j^{(n)}$ denote the probability that the environment starts from the initial state j and returns to state j for the first time after n transitions. The sum $f_j := \sum_{n=1}^{\infty} f_j^{(n)}$ is the probability that the environment ever returns to state j. The state j is *recurrent* (or *persistent*) if $f_j = \sum_{n=1}^{+\infty} f_j^{(n)} = 1$, and *transient* if $f_j = \sum_{n=1}^{+\infty} f_j^{(n)} < 1$. Therefore, if j is transient, then there is a positive probability then the environment will never return to j.

Let $\tau_j := \sum_{n=1}^{\infty} n f_j^{(n)}$ be the *mean recurrence time* for j. If there exists no n such that $f_j^{(n)} = 0$, we let $\tau_j = +\infty$. When τ_j is finite, j is said to be *non-null*; when τ_j is infinite, j is said to be *null.*

To illustrate the above notions, consider the following Markov chain: the states are partitioned into three *communicating classes*, (0), (1, 2, 3) and (4, 5, 6). Two of these classes are *closed*, meaning that one cannot escape from either (1, 2, 3) or (4, 5, 6). These closed classes are *recurrent.* The class (0) is *transient*, and the state (0) is null.

We now discuss the *periodicity.* We say that a recurrent state j has period $\mathbf{T}$ if a return to j is impossible except possibly after $\mathbf{T}$, $2\mathbf{T}$, $3\mathbf{T}\ldots$ transitions and $\mathbf{T}$ is the largest such integer satisfying this property. When $\mathbf{T}$=1, j is said to be *aperiodic.*

A recurrent non-null state which is aperiodic is said to be *ergodic.*

The following results hold for a discrete-time Markov chain:

(P1) j is a transient state if $\sum_{n=1}^{\infty} p_{jj}^{(n)}$ is convergent;

(P2) j is a recurrent state if $\sum_{n=1}^{\infty} p_{jj}^{(n)}$ is divergent;

(P3) $p_{jj}^{(n)} \to \frac{1}{\tau_j}$, if j is ergodic;

(P4) $p_{jj}^{(n)} \to 0$ if j is null;

(P5) $p_{jj}^{(n\mathbf{T})} \to \frac{\mathbf{T}}{\tau_j}$ if j is recurrent, non-null, and with period $\mathbf{T}$;

(P6) In an irreducible Markov chain, all states belong to the same class and they are all transient, they are either all recurrent null or all recurrent non-null, and have the same periodicity.

For a recurrent Markov chain with a countably generated σ-algebra $\mathcal{Y}$ on the phase space Y, there exists a unique (up to a constant factor) distribution $(p_i; i = 1, 2, \ldots)$ such that

$$p_i = \sum_{j=1}^{\infty} p_{ij} p_j.$$

Distribution $(p_j)_{j \geq 1}$ defines the *stationary distributions.*

Many long-time properties of a Markov chain are connected with the notion of an *invariant distribution* or *invariant measure.* We say a distribution p is *invariant* if $p\mathbf{P} = p$. The terms *equilibrium* and *stationary* are also used for invariant measures.

The most general two-state chain has the transition matrix

$$\mathbf{P} = \begin{pmatrix} 1-\alpha & \alpha \\ \beta & 1-\beta \end{pmatrix},$$

with $0 < \alpha, \beta \leq 1$.

From the relation $\mathbf{P}^{n+1} = \mathbf{P}^n \mathbf{P}$, we get

$$p_{11}^{(n+1)} = p_{12}^{(n)} \beta + p_{11}^{(n)} (1 - \alpha).$$

As $p_{11}^{(n)} + p_{12}^{(n)} = 1$, we can eliminate $p_{12}^{(n)}$ and get a recurrence relation for $p_{11}^{(n)}$ as follows

$$p_{11}^{(n)} = (1 - \alpha - \beta) p_{11}^{(n)} + \beta, \qquad p_{11}^{(0)} = 1.$$

This has a unique solution given by

$$p_{11}^{(n)} = \beta/(\alpha + \beta) + (\alpha/(\alpha + \beta))(1 - \alpha - \beta)^{(n)}, \qquad \text{if } \alpha + \beta > 0,$$

or

$$p_{11}^{(n)} = 1 \qquad \text{if } \alpha + \beta = 0.$$

In particular,

$$\mathbf{P}^n \to \begin{pmatrix} \beta/(\alpha+\beta) & \alpha/(\alpha+\beta) \\ \beta/(\alpha+\beta) & \alpha/(\alpha+\beta) \end{pmatrix} \qquad \text{as } n \to +\infty,$$

so the distribution $p = (\beta/(\alpha+\beta), \alpha/(\alpha+\beta))$ is invariant, namely, $p\mathbf{P} = p$.

General Case of Discrete-time Markov Chain. In a general case with a measurable phase space of states $(Y, \mathcal{Y})$, for a recurrent Markov chain $(y_n)_{n\in Z^+}$ in Y with a countably generated σ-algebra $\mathcal{Y}$, there exists a *unique invariant measure* $p(B)$ such that

$$p(B) = \int_Y p(dy)P(y,B), \forall B \in \mathcal{Y}. \tag{A.1}$$

If the invariant measure is finite, then we can assume that it is normalized, i.e., $p(Y) = 1$. It follows from (A.1) that $\mathcal{P}\{y_n \in B\} = p(B)$ for all $n \geq 1$ and $\mathcal{P}\{y_0 \in B\} = p(B)$. Hence, the measure $p(B)$ defines the *stationary distribution* of a Markov chain in $(Y, \mathcal{Y})$. Using (A.1), we can get an equivalent definition of an *ergodic Markov chain*: it is an aperiodic Markov chain with a stationary distribution p, which is defined by relation (A.1) and satisfies the condition $p(Y) = 1$.

For an ergodic Markov chain, we have

$$\begin{aligned} \lim_{n\to+\infty} P^n(y,A) &= \lim_{n\to+\infty} \mathcal{P}\{y_n \in A | y_0 = y\} \\ &= p(A), \quad \forall A \in \mathcal{Y}, \forall y \in Y. \end{aligned}$$

The class structure of a continuous-time Markov chain $(y_t)_{t\in R^+}$ with the associated Q-matrix Q and $P(t) := \exp(tQ)$ is the same as the case of the above described discrete-time Markov chains. We say that i *leads* to j and write $i \to j$ if $\mathcal{P}_i(y(t) = j) := \mathcal{P}\{y(t) = j | y(0) = i\} > 0$ for some $t \geq 0$. The notions of *communication*, *communicating class*, *clossed class*, *absorbing state*, and *irreducibility* are analogous to those for discrete-time Markov chains. We say a state i is *recurrent* if $\mathcal{P}_i([t \geq 0 : y(t) = i] \quad \text{is unbounded}) = 1$. A state i is recurrent if $q_i = 0$ or $\mathcal{P}(T_i < \infty) = 1$, where $T_i(\omega) \equiv T_i := \inf(t \geq \tau_1(\omega) : y(t) = i)$ is the *first passage time* of $y(t)$ to state i, and $\tau_1 := \inf(t \geq 0 : y(t) \neq y(0))$. Therefore, a state i is *transient* if $\mathcal{P}_i([t \geq 0 : y(t) = i] \quad \text{is unbounded}) = 0$.

The following dichotomy holds: a) if $q_i = 0$, then i is recurrent and $\int_0^\infty p_{ii}(t)dt = \infty$; b) if $q_i > 0$, then i is transient and $\int_0^\infty p_{ii}(t)dt < \infty$.

The notions of invariant distributions and measures play an important role in the study of continuous-time Markov chains $(y_t)_{t\in R^+}$. We say that p is *invariant* if $pQ = 0$.

It is known that if Q is irreducible and recurrent, then Q has an invariant measure p which is unique up to scalar multiplies. Moreover, for all states i, j we have $p_{ij}(t) \to p_j$ as $t \to \infty$. Finally, if Q is irreducible and recurrent, and if p is a measure, then for each fixed $s > 0$ $pQ = 0$ if and only if $pP(s) = p$.

Ergodic Theorem for continuous-time Markov chains asserts that

$$\mathcal{P}((1/t)\int_0^t \mathbf{1}_{(y(s)=i)} ds \to 1/(m_i q_i)) = 1 \qquad \text{as } t \to +\infty,$$

where $m_i := E_i(T_i)$ is the expected return time to state i, and $\mathbf{1}_A$ is the characteristic function of a set A.. Moreover, if $q_i = 0$, then for any bounded function $f : Y \to R$ we have

$$\mathcal{P}((1/t)\int_0^t f(y(s))ds \to \hat{f}) = 1 \qquad \text{as } t \to \infty,$$

where $\hat{f} := \sum_{i\in Y} p_i f_i$ and where $(p_i : i \in Y)$ is the unique invariant distribution.

Operator of Transition Probabilities and Potential of a Markov Chains. Let $\mathbf{B}(Y)$ be *a normed space* of Y-measurable bounded functions $f : Y \to R$ with norm $||f|| := \sup_{y\in Y} |f(y)|$. An operator of transition probabilities $\mathbf{P}$ in the space $\mathbf{B}(Y)$ is defined by the stochastic kernel $P(y, A)$ as

$$\mathbf{P}f(y) := \int_Y P(y, dz)f(z).$$

A stationary projector Π in $\mathcal{B}(Y)$ is defined by the stationary distribution $p(A)$ of the ergodic irreducible Markov chain as

$$\Pi f(y) := \int_Y p(dz)f(z) =: \hat{f} \cdot \mathbf{1}(y). \tag{A.2}$$

Here,

$$\mathbf{1}(y) \equiv 1 \quad \text{for all } y\epsilon Y, \quad \text{and } \hat{f} := \int_Y p(dz)f(z).$$

The operator Π defined by (A.2) possesses the projection property, i.e., $\Pi^2 = \Pi$.

An ergodic Markov chain with the operator of transition probabilities $\mathbf{P}$ and the stationary projector Π is called *uniformly ergodic* if

$$\lim_{n\to+\infty} \sup_{||f||\leq 1} ||(\mathbf{P}^n - \Pi)f|| = 0 \qquad \forall f \in \mathbf{B}(Y).$$

For a uniformly ergodic Markov chain, the operator $Q := \mathbf{P} - \mathbf{I}$ is invertibly reducible, namely,

$$\mathbf{B}(Y) = N(Q) \oplus R(Q), \dim N(Q) = 1,$$

where $N(Q)$ is the null-space of Q, which consists of all constant functions, and $R(Q)$ is the range of the operator Q, which is closed, dim $N(Q)$ is the dimension of the space $N(Q)$. The stationary projector is a projector onto the null-space $N(Q)$.

For a uniformly ergodic Markov chain the following operator, called *the potential*

$$\mathbf{R}_0 = (\mathbf{P} - \mathbf{I} + \Pi)^{-1} - \Pi = \sum_{n=1}^{+\infty} (\mathbf{P}^n - \Pi)$$

is well-defined. The boundness of the linear operator $\mathbf{R}_0$ follows from the uniform convergence of the series

$$\sum_{n=1}^{+\infty} ||(\mathbf{P}^n - \Pi)f|| < +\infty \qquad \forall f \epsilon \mathbf{B}(Y)$$

with $||f|| \leq 1$.

We note, that if $(y_n)_{n \in Z^+}$ is a stationary ergodic Markov chain with ergodic distribution $p(dy)$, then for every $f \in \mathbf{B}(Y)$ we have that

$$\mathcal{P}\{\lim_{n \to +\infty} \frac{1}{n} \sum_{k=1}^{n} f(y_k) = \int_Y f(y)p(dy)\} = 1.$$

Reducible Markov Chains. For *a reducible Markov chain* $(y_n)_{n \in Z^+}$ in a phase space $(Y, \mathcal{Y})$, we can define a stochastic kernel $P(y, A)$ associated with the *decomposition* of the phase space

$$Y = \bigcup_{v \epsilon V} Y_v \tag{A.3}$$

into disjoint classes Y_v of closed sets of states in the following manner:

$$P(y, Y_v) = \mathbf{1}_v(y) = \begin{cases} 1, & y \in Y_v, \\ 0, & y \notin Y_v \end{cases}$$

If $V = \{1, 2, ..., N\}$, for example, then such decomposition defines a merging function $v(y)$ by the relation

$$v(y) = v,$$

if $y \in Y_v, \qquad v = 1, ..., N.$

It means that for every class Y_v with $y \in Y_v$ corresponds one merged state $v \in V$.

We conclude this section with some remarks about reducible ergodic Markov chains whose closed sets of states $Y_v, v \in V$, are ergodic, so that the stationary distributions $p_v(A), A \in \mathcal{Y}_v (\mathcal{Y}_v$ is a σ-algebra in $Y_v)$

$$p_v(A) = \int_Y p_v(dz)P(z, A), A \in \mathcal{Y}_v, v \in V,$$

exist with the normalization property $p_v(Y_v) = 1$. Essentially, a reducible ergodic Markov chain with the decomposition of the phase space (A.3) consists of different irreducible ergodic Markov chains defined by the stochastic kernels

$$P_v(y, B) = P(y, B), \qquad y \in Y_v, \qquad B \in \mathcal{Y}_v.$$

For a *reducible Markov chain*, defined on the phase space $(Y, \mathcal{Y})$ with the decomposition (A.3) the definition of uniform ergodicity remains valid, where the stationary projector Π is defined by the relation

$$\Pi f(y) := (\hat{f}_v)_{v \in V},$$

$$\hat{f}_v := \int_{Y_v} p_v(dy) f(y), \qquad \forall y \in Y_v, \qquad \forall v \in V$$

Another definition of a Markov chain can be formulated as follows. Let $(\mathcal{F}_u), \mathcal{F}_0 \subseteq \mathcal{F}_1 \subseteq \ldots \subseteq \mathcal{F}$, be a non-decreasing family of σ-algebras and $(\Omega, \mathcal{F}, \mathcal{P})$ is a probability space. A stochastic sequence $(y_n)_{n \in Z^+}$ is called a *Markov chain with respect to the measure* $\mathcal{P}$, if for any $n \geq m \geq 0$ and any $A \in \mathcal{Y}$

$$\mathcal{P}\{y_n \in A | \mathcal{F}_m\} = \mathcal{P}\{y_n \in A | y_m\}.$$

In the particular case, where $\mathcal{F}_n = \mathcal{F}_n^Y := \sigma\{\omega : y_0, \ldots y_n\}$ and the stochastic sequence $(y_n)_{n \in Z^+}$ satisfies (1.16), the sequence $(y_n)_{n \in Z^+}$ is the Markov chain defined in the previous sections.

A.2 Markov Renewal Processes

Let us start with an ordinary *renewal process.* This is a sequence of independent identically distributed non-negative random variables $(\theta_n)_{n \in Z^+}$ with a common *distribution function* $F(x)$, where $F(x) := \mathcal{P}\{\omega : \theta_n(\omega) \leq x\}$. The renewal process counts events, and the random variables θ_n can be interpreted as *lifetimes (operating periods, holding times, renewal periods)* of a certain system in a random environment. In particular, θ_n can be regarded as the duration between the $(n-1)th$ and the nth event. From this renewal process we can construct another renewal process $(\tau_n)_{n \in Z^+}$ by $\tau_n := \sum_{k=0}^{n} \theta_k$. The random variables τ_n are called *renewal times* (or *jump times*).

Let $\nu(t) := \sup\{n : \tau_n \leq t\} := \sum_{n=0}^{\infty} \mathbf{1}_{[0,t]}(\tau_n)$. This is called the *counting process.* In the particular case where $F(t) = 1 - \exp\{-\lambda t\}$, $t \geq 0$, the counting process $\nu(t)$ is the homogeneous Poisson process, namly, $E[\nu(t)] = \lambda t$. From the above definitions it follows that $\tau_n \to +\infty$ a.s. as $n \to +\infty$ and $\nu(t) \to +\infty$ a.s. $t \to +\infty$. Also, since $\theta_{\nu(t)} < \infty$ (a.s.), we have that $\theta_{\nu(t)}/\nu(t) \to 0$ a.s. as $t \to +\infty$.

The so-called *renewal function* $m(t) := E[\nu(t)]$ has a number of important properties. First of all, the *Elementary Renewal Theorem* asserts that

$$m(t)/t \to 1/\mu \qquad \text{as } t \to +\infty,$$

where $\mu := E(\theta_1)$ and we regard $1/\infty$ as 0.

Let $\mu = E(\theta_1) < \infty$. Then the *Strong Law of Large Numbers* asserts that

$$\nu(t)/t \to 1/\mu, a.s. \qquad \text{as } t \to +\infty$$

Furthermore, if $0 < \sigma^2 := Var(\theta_1) < \infty$, then the *Central Limit Theorem* asserts that

$$\frac{\nu(t) - t/\mu}{\sqrt{t\sigma^2/\mu^3}} \to N(0,1) \qquad \text{as } t \to +\infty,$$

where $N(0,1)$ is a *normally distributed random variable* with mean value 0 and variance 1, namely, the *density function* $f(x) := F'(x)$ of this variable is equal to $(1/\sqrt{2\pi})\exp\{-x^2/2\}$.

To introduce Markov renewal processes, we need to introduce the notion of a semi-Markov kernel. A function $Q(y,A,t), y \in Y, A \in \mathcal{Y}, t \geq 0$, is called *a semi-Markov kernal* in the measurable space $(Y,\mathcal{Y})$ if it satisfies the following conditions:

(i) $Q(y,A,t)$ is measureable with respect to (y,t);

(ii) for fixed $t > 0, Q(y,A,t)$ is a semi-stochastic kernel in (y,A) and $Q(y,A,t) \leq 1$;

(iii) for fixed $(y,A), Q(y,A,t)$ is a nondecreasing function continuous from the right with respect to $t \geq 0$ and $Q(y,A,0) = 0$;

(iv) $Q(y,A,+\infty) =: P(y,A)$ is a stochastic kernel;

(v)) for fixed $y \in Y$ and $A \in \mathcal{Y}$, the function $Q(y,Y,t) =: G_y(t)$ is a distribution function with respect to $t \geq 0$.

In a discrete phase space $Y = \{1,2,3,\ldots\}$, a semi-Markov kernel is defined by a *semi-Markov matrix* $Q(t) := [Q_{ij}(t); i, j\epsilon Y]$, where $Q_{ij}(t)$ are nondecreasing functions in $t, \sum_{j\in Y} Q_{ij}(t) = G_j(t)$ are distribution functions of t and $[Q_{ij}(+\infty) = p_{ij}; i,j \in Y] =: P$ is a *stochastic matrix.*

A homogeneous two-dimensional Markov chain $(y_n;\theta_n)_{n\in Z^+}$ with values on $Y \times R_+$ is called *a Markov renewal process* (MRP) if its transition probabilities are given by the semi-Markov kernel

$$Q(y,A,t) = \mathcal{P}\{y_{n+1} \in A, \theta_{n+1} \leq t | y_n = y\}, \forall y \in Y, A \in \mathcal{Y}, t \in R_+.$$

It follows that the transition probabilities are independent of the second component. This distinguishes a Markov renewal processes from an arbitrary Markov chain with a non-negative second component. The first component $(y_n)_{n\in Z^+}$ of the Markov renewal process forms a Markov chain which is called an *imbedded Markov chain (IMC).*

The transition probabilities of the IMC can be obtained by setting $t = +\infty$, resulting

$$P(y,A) := Q(y,A,+\infty) = \mathcal{P}\{y_{n+1}\epsilon A | y_n = y\}.$$

In a MRP, the non-negative random variables $(\theta_n; n \geq 1)$ define the intervals between the Markov renewal times $\tau_0 := 0$ and $\tau_n := \sum_{k=1}^n \theta_k$, for $n \geq 1$.

The distribution functions of the renewal times depend on the states of the imbedded Markov chain

$$G_y(t) := \mathcal{P}\{\theta_{n+1} < t | y_n = y\} = Q(y, Y, t). \tag{A.4}$$

For a right-continuous process $(y(t))_{t \in R^+}$, the jump times τ_n and the holding times θ_n are obtained by

$$\tau_0 = 0, \tau_{n+1} := \inf\{t \geq \tau_n : y(t) \neq y(\tau_n)\},$$

for $n \geq 0$ and $\theta_n := \tau_n - \tau_{n-1}$ if $\tau_n < \infty$, or $\theta := 0$ if otherwise.

The discrete-time process $(y_n)_{n \in Z^+}$ given by $y_n := y(\tau_n)$ is called the *imbedded Markov chain* or *jump chain.*

We can now state the *ergodic theorem* for a Markov renewal process. Let $\alpha(y)$ be a measurable and bounded function of $y \in Y$. Then

$$t^{-1} \sum_{k=1}^{\nu(t)} \alpha(y_k) \to \hat{\alpha} \qquad \text{as } t \to +\infty,$$

where

$$\hat{\alpha} := \int_Y p(dy) \alpha(y)/m$$

and $p(y)$ is a stationary distribution of the Markov chain $(y_n)_{n \in Z^+}$.

A.3 Semi-Markov Processes

The MRP $(y_n, \theta_n)_{n \in Z^+}$ considered in previous section may serve as a convenient constructive tool to define a *semi-Markov process.*

A semi-Markov process(SMP) $(y(t))_{t \in R^+}$ is defined by the following relations:

$$y(t) := y_{\nu(t)}, \tag{A.5}$$

where

$$\nu(t) := \sup \{n : \tau_n \leq t\}. \tag{A.6}$$

The process $\nu(t)$ in (A.6) is called a *counting process.* It determines the number of renewal times on the segment $[0, t]$. Since the counting process $\nu(t)$ assumes constant values on the intervals $[\tau_n, \tau_{n+1})$ and is continuous from the right, that is,

$$\nu(t) = n, \tau_n \leq t < \tau_{n+1},$$

we conclude that the SMP $(y(t))_{t \in R^+}$ also assumes constant values on the same intervals and is continuous from the right. Namely,

$$y(t) = y_n, \tau_n \leq t < \tau_{n+1}.$$

Moreover,

$$y(\tau_n) = y_n, n \geq 0. \tag{A.7}$$

Relation (A.7) illustrates the concept of the *imbedded Markov chain* $(y_n)_{n \in Z^+}$.

For the SMP $(y(t))_{t \in R^+}$ in (A.5), the *renewal times (periods)* $\theta_n := \tau_{n+1} - \tau_n$ can be naturally interpreted as the *occupation times (life-times)* in the states $(y_n)_{n \in Z^+}$.

In what follows, we consider only a *regular SMP*.This is a SMP that has finite numbers of renewals on a finite period of time with probability 1. That is,

$$\mathcal{P}\{\nu(t) < +\infty\} = 1 \text{ for all } t > 0.$$

The renewal times $(\tau_n)_{n \in Z^+}$ and the IMC $(y_n)_{n \in Z^+}$ form a two-dimensional Markov chain $(y_n, \tau_n)_{n \in Z^+}$ homogeneous with respect to the second component, with transition probabilities $P(y, A), y \in Y, A \in \mathcal{Y}$. The two-dimensional Markov chain $(y_n; \tau_n)_{n \in Z^+}$ is also called a *MRP generating the semi-Markov process* $(y(t))_{t \in R_+}$ in (A.5). The process $\gamma(t) := t - \tau_{\nu(t)}$ is sometimes called an *age process* and the two-component process $(y_n, \theta_n)_{n \in Z_+}$ is also called a generating *semi-Markov chain,* although this is actually a Markov chain and the joint distribution of (y_{n+1}, θ_n) depends only on y_n.

Just as in the right-continuous Markov process, the epochs of jumps are regeneration points erasing the influence of the past. The only difference is that sojourn time at a point $y \in Y$ has an arbitrary distribution $G_y(t), y \in Y, t \in R^+$, which depends on the terminal state $y \in Y$.

Let $\alpha(y)$ be a measurable and bounded function on Y. The *ergodic theorem* for a semi-Markov process $(y(t))_{t \in R^+}$ then states

$$t^{-1} \int_0^t \alpha(y(s))ds \to \tilde{\alpha} \qquad t \to +\infty,$$

where

$$\tilde{\alpha} := \int_Y p(dy)m(y)\alpha(y)/m,$$

$$m := \int_Y p(dy)m(y),$$

$$m(y) := \int_0^{+\infty} tG_y(dt),$$

and $p(dy)$ is the stationary distribution of the Markov chain $(y_n)_{n \geq 1}$, $G_y(t)$ is defined in (A.4).

A Markov renewal process $(y_n; \tau_n)_{n \in Z^+}$, its counting process $\nu(t) := \max\{n : \tau_n \leq t\}$, and the associated semi-Markov process $y(t) = y_{\nu(t)}$ together generate several so-called auxiliary processes described below which play an important role in the theory of stochastic evolutionary systems.

The aforementioned auxiliary processes are defined as follows: *A point process* $\tau(t)$ is defined by:

$$\tau(t) := \tau_{\nu(t)}, t \geq 0;$$

an occupation time process $\theta(t)$ is defined by:

$$\theta(t) := \theta_{\nu(t)}, t \geq 0;$$

a running occupation time (defect process, age process) $\gamma(t)$ is given by:

$$\gamma(t) := t - \tau(t), t \geq 0;$$

and *a residual occupation time (excess process)* is given by

$$\gamma^+(t) := \tau_{\nu(t)+1} - t = \tau(t) + \theta(t+1) - t.$$

To describe some properties of these processes, let us first fix a Markov chain $(y_n)_{n\in Z^+}$ and write $\mathcal{F}_n^Y$ for the collection of all sets depending only on $y_0, y_1, y_2, ..., y_n$. The sequence $\mathcal{F}_n^Y$ is called the *filtration* of $(y_n)_{n\in Z^+}$ and we think of $\mathcal{F}_n^Y$ as representing the state of knowledge, or history, of the chain up to time n. We denote this by $\mathcal{F}_n^Y := \sigma\{y_k; 0 \leq k \leq n\}$ and call it the *σ-algebra generated by* chains $y_0, y_1, y_2, ..., y_n$. Let us also fix a continuous-time process $(y(t))_{t\in R^+}$ and write $\mathcal{F}_t^Y$ for the collection of all sets depending only on $y(s)$ for all $0 \leq s \leq t$. The sequence $\mathcal{F}_t^Y$ is called the filtration of $y(t)$ and we can think of $\mathcal{F}_t^Y$ as representing the state of knowledge, or history, of the process up to time t. We denote this by $\mathcal{F}_t^Y := \sigma\{y_s; 0 \leq s \leq t\}$ and call it the *σ-algebra generated by* the process $(y_s; 0 \leq s \leq t)$.

We can then define the following flows of σ-algebras:

$$\mathcal{F}_n^Y := \sigma\{y_0, y_1, \ldots, y_n\} := \sigma\{y_k; 0 \leq k \leq n\},$$

$$\mathcal{H}_t := \sigma\{\nu(t); \tau_n, \ldots, \tau_{\nu(t)}, y_0, y_1, \ldots, y_{\nu(t)}\};$$

$$\mathcal{F}_t^\theta := \sigma\{y(s); \theta(s); 0 \leq s \leq t\};$$

$$\mathcal{F}_t^\gamma := \sigma\{y(s); \gamma(s); 0 \leq s \leq t\};$$

$$\mathcal{F}_t^{\gamma+} := \sigma\{y(s); \gamma^+(s); 0 \leq s \leq t\}.$$

The most important property of the above auxiliary processes is that they complement a semi-Markov process $(y(t))_{t\in R^+}$ to a Markov process with respect to the corresponding flow. That is, each two-component process, $(y(t), \theta(t)), (y(t), \gamma(t))$ or $(y(t), \gamma^+(t))$ is a Markov process on the phase space $Y \times R^+$ with respect to the σ-algebra $\mathcal{F}_t^\theta, \mathcal{F}_t^\gamma$ or $\mathcal{F}_t^{\gamma+}$, respectively.

A.4 Jump Markov Processes

A regular homogeneous jump Markov process $(y(t))_{t\in R^+}$ is defined in terms of the MRP $(y_n;\theta_n)_{n\in Z^+}$ with the semi-Markov kernel

$$Q(y,A,t) = P(y,A)(1-e^{-\lambda(y)t})$$

as follows

$$y(t) = y_{\nu(t)}, \nu(t) := \max\{n : \tau_n \leq t\}.$$

Thus, a regular jump Markov process is a SMP with exponentially distributed occupation times with the parameters $\lambda(y) \geq 0, y \in Y$, depending on the states of the IMC $(y_n)_{n\in Z^+}$.

To define a jump Markov process constructively, it we need to have two functions: a stochastic kernel $P(y,A), y \in Y, A \in \mathcal{Y}$, determining the probabilities of jumps of the process (transition probabilities of the imbedded Markov chain), and a non-negative function $\lambda(y) \geq 0, y \epsilon Y$, which fixes the parameters of the exponential distributions of occupation times $\theta_n, n \geq 0$.

In the discrete phase space of states $Y = \{1,2,3,\ldots\}$, a jump Markov process is defined by a semi-Markov matrix

$$Q(t) = [Q_{ij}(t); ij \in Y],$$

where

$$Q_{ij}(t) = p_{ij}(1-e^{-\lambda(i)t}), i,j \in Y, \lambda_i \geq 0.$$

We now define *A general Markov process*, which we will use for martingale characterization and definition of semigroups of operators. Let $(y(t))_{t\in R^+}$ be a stochastic process defined on a probability space $(\Omega, \mathcal{F}, \mathcal{F}_t, P)$, where $\mathcal{F}_t$ is a nondecreasing family of σ-algebras: $\mathcal{F}_s \subseteq \mathcal{F}_t \subseteq \mathcal{F}$ for all $0 \leq s \leq t < +\infty$. The *Markov property* of the process $(y(t))_{t\in R_+}$, with respect to $\mathcal{F}_t$ is defined by

$$E[f(y(t))/\mathcal{F}_s] = E[f(y(t))/y(s)], s \leq t,$$

for any $f \epsilon \mathbf{B}(Y)$. A Markov process is a stochastic process $(y(t))_{t\in R^+}$ satisfying the Markov property. The Markov process $(y(t))_{t\in R^+}$ is uniquely determined by the *transition probabilities*

$$P(s,y;t,A) := \mathcal{P}\{y(t)\epsilon A | y(s) = y\}, s \leq t,$$

and the initial distribution $p(A) := \mathcal{P}\{y(0) \in A\}$.

The Markov property (1.28) yields the *Kolmogorov-Chapman equation* for the transition probability

$$P(s,y;t,A) = \int_Y P(s,y;u,dz)P(u,z,t,A),$$

where $s \leq u \leq t, y \in Y, A \in \mathcal{Y}$.

In the particular case of *time-homogeneous* Markov processes, the transition probabilities are determined by the family of *stochastic kernels*

$$P_t(y, A) := \mathcal{P}\{y(s+t) \in A/y(s) = y\}, s \leq t, y \in Y, A \in \mathcal{Y}.$$

The corresponding *Kolmogorov-Chapman equation* becomes

$$P_{t+s}(y, A) = \int_Y P_t(y, dz) P_s(z, A).$$

A.5 Wiener Processes and Diffusion Processes

A simple example of a *Brownian motion* is a symmetric random walk, choosing the directions with the same probability, in an Euclidean space which takes infinitesimal jumps with infinite frequency. It is named after a botanist who observed such a motion when looking at pollen grains under a microscope. The mathematical object, now called Brownian motion, was actually discovered by N. Wiener, and is thus called the *Wiener process.*

To give the definition of a Wiener process, we start with the definition of *Gaussian distribution with mean* 0 *and variance* t for a real-valued random variable. That is, the random variableb has density function

$$\phi_t(x) = (2\pi t)^{-1/2} \exp(-x^2/2t).$$

A real-valued process $(y(t))_{t \in R_+}$ is said to be *continuous* if

$$\mathcal{P}(\omega : t \to y(t, \omega) \quad \text{is continuous}) = 1.$$

A continuous real-valued process $(w(t))_{t \in R_+}$ is called a *Wiener process* if $w_0 = 0$ and if for all $0 = t_0 < t_1 < ... < t_n$ the increments

$$w_{t_1} - w_{t_0}, ..., w_{t_n} - w_{t_{n-1}}$$

are independent Gaussian random variables of mean 0 and variance $t_1 - t_0, ..., t_n - t_{n-1}$. We note that a Wiener process is a Markov process.

One important example of Markov process is the *diffusion process.*

By a *diffusion process* we mean a continuous Markov process $(y(t)_{t \in R_+}$ in $R(Y = R)$ with transition probabilities $P(s, y; t, A), s \leq 5, y \epsilon R, A \epsilon \mathcal{R}$, satisfying the following conditions:

(i)

$$\lim_{\Delta t \to 0} \frac{1}{\Delta t} \int_{|z-y|>\varepsilon} P(s, y, s+\Delta t, dz) = 0,$$

(ii)

$$\lim_{\Delta t \to 0} \frac{1}{\Delta t} \int_{|z-y|\leq\varepsilon} (z-y) P(s, y, s+\Delta t, dz) = a(s, y);$$

(iii)

$$\lim_{\triangle t \to 0} \frac{1}{\triangle t} \int_{|z-y| \leq \varepsilon}^{2} (z-y)^2 P(s,y,s+\triangle t, dz) = \sigma^2(s,y), \forall \epsilon > 0$$

The value $a(s,y)$ characterizes a mean trend in the evolution of the random process $(y(t))_{t\in R_+}$ for small interval of time from s to $s+\triangle t$ provided that $y(s) = y$, and is called the *drift coefficient.* The value $\sigma(s,y)$ characterizes a mean square deviation of the process $(y(t))_{t\in R^+}$ from its mean value and is called the *diffusion coefficient.* We have

$$y(s+\triangle t) \approx y(s) + a(s,y(s))\triangle t + \sigma(s,y(s))\triangle w(s),$$

where $\triangle w(s)$ is a random variable such that

$$E\triangle w(s) \sim 0 \text{ and } E(\triangle w(s))^2 \sim \triangle t. \tag{A.8}$$

Usually, diffusion processes are represented in differential form as a *stochastic differential equation*

$$dy(t) = a(t,y(t))dt + \sigma(t,y)dw(t)$$

where $(w(t))_{t\in R^+}$ is a standard Wiener process satisfying (A.8).

In such a way, a standard Wiener process is also a diffusion process with drift $a(t,y) \equiv 0$ and diffusion $\sigma(t,y) \equiv 1$.

A.6 Counting and Poisson Process

A.6.1 Counting Process (CP)

A stochastic process $N(t), t \geq 0$, is said to be a *counting process* if $N(t)$ represents the total number of "events" that occur by time t.

From this definition it follows that $N(t)$ satisfies the following properties:

1. $N(t) \geq 0$.
2. $N(t)$ is integer values.
3. If $s < t$, then $N(s) \leq N(t)$.
4. For $s < t$, $N(t) - N(s)$ equals the number of events that occur in the interval $(s,t]$.

Examples of CP:

a) The number of persons who enter a particular store at or prior to time t;

b) Total number of people who were born by time t;

c) The number of goals that a given soccer player scores at time t.

A.6.2 Poisson Process (PP)

One of the most important counting process is the Poisson process.

A CP is said to possess *independent increments* if the numbers of events that occur in disjoint intervals are independent.

E.g., for CP $N(10)$ must be independent of $N(15) - N(10)$.

A CP is said to possess *stationary increments* if the distribution of the number of events that occur in any interval of time depends only on the length of the time interval.

It means that the number of events in the interval $(s, s+t]$ has the same distribution for all s.

The counting process $N(t)$ is said to be a *Poisson process* having rate $\lambda > 0$, if

i) $N(0) = 0$;

ii) The process has independent and stationary increments;

iii)

$$P(N(t+s) - N(s) = n) = e^{-\lambda t}\frac{(\lambda t)^n}{n!}, \quad n = 0, 1, 2, \ldots.$$

We note, that from iii) it follows that PP has stationary increments and

$$E[N(t)] = \lambda t.$$

Alternate definition of a PP is the following which is equaivalent to Def. 4.

The counting process $N(t)$ is a PP with rate $\lambda > 0$, if

i) $N(0) = 0$;

ii) The process has independent and stationary increments;

iii) $P(N(h) = 1) = \lambda h + o(h)$;

iv) $P(N(h) \geq 2) = o(h)$.

Here, $o(h)$ is the function of h that satisfies the condition:

$$\lim_{h \to 0} \frac{o(h)}{h} = 0.$$

For example, x^2 is $o(h)$. However, x is not $o(h)$, as $o(h)/h = h/h = 1$, not 0.

A.6.3 Compound Poisson Process (CPP)

A stochastic process $X(t)$ is said to be a *compound Poisson process* if it can be represented as

$$X(t) = \sum_{i=1}^{N(t)} Y_i,$$

where $N(t)$ is a Poisson process, and $Y_i, i \geq 1$ is a family of idependent and identically distributed r.v.'s that is also independent of $N(t)$.

We note, that mean value of $X(t)$ is

$$E[X(t)] = E\sum_{i=1}^{N(t)} Y_i = EN(t)EY_i = \lambda tEY_1,$$

and variance (use law of total variance $Var(X) = E[Var(X/Y)] + Var[E(X/Y)]$)

$$Var[X(t)] = Var[\sum_{i=1}^{N(t)} Y_i] = \lambda tE[Y_1^2].$$

Moment generating function (MGF) of CPP is (use law of total probability)

$$M_X(s) := Ee^{sX(t)} = e^{\lambda t(M_Y(s)-1)},$$

where $M_Y(s)$ is the MGF of Y_i.

Examples of CPP

1. If $Y_1 = 1$, then $X(t) = N(t)$-it is usual PP.

2. Suppose that customers leave a supermarket in accordance with a PP. If the Y_i, the amount of money spent by the ith customer, $i = 1, 2, ...$, are i.i.d.r.v.'s, then $X(t)$ is a CPP, where $X(t)$ is the total amount of money spent by time t.

3. Total amount of claims $X(t)$ to an insurance company follows a CPP:

$$X(t) := \sum_{k=1}^{N(t)} X_k,$$

where $N(t)$ is a Poisson process describing a number of claims to an insurance company, and X_k is a size of kth claim.

A.7 Hawkes Processes

(One-dimensional Hawkes Process) (Hawkes (1971)). The one-dimensional Hawkes process is a point point process $N(t)$ which is characterized by its intensity $\lambda(t)$ with respect to its natural filtration:

$$\lambda(t) = \lambda + \int_0^t \mu(t-s)dN(s),$$

where $\lambda > 0$, and the response function $\mu(t)$ is a positive function and satisfies $\int_0^{+\infty} \mu(s)ds < 1$.

The constant λ is called the background intensity and the function $\mu(t)$ is sometimes also called the excitation function. We suppose that $\mu(t) \neq 0$

to avoid the trivial case, which is, a homogeneous Poisson process. Thus, the Hawkes process is a non-Markovian extension of the Poisson process.

The interpretation of equation above is that the events occur according to an intensity with a background intensity λ which increases by $\mu(0)$ at each new event then decays back to the background intensity value according to the function $\mu(t)$. Choosing $\mu(0) > 0$ leads to a jolt in the intensity at each new event, and this feature is often called a self-exciting feature, in other words, because an arrival causes the conditional intensity function $\lambda(t)$ in (1)–(2) to increase then the process is said to be self-exciting.

With respect to definitions of $\lambda(t)$ and $N(t)$, it follows that

$$P(N(t+h) - N(t) = m|\mathcal{F}^N(t)) = \begin{cases} \lambda(t)h + o(h), m = 1 \\ o(h), m > 1 \\ 1 - \lambda(t)h + o(h), m = 0. \end{cases}$$

We should mention that the conditional intensity function $\lambda(t)$ can be associated with the compensator $\Lambda(t)$ of the counting process $N(t)$, that is:

$$\Lambda(t) = \int_0^t \lambda(s)ds.$$

Thus, $\Lambda(t)$ is the unique $\mathcal{F}^N(t), t \geq 0$, predictable function, with $\Lambda(0) = 0$, and is non-decreasing, such that

$$N(t) = M(t) + \Lambda(t) \quad a.s.,$$

where $M(t)$ is an $\mathcal{F}^N(t), t \geq 0$, local martingale (This is the Doob-Meyer decomposition of N.)

A common choice for the function $\mu(t)$ in (2) is one of exponential decay:

$$\mu(t) = \alpha e^{-\beta t},$$

with parameters $\alpha, \beta > 0$. In this case the Hawkes process is called the Hawkes process with exponentially decaying intensity.

Thus, the equation above for $\lambda(t)$ becomes

$$\lambda(t) = \lambda + \int_0^t \alpha e^{-\beta(t-s)} dN(s),$$

We note, that in the case of (4), the process $(N(t), \lambda(t))$ is a continuous-time Markov process, which is not the case for the $\lambda(t)$.

With some initial condition $\lambda(0) = \lambda_0$, the conditional density $\lambda(t)$ in (5) with the exponential decay in (4) satisfies the SDE

$$d\lambda(t) = \beta(\lambda - \lambda(t))dt + \alpha dN(t), \quad t \geq 0,$$

which can be solved (using stochastic calculus) as

$$\lambda(t) = e^{-\beta t}(\lambda_0 - \lambda) + \lambda + \int_0^t \alpha e^{-\beta(t-s)} dN(s).$$

Another choice for $\mu(t)$ is a power law function:

$$\lambda(t) = \lambda + \int_0^t \frac{k}{(c + (t-s))^p} dN(s)$$

for some positive parameters c, k, p.

This power law form for $\lambda(t)$ in (6) was applied in the geological model called Omori's law, and used to predict the rate of aftershocks caused by an earthquake.

Many generalizations of Hawkes processes have been proposed. They include, in particular, multi-dimensional Hawkes processes, non-linear Hawkes processes, mixed diffusion-Hawkes models, Hawkes models with shot noise exogenous events, Hawkes processes with generation dependent kernels.

A.8 Martingales

Let $(\Omega, \mathcal{F}, \mathcal{F}_t, \mathcal{P})$ be a probability space with a non-decreasing flow of σ-algebras

$$\mathcal{F}_s \subseteq \mathcal{F}_t \subseteq \mathcal{F}, \quad \forall s \leq t.$$

An *adapted* (e.g., $\mathcal{F}_t$-measurable) *integrable* (e.g., $E|m(t)| < +\infty$ for all $t \in R_+$) collection $(m(t), \mathcal{F}_t, \mathcal{P})$ is called a *martingale*, if for any $s < t$ and $t \epsilon R^+$, we have

$$E[m(t)|\mathcal{F}_s] = m(s) \qquad \text{a.s.},$$

where $E[\cdot|\mathcal{F}_s]$ is a conditional expectation with respect to σ-algebra $\mathcal{F}_s$. This collection is called *submartingale (supermartingale)* if the relation (1.36) is fulfilled with sign $\geq$ $(\leq)$. Submartingales and supermartingales are called *semimartingales.* Later on, we will give martingale characterizations of Markov and semi-Markov processes, respectively.

Similarly, we can define discrete time martingales $(m_n)_{n \in Z^+}$. An *adapted* (e.g., $\mathcal{F}_n$-measurable) *integrable* (e.g., $E|m_n| < +\infty$) process $(m_n)_{n \in Z^+}$ is called a *martingale* if

$$E(m_{n+1}|\mathcal{F}_n) = m_n,$$

where $\mathcal{F}_n$ is the filtration: $\mathcal{F}_n \subseteq \mathcal{F}_m \subseteq \mathcal{F}$, where $n \leq m$. This collection is called *submartingale (supermartingale)* if this relation is fulfilled with sign $\geq (\leq)$.

If $(\xi_n)_{n \in Z^+}$ are the sequence of independent random variables such that $E\xi_n = 0$ for all $n \geq 0$ and $\mathcal{F}_n := \sigma(\xi_k; 0 \leq k \leq n)$, then the process $m_n := \sum_{k=0}^n \xi_k$ is martingale with respect to the $\mathcal{F}_n$.

If $E\xi_n = 1$ for all $n \geq 0$, then process $m_n := \prod_{k=1}^n \xi_k$ is also martingale.

Later on, we will give martingale characterization of discrete-time Markov chains. There is a short list of martingales:

(i) A standard Wiener process $(w(t))_{t \in R^+}$ is a martingale with respect to the natural σ-algebra

$$\mathcal{F}_t^w := \sigma\{w(s); 0 \leq s \leq t\}.$$

This is because $Ew(t) = 0$ and because it is a process with independent increments.

(ii) Process $w^2(t) - t$ is also a martingale with respect to $\mathcal{F}_t^w$.

(iii) Process $|w(t)|$ is a submartingale with respect to $\mathcal{F}_t^w$, hence, process $(-|w(t)|)$ is a supermartingale.

We will need *Kolmogorov-Doob inequalities* for semimartingales. Let $(m_n)_{n \in Z^+}$ be a nonnegative submartingale. Then for every $\lambda > 0$ and every $n \geq 0$, we have

$$\mathcal{P}\{\max_{0 \leq k \leq n} |m_k| \geq \lambda\} \leq \frac{Em_n}{\lambda}.$$

If $(m_n)_{n \in Z^+}$ is a supermartingale, then for every $\lambda > 0$ and every $n \geq 0$, we have

$$\mathcal{P}\{\max_{0 \leq k \leq n} |m_k| \geq \lambda\} \leq \frac{Em_0}{\lambda}.$$

The same inequalities are true for continuous time martingales.

A.9 Martingale Characterization of Markov and Semi-Markov Processes

We now describe martingale properties of Markov chains, Markov processes and semi-Markov processes.

A.9.1 Martingale Characterization of Markov Chains

Let $(y_n)_{n \in Z^+}$ be a homogeneous Markov chain on a measurable phase space $(Y, \mathcal{Y})$ with stochastic kernel $P(y, A), y \epsilon Y, A \epsilon \mathcal{Y}$. Let P be the operator on the space $\mathbf{B}(Y)$

$$\begin{aligned} Pf(y) &:= \int_Y P(y, dz) f(z) = E[f(y_n) | y_{n-1} = y] \\ &:= E_y[f(y_n)], \end{aligned} \tag{A.9}$$

generated by $P(y, A)$, and $\mathcal{F}_n^Y := \sigma\{y_k; 0 \leq k \leq n\}$ be a natural filtration generated by $(y_n)_{n \in Z^+}$. The Markov property can be described by

$$E[f(y_n) | \mathcal{F}_{n-1}^Y] = Pf(y_{n-1}).$$

This is because (A.9) and the Markov property of the chain $(y_n)_{n\in Z_+}$

$$E[f(y_n)|\mathcal{F}_{n-1}^Y] = E[f(y_n)|y_{n-1} = y\} = Pf(y_{n-1}).$$

We note that

$$Pf(y_n) - f(y) = \sum_{k=0}^{n}[P - I]f(y_n), y_0 = y.$$

Hence, from the last two expressions it follows that

$$E(f(y_n) - f(y) - \sum_{k=0}^{n}[P - I]f(y_n)|y_{n-1} = y) = 0,$$

or

$$E(f(y_n) - f(y) - \sum_{k=0}^{n}[P - I]f(y_n)|\mathcal{F}_{n-1}^Y) = 0.$$

Consequently,

$$M_n := f(y_n) - f(y) - \sum_{k=0}^{n}[P - I]f(y_k)$$

is an $\mathcal{F}_n^Y$-martingale.

The *quadratic variation*

$$< m_n >:= \sum_{k=1}^{n} E[(m_k - m_{k-1})^2|\mathcal{F}_{k-1}^Y]$$

of the martingale m_n in (54) is given by

$$< m_n >= \sum_{k=0}^{n}[Pf^2(y_k) - (Pf(y_k))^2].$$

A.9.2 Martingale Characterization of Markov Processes

Let $(y(t))_{t\in R^+}$ be a homogeneous Markov process on a measurable phase space $(Y, \mathcal{Y})$ with transition probabilities $P(t, y, A), t \in R^+$, $y \in Y$, $A \in \mathcal{Y}$. The transition probabilities $P(t, y, A)$ generate the contraction semigroup $\Gamma(t)$ on the Banach space $\mathbf{B}(Y)$ by the following formula

$$\Gamma(t)f(y) := \int_Y P(t, y, dz)f(z) = E[f(y(t))|y(0) = y].$$

Let Q be the infinitesimal operator of the Markov process $(y(t))_{t\in R_+}$. Then

$$\begin{aligned}\Gamma(t)f(y) - f(y) &= \int_0^t Q\Gamma(s)f(y(s))ds \\ &= \int_0^t \Gamma(s)Qf(y(s))ds.\end{aligned}$$

From this and the Markov property it follows that

$$E[f(y(t)) - f(y) - \int_0^t Qf(y(s))ds|\mathcal{F}_s^Y] = 0,$$

where

$$\mathcal{F}_s^Y := \sigma\{(y(u)); 0 \le u \le s\}.$$

Therefore,

$$m(t) := f(y(t)) - f(y) - \int_0^t Qf(y(s))ds$$

is $\mathcal{F}_t^Y$-martingale. The quadratic variation of $m(t)$ is

$$< m(t) >:= \int_0^t [Qf^2(y(s)) - 2f(y(s))Qf(y(s))]ds.$$

A.9.3 Martingale Characterization of Semi-Markov Processes

Let $y(t) := y_{\nu(t)}$ be a semi-Markov process, constructed by Markov renewal process $(y_n, \theta_n)_{n \in Z^+}$ on $Y \times R^+$ and and $\gamma(t) := t - \tau_{\nu(t)}$ be a defect process. Then the process $(y(t), \gamma(t))_{t \in R^+}$ on $Y \times R^+$ is a homogeneous Markov process with the generator

$$\hat{Q}f(t, y) = \frac{d}{dt} f(t, y) + \frac{g_y(t)}{\bar{G}_y(t)} [Pf(0, y) - f(t, y)],$$

where $f \in C^1(R_+) \times C(Y)$.

Let $\hat{\Gamma}(t)$ be the semigroup generated by transition probabilities $\hat{P}(t, (s, y), \cdot)$ of the process $(y(t), \gamma(t))_{t \in R^+}$. Then

$$\hat{\Gamma}(t)f(0, y) - f(0, y) = \int_0^t \hat{Q}\hat{\Gamma}(s)f(\gamma(s), y(s))ds.$$

This, together with the Markov property, implies

$$E[f(\gamma(t), y(t)) - f(0, y) - \int_0^t \hat{Q}f(\gamma(s), y(s))ds|\hat{\mathcal{F}}_s] = 0,$$

where

$$\hat{\mathcal{F}}_s := \sigma\{\gamma(u), y(u); 0 \le u \le s\}.$$

Consequently,

$$\hat{m}(t) := f(\gamma(t), y(t)) - f(0, y) - \int_0^t \hat{Q}f(\gamma(u), y(u))du$$

is an $\hat{\mathcal{F}}_t$- martingale.

The quadratic variation $< \hat{m}(t) >$ of the martingale $\hat{m}(t)$ is given by

$$\begin{aligned} &< \hat{m}(t) > \\ = \ & \textstyle\int_0^t [\hat{Q}f^2(\gamma(u), y(u)] - 2f(\gamma(u), y(u))\hat{Q}f(\gamma(u), y(u))du. \end{aligned}$$

A.10 Conclusion

This Appendix presented all necessary basic definitions and facts in stochastic processes which we used in this book.

It includes discrete- and continuous-time Markov chains, Wiener, Poisson, compound Poisson, Hawkes, Markov, Markov renewal and semi-Markov processes, and martingales. We refered to [1]–[7] for more information and results.

Bibliography

[1] Hawkes, A. (1971): Spectra of some self-exciting and mutually exciting point processes. *Biometrica*, 58, 83–90.

[2] Karlin, S. and Taylor, H. (1975): *A First Course in Stochastic Processes*, Academic Press, 2nd Ed.

[3] Laub, P., Taimre, T. and Pollett, P. (2015): Hawkes Processes.arXiv: 1507.02822v1[math.PR]10 Jul 2015.

[4] Norris J. R. (1997): *Markov Chains.* Cambridge Series in Statistical and Probabilistic Mathematics.

[5] Shreve, S. (2004): *Stochastic Calculus for Finance*, Springer, V.1–2.

[6] Swishchuk, A. and Wu, J. (2003): *Evolutionary Biological Systems in Random Media: Limit Theorems and Stability.* Kluwer AP.

[7] Williams, D. (1991): *Probability with Martingales*, University of Cambridge.

Index